Edited by
Samrath L. Chaplot, Ranjan Mittal,
and Narayani Choudhury

Thermodynamic Properties of Solids

Related Titles

Owens, F. J., Poole, Jr., C. P.

The Physics and Chemistry of Nanosolids

2008

ISBN: 978-0-470-06740-6

Haussühl, S.

Physical Properties of Crystals

An Introduction

2007

ISBN: 978-3-527-40543-5

Sandler, S. I.

Chemical, Biochemical, and Engineering Thermodynamics

2006

ISBN: 978-0-471-66174-0

Balluffi, R. W., Allen, S. M., Carter, W. C.

Kinetics of Materials

2005

ISBN: 978-0-471-24689-3

Adachi, S.

Properties of Group-IV, III-V and II-VI Semiconductors

2005

ISBN: 978-0-470-09033-6

Stolen, S., Grande, T.

Chemical Thermodynamics of Materials

Macroscopic and Microscopic Aspects

2004

ISBN: 978-0-471-49230-6

Trebin, H.-R. (ed.)

Quasicrystals

Structure and Physical Properties

2003

ISBN: 978-3-527-40399-8

Edited by
Samrath L. Chaplot, Ranjan Mittal, and Narayani Choudhury

Thermodynamic Properties of Solids

Experiment and Modeling

WILEY-VCH
WILEY-VCH Verlag GmbH & Co. KGaA

The Editors

Dr. Samrath L. Chaplot
Bhabha Atomic Research Centre
Solid State Physics Division
Trombay, Mumbai 400 085
India
chaplot@barc.gov.in

Dr. Ranjan Mittal
Bhabha Atomic Research Centre
Solid State Physics Division
Trombay, Mumbai 400 085
India
rmittal@barc.gov.in

Dr. Narayani Choudhury
Bhabha Atomic Research Centre
Solid State Physics Division
Trombay, Mumbai 400 085
India
narayani@gmail.com
narayani@uark.edu

Cover Illustration
Spieszdesign, Neu-Ulm, Germany

Library of Congress Card No.: applied for

British Library Cataloguing-in-Publication Data
A catalogue record for this book is available from the British Library.

Bibliographic information published by the Deutsche Nationalbibliothek
The Deutsche Nationalbibliothek lists this publication in the Deutsche Nationalbibliografie; detailed bibliographic data are available on the Internet at http://dnb.d-nb.de.

Cover Formgeber, Eppelheim
Typesetting Thomson Digital, Noida, India
Printing and Binding Bell & Bain Ltd., Glasgow

Printed in Great Britain
Printed on acid-free paper

ISBN: 978-3-527-40812-2

We dedicate this book to the memory of the late Dr. Krishnarao Raghavendra Rao, who was much more than a teacher and a mentor to us.

Contents

Thermodynamic Properties of Solids: Experiment and Modeling
Edited by S. L. Chaplot, R. Mittal, and N. Choudhury

ISBN: 928-3-527-40812-2

Preface

We are happy to present this book with contributions from leading researchers around the world. We were invited to take on this project by Dr. Edmund H. Immergut, Consulting Editor, who discussed it at the meetings of the German Physical Society with a number of scientists and the publisher. We thank Dr. Immergut for the invitation.

In the recent years, there has been a growing interest in the field of thermodynamic properties of solids due to the development of advanced experimental and modeling tools. Predicting structural phase transitions and thermodynamic properties find important applications in condensed matter and material science research, as well as in interdisciplinary research involving geophysics and earth sciences. Contributed by the experts in their respective fields, each topic of this book aims at meeting the need of the academic and industrial researchers, graduate students, and nonspecialists working in these fields. The book covers various experimental and theoretical techniques relevant to the subject.

The first few chapters give details about the experimental techniques used to determine the thermodynamic properties of solids at high pressures and temperatures. These include spectroscopic techniques of Raman, infrared, Brillouin, neutron and X-ray scattering, neutron and X-ray diffraction, calorimetry, and so on and interpretations of the experimental data to understand the thermodynamic behavior of solids.

The modeling of the thermodynamic properties can be carried out using a combination of classical and quantum mechanical approaches. Semiempirical methods are quite successful in spite of their approximate nature. The techniques of lattice dynamics, molecular dynamics simulations, and *ab initio* quantum mechanical simulations that play an important role in the modeling are discussed extensively.

Numerous applications are presented that cover a wide variety of technologically relevant materials as well as geophysically important minerals. These include, for example, novel negative thermal expansion materials, ferroelectrics, oxides, silicates, and garnets. The results obtained using a combination of experiments and theory are analyzed. The theoretical studies enable the planning, analysis, and

Thermodynamic Properties of Solids: Experiment and Modeling
Edited by S. L. Chaplot, R. Mittal, and N. Choudhury

ISBN: 928-3-527-40812-2

interpretation of various experiments as well as provide fundamental insights into the origins of observed anomalous properties.

The book is expected to be a useful reference tool for the academic and industrial researchers and graduate students of physics, chemistry, and material science. We are grateful to the various authors for their contributions that form the backbone of this book.

The publishers, Wiley-VCH, especially Anja Tschörtner and her team, have been very cooperative and understanding, and we are indeed grateful for their untiring efforts in bringing out the excellent publication.

Solid State Physics Division
Bhabha Atomic Research Centre
Trombay, Mumbai 400 085, India

Samrath L. Chaplot,
Ranjan Mittal, and
Narayani Choudhury

List of Contributors

S.N. Achary
Solid State Chemistry Section
Chemistry Division
Government of India
Bhabha Atomic Research Centre
Mumbai – 400 085
India

Rajeev Ahuja
Condensed Matter Theory Group
Department of Physics and Materials
Science Uppsala University
Box-530
75121 Uppsala
Sweden

C. Moysés Araújo
Condensed Matter Theory Group
Department of Physics and Materials
Science Uppsala University
Box-530
75121 Uppsala
Sweden

Alexei Bosak
European Synchrotron Radiation
Facility
6, rue Jules Horowitz BP 220
38043 Grenoble Cedex 9
France

Razvan Caracas
CNRS
Laboratoire de Sciences de la Terre
Ecole Normale Supérieure de Lyon
Université de Lyon
46, allée d'Italie
69364 Lyon
France

Samrath L. Chaplot
Solid State Physics Division
Bhabha Atomic Research Centre
Mumbai 400 085
India

Narayani Choudhury
Solid State Physics Division
Bhabha Atomic Research Centre
Mumbai 400 085
India

Irmengard Fischer
European Synchrotron Radiation
Facility
6, rue Jules Horowitz BP 220
38043 Grenoble Cedex 9
France

Thermodynamic Properties of Solids: Experiment and Modeling
Edited by S. L. Chaplot, R. Mittal, and N. Choudhury

ISBN: 928-3-527-40812-2

Xavier Gonze
Unité Physico-Chimie et de Physique des Matériaux
Université catholique de Louvain
Place Croix du Sud, 1
1348 Louvain-la-Neuve
Belgique

Michael Krisch
European Synchrotron Radiation Facility
6, rue Jules Horowitz BP 220
38043 Grenoble Cedex 9
France

Ranjan Mittal
Solid State Physics Division
Bhabha Atomic Research Centre
Mumbai 400 085
India

Olivier Pagès
Laboratoire de Physique des Milieux Denses
Institut Jean Barriol
Université Paul Verlaine – Metz
1, Bd. Arago
57078 Metz
France

Alain Polian
Physique des Milieux Denses
Institut de Minéralogie et Physique des Milieux Condensés
CNRS UMR 7590
Université Pierre et Marie Curie – Paris 6
140, rue de Lourmel
75015 Paris
France

Patrick Simon
Conditions Extrêmes et Matériaux: Haute Température et Irradiation
CNRS UPR 3079
45071 Orléans Cedex 2
France

and

Université d'Orléans
45071 Orléans Cedex 2
France

Toshihide Tsuji
School of Materials Science
Japan Advanced Institute of Science and Technology
1-1 Asahidai, Nomi
Ishikawa 923-1292
Japan

A.K. Tyagi
Solid State Chemistry Section
Chemistry Division
Government of India
Bhabha Atomic Research Centre
Mumbai – 400 085
India

Abbreviations

AFE	antiferroelectric
AHWR	Advanced Heavy Water Reactor
APS	Advanced Photon Source
APW	augmented plane wave method
ASA	atomic sphere approximation
ASC	adiabatic scanning calorimeter
BCC	body-centered cubic structure
BLS	Brillouin light scattering
BO	Born–Oppenheimer
BZ	Brillouin zone
CPA	coherent potential approximation
DAC	diamond–anvil cell
DFPT	density functional perturbation theory
DFT	density-functional theory
DOS	density of state
EOS	equation of state
e–ph	electron–phonon
ESRF	European Synchrotron Radiation Facility
EXAFS	extended X-ray absorption fine structure
FCC	face-centered cubic structure
FE	ferroelectric
fef	free energy function
FP	full potential
FPLMTO	full potential linearized muffin-tin orbital
FSR	free spectral range
GGA	generalized gradient approximation
HDA	high-density amorphous
HK	Hohenberg and Kohn
HT	high-temperature
ICCD	intensified CCD
IFC	*interatomic force constant*
ILL	Institut Laue Langevin
INS	inelastic neutron scattering

Thermodynamic Properties of Solids: Experiment and Modeling
Edited by S. L. Chaplot, R. Mittal, and N. Choudhury

ISBN: 928-3-527-40812-2

IR	infrared
IXS	inelastic X-ray scattering
KS	Kohn and Sham
LA	longitudinal acoustic
LAPW	linearized augmented plane wave
LDA	low-density amorphous
LMTO	linearized muffin-tin orbital
LO	longitudinal optic
LVDT	linear variable differential transformer
MD	molecular dynamics
MDS	Molecular dynamics simulation
MREI	modified random element isodisplacement
MT	muffin-tin
NN	next-neighbor
NTE	negative thermal expansion
PAW	projector augmented wave
PBE	Perdew, Burke, and Enzerhof
PDA	photodiode array detector
PDF	pair distribution function
PDOS	phonon density of state
PREM	preliminary reference earth model
PW	plane wave
qTA	quasitransverse acoustic
TA	transverse acoustic
TO	transverse optic
TOF	time-of-flight
US	ultrasonic
VDOS	vibrational density of states

1 Thermodynamic Properties of Solids: Experiment and Modeling

Samrath L. Chaplot, Ranjan Mittal, and Narayani Choudhury

1.1 Introduction

While thinking of thermodynamic properties of solids, a wide variety of properties and phenomena come to mind. Perhaps the most notable are specific heat, phase transitions, thermal expansion, thermal conductivity, melting, and so on. The macroscopic thermodynamic properties [1–11] are determined by microscopic crystalline and electronic structure and atomic vibrations, and these are determined by the nature of bonding between the atoms. In this book, we focus on the understanding and modeling of these microscopic and macroscopic properties and the experimental techniques [12–21] used in their investigation.

The modeling of the structure, dynamics, and various thermodynamic properties is done either by the first-principles quantum mechanical methods [6, 7] or by the semiempirical methods [8–11] largely based on models of interatomic interactions. The former is computationally far more intensive; therefore, its application to complex structures has been more recent and somewhat limited because of the available computational resources. The latter has been more widely used. Both of these techniques are extensively covered in this book.

On the experimental side, a variety of microscopic and macroscopic techniques are in use. The visible light, infrared and X-ray photons, and thermal neutrons are most widely used microscopic probes. These spectroscopic techniques [12–17] generate a rich amount of complex data of all kinds of vibrational modes of various polarizations and symmetry. Theoretical lattice dynamical calculations [7–10] are necessary for optimal planning of the experiments and for the microscopic interpretation of complex experimental data. Macroscopic measurements of specific heat [18] and thermal expansion [19, 20] and use of high-pressure, high-temperature devices [21] are also particularly important for thermodynamic investigations. These experimental techniques and the interpretations of their results by theoretical techniques are presented in individual chapters.

Thermodynamic Properties of Solids: Experiment and Modeling
Edited by S. L. Chaplot, R. Mittal, and N. Choudhury

ISBN: 928-3-527-40812-2

1.2
Spectroscopic Techniques and Semiempirical Theoretical Methods

The macroscopic thermodynamic properties are closely related to the microscopic dynamics of atoms. The collective vibrations of atoms in solids, which are called lattice vibrations, occur in discrete energies. These quanta of lattice vibrations are known as phonons.

Phonons are one of the fundamental excitations in a solid, and along with electrons they determine the thermodynamic properties of a material. In insulators and semiconductors, phonons play a prominent role. They directly contribute to a number of phenomena such as the thermal expansion, temperature dependence of mechanical properties, phase transitions, and phase diagrams. The understanding of phonon spectra, especially that of new materials, is essential for future technological developments.

The theoretical methods of lattice dynamics and the calculation of thermodynamic properties dealing with semiempirical and *ab initio* approaches are covered in Chapters 3 and 8, respectively. The concept of phonons assumes that the atomic vibrations are harmonic in nature, which is strictly valid at low temperatures, typically below the Debye temperature of the solid. As a complementary tool, molecular dynamics simulation is especially useful in studying the dynamics at high temperatures and in understanding the mechanisms of solid–solid phase transitions and melting, and so on. The simulations are also discussed in Chapter 3.

Spectroscopic techniques aim to determine the characteristics of these phonons. There are three major spectroscopic techniques, namely, optical spectroscopy (Chapter 2), inelastic neutron scattering (Chapter 3), and inelastic X-ray scattering (Chapter 4), which are complementary to each other.

Recent developments in lasers, optics, and electronics have made a significant impact on the modern optical spectroscopic methods and instrumentations. The optical techniques of Raman, infrared, and Brillouin scattering are reviewed in Chapter 2. Here the authors present a detailed comparison among the three techniques and provide a theoretical and experimental methodology. The chapter also gives a detailed account of the contribution of optical spectroscopy methods for studying the vibrational properties of materials under extreme conditions of high pressures and high temperatures. The uses of the spectroscopic methods are illustrated by examples taken from recent literature.

Inelastic neutron scattering is a spectroscopic technique in which neutrons are used to probe the dynamics of atoms and molecules in solids. It is the main experimental technique for determining the phonon dispersion curves as it offers both energy and momentum well matched for studies of various materials. Chapter 3 describes the principles and recent applications of inelastic neutron scattering, theoretical lattice dynamics, and molecular dynamics simulations. The calculations provide microscopic insights into a variety of novel phenomena like high-pressure phonon softening, structural phase transitions, melting, and so on.

Inelastic neutron scattering experiments take long data acquisition time and require large sample volume. This situation has changed with the advances in synchrotron radiation instrumentation. The improvement in X-ray photon flux combined with advances in instrumentation capabilities has made it possible to carry out detailed phonon studies for small quantities of samples, which also enables measurements under very high pressure (Chapter 4). The X-ray beam can be focused down to micron size, which allows investigation of very small crystals. The technique of inelastic X-ray scattering has opened new and important opportunities in the investigation of dynamics of atoms.

1.3 Thermal Measurement Techniques

The specific heat of a material is one of the most important thermodynamic properties indicating its heat retention or loss ability. Specific heat measurements also reveal signatures of crystalline or magnetic phase transitions. The major factors contributing to the specific heat are the atomic vibrations and electrons, which are called the lattice and electronic specific heat, respectively. The relation of the specific heat to these important physical properties and phase transitions makes this subject especially interesting for experimental and theoretical investigations. Chapter 5 gives a detailed theory of specific heat and presents a number of theoretical models. The principles and experimental methods of calorimetry to measure heat capacity of solids are also described.

Thermal expansion, an important thermophysical property of materials, has been of considerable interest for research and development of technology. This is an important material property considered for any structural materials experiencing a temperature gradient. Examples range from metals and ceramic parts of cooking wares to highly sophisticated engineering mechanical structures, such as buildings, bridges, air/spacecraft bodies, vessels, kiln, furnaces, and so on. Thermal expansion data of ceramics have also been a prime consideration for the design of electrolytes and electrodes of solid oxide fuel cells as well as reactor technology.

The bulk thermal expansion is generally measured by techniques like dilatometer and interferometers, while lattice thermal expansion is generally determined by the diffraction methods like variable-temperature X-ray or neutron diffraction. The nature and type of interatomic bonding, polyhedra around the cations, and packing of atoms in the unit cell and so on are key features in governing the magnitude and anisotropy of thermal expansion behavior. The salient details of the diffraction method and its importance for thermal expansion measurements are discussed in Chapter 6.

1.4 First-Principles Quantum Mechanical Methods

The study of electronic structure in solids provides important insights into the atomic structure and electronic properties of materials. Chapter 7 describes the main theoretical approaches and computational techniques for electronic structure calculations. Various practical methods used to solve the electronic structure problem are discussed. Finally, the electronic structure and crystallographic phase transformations of elemental compounds under high external pressure are described by means of the first principles theory.

The *ab initio* quantum mechanical computation methods have now become very useful research tool in the condensed matter physics. It is now possible to calculate a wide variety of material properties like electronic structure, elastic constants, phonon dispersion relations, and so on at high pressures and temperatures. Because of this, *ab initio* methods are now quite useful in earth science since they provide data that are complementary to the experiments on material properties at extreme pressure and temperature conditions found in the earth's interior. The ab initio lattice dynamics is studied by means of a perturbative approach to the density functional theory (Chapter 8). The study of phonon dynamics in multilayers, surfaces, crystals with defects and impurities, and so on is very limited. *Ab initio* phonon calculation is expected to provide new insights into such systems.

1.5 Outlook

We have discussed above a variety of experimental and theoretical methods used in the study of thermodynamic properties and related topics. Two major developments seem to stand out in the immediate future. One is the availability of the next generation of neutron and synchrotron sources that would bring a revolution in the way experiments are carried out and the nature of information and knowledge that is derived. We may expect new physics about the local short-range structures and ordering in complex and mixed solids; real-time analysis of dynamical phenomena including phase transitions and growth of novel structures of low dimensions, especially nanostructures and multilayers; new applications in energy systems and environment and earth sciences [22]; and so on. The second major development is the spurt in massively parallel computing that would enable the modeling of the structures and dynamical phenomena just noted above in the context of next-generation experimental facilities. We hope and believe that the contents of various topics covered in this book would prove extremely valuable in these new developments.

References

1 Born, M. and Huang, K. (1954) *Dynamical Theory of Crystal Lattices*, Oxford University Press, London.

2 Venkataraman, G., Feldkamp, L., and Sahni, V.C. (1975) *Dynamics of Perfect Crystals*, MIT Press, Cambridge.

3 Bruesch, P. (1982) *Phonons: Theory and Experiments I*, Springer-Verlag, Berlin.

4 Bruesch, P. (1986) *Phonons: Theory and Experiments II*, Springer-Verlag, Berlin.

5 Dove, M.T. (1993) *Introduction to Lattice Dynamics*, Cambridge University Press, Cambridge.

6 Martin, R.M. (2004) *Electronic Structure: Basic Theory and Practical Methods*, Cambridge University Press.

7 Baroni, S., de Gironcoli, S., Corso, A.D., and Giannozzi, P. (2001) Phonons and related crystal properties from density functional perturbation theory. *Rev. Mod. Phys.*, **73**, 515–562.

8 Chaplot, S.L., Choudhury, N., Ghose, S., Rao, M.N., Mittal, R., and Goel, P. (2002) Inelastic neutron scattering and lattice dynamics of minerals. *Eur. J. Mineral*, **14**, 291–329.

9 Mittal, R., Chaplot, S.L., and Choudhury, N. (2006) Modeling of anomalous thermodynamic properties using lattice dynamics and inelastic neutron scattering. *Prog. Mater. Sci.*, **51**, 211–286.

10 Choudhury, N. and Chaplot, S.L. (2009) Inelastic neutron scattering and lattice dynamics: perspectives and challenges in mineral physics, in *Neutron Applications in Earth, Energy and Environmental Sciences* (eds L. Liang, H. Schober, and R. Rinaldi), Springer, New York, pp. 145–188.

11 Allen, M.P. and Tildesley, D.J. (1987) *Computer Simulation of Liquids*, Clarendon, Oxford.

12 Cardona, M. and Güntherodt, G. (eds) (1982) *Topics in Applied Physics: Light Scattering in Solids II*, Springer-Verlag.

13 Bacon, G.E. (1975) *Neutron Diffraction*, Oxford University Press, Oxford.

14 Dorner, B. (1982) *Coherent Inelastic Neutron Scattering in Lattice Dynamics*, Springer-Verlag, Berlin.

15 Price, D.L. and Skold, K. (1986) in *Methods of Experimental Physics: Neutron Scattering Part A* (eds K. Skold and D.L. Price), Academic Press, Orlando.

16 Burkel, E. (1991) *Inelastic Scattering of X-Rays with Very High Energy Resolution*, vol. 125, Springer Tracts in Modern Physics, Springer-Verlag, Berlin.

17 Krisch, M. and Sette, F. (2007) Inelastic X-ray scattering from phonons, in *Light Scattering in Solids, Novel Materials and Techniques*, vol. 108, Topics in Applied Physics, Springer-Verlag, Berlin.

18 Sorai, M. (ed.) (2004) *Comprehensive Handbook of Calorimetry and Thermal Analysis*, John Wiley & Sons, Ltd.

19 Yates, B. (1972) *Thermal Expansion*, Plenum Press, New York.

20 Krishnan, R.S., Srinivasan, R., and Devanarayann, S. (1979) *Thermal Expansion of Crystal*, Pergamon Press, Oxford.

21 Eremets, M. (1996) *High Pressure Experimental Methods*, Oxford Science Publications, Oxford.

22 Liang, L., Schober, H., and Rinaldi, R. (eds) (2009) *Neutron Applications in Earth, Energy and Environmental Sciences*, Springer, New York.

2
Optical Spectroscopy Methods and High-Pressure–High-Temperature Studies

Alain Polian, Patrick Simon, and Olivier Pagès

This chapter aims to show how optical spectroscopies contribute to the study of vibrational properties under extreme conditions. Now, what do we mean by extreme conditions? We can take it in the other way and mention that ambient conditions have nothing more special as to permit life on earth – that is already not bad! From the point of view of physics, no particular property occurs at ambient conditions. On the contrary, most of the matter in the universe is under extreme pressure (P) and temperature (T), as it is in planets, stars, and more exotic objects. Studies at high pressure and high temperature are hence only the study of matter in its "normal" conditions. The reason why geoscientists are much interested in such studies is self-explanatory, but other fields are widely studied. The application of high pressure and high temperature enables to explore the repulsive part of the potential energy (fundamental physics), to provoke phase transformations thereby leading to new structures eventually metastable at ambient conditions (materials sciences), to orient chemical reactions in requested directions (chemistry), and even to crystallize proteins, that otherwise would not happen by using other techniques (biology). Another interest to work under variable conditions is that, obviously, more insight is given into the considered phenomenon, in that the pressure and/or temperature derivatives of the phenomenon become available, thus offering a more complete picture to achieve optimum understanding and/or modeling.

There are mainly five experimental methods to probe the vibrational properties of matter: optical spectroscopies (of main interest here), ultrasonics (US), inelastic neutron scattering (INS, cf. Chapter 3), more recently inelastic X-ray scattering (IXS, cf. Chapter 4) – that was made possible due to the introduction of third generation synchrotron radiation facilities – and picosecond (PS) acoustics.

INS and IXS are used to determine the dispersion of vibration modes, that is, acoustical as well as optical ones, over the whole Brillouin zone (BZ). The most stringent limitations are the needed sources (national-size ones, i.e., a nuclear reactor for INS and a monochromatized X-ray beam as delivered by a synchrotron for IXS), and a limitation to small momentum transfer ($q < 0.1q_{\mathrm{BZB}}$, where q is the magnitude of the wavevector and BZB is the BZ boundary). An additional limitation for INS is

Thermodynamic Properties of Solids: Experiment and Modeling
Edited by S. L. Chaplot, R. Mittal, and N. Choudhury

ISBN: 928-3-527-40812-2

that the sample should be large ($\sim$1 cm^3). For an insight into acoustical modes close to the center of the BZ ($q \sim 0$), the commonly used and the precise technique, at least at ambient conditions, is US, in which the sound velocity is deduced from the measurement of the transit time throughout the sample. Again, a large sample is needed, that is, some millimeters long. The latest technique, a pump-probe one, is PS. With this, a pump laser pulse is sent onto an absorbing sample (a metal), creating a heat pulse that propagates throughout the sample. Some picoseconds after the pump, a probe laser beam is sent to detect the arrival of the heat pulse at the other end of the sample, and measures the transit time. Unfortunately, for the moment, only the longitudinal acoustical modes can be detected by this technique. Decisive advantages are that small samples can be used, transparent or not (it is easy to evaporate a thin metal layer on a transparent sample), and the technique can be implemented at the laboratory scale. A synthetic presentation of the different q-domains addressed by the different techniques in relation to acoustical modes is given in Figure 2.1.

Currently, the versatile techniques at the laboratory scale are certainly optical spectroscopies, that is, infrared (IR) absorption, a generic terminology covering reflection/transmission/emission, and visible scattering, under the Raman and Brillouin variants. These provide complementary insights into acoustical (Brillouin) and optical (Raman/IR) modes. Again, optical spectroscopies cover a limited q-domain, that is, they operate close to the BZ-center, but with an unequalled resolution (less than 1 cm^{-1}, typically). Also, very small samples can be used (from several cubic millimeter with standards setups, down to several cubic micrometer with the Raman microprobe), transparent ones being preferable for the determination of bulk properties.

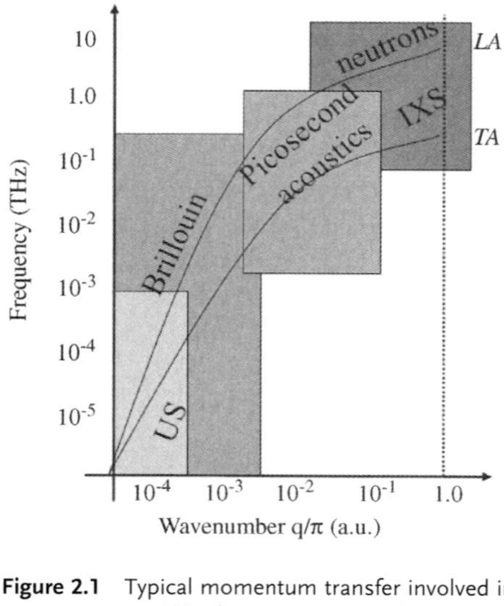

Figure 2.1 Typical momentum transfer involved in the various techniques of elastic properties measurements. US: ultrasonics; IXS: inelastic X-ray scattering and LA (TA): longitudinal (transversal) acoustic mode.

2.1 Methods and Principles: Ambient Conditions

To introduce the basic principles of optical spectroscopies, we limit ourselves to ambient conditions – the pressure and temperature aspects making the objects of specific developments in Section 2.2, and focus on a model system, that is, a semiconductor. With semiconductors, nature seems to offer to scientists "objects" with a quasimathematical perfection (simple atoms arranged on a regular lattice), where models can be developed at the ultimate scale of the atoms themselves. Here, vibrational properties reveal the lattice dynamics, that is, the vibration modes are collective, referred to as phonons. In this back-to-basics section, the main sources in the literature are quoted directly in the title of each paragraph – even though the present formulation might be different – while more regular insight is indicated by in-text references.

2.1.1 Semiconductors

Among semiconductors, the leading system is silicon (Si). Silicon belongs to column IV in the periodic table, and thereby has an electronic configuration of the sp^3 type. Four electrons are available per atom for the chemical bonding, which ends up in a tetrahedral environment, and a crystal structure of the diamond type. The intrinsic symmetry leads to a purely covalent bond, as represented by an electronic charge accumulated at intermediary distance between the atoms. Now, such monoatomic crystals bring only a limited insight into the phonon properties of semiconductors; the full picture emerges out by considering AB semiconductor compounds (see details below). With these, the same average sp^3 electronic configuration is achieved by combining two elements taken symmetrically on each side of column IV. A deficit of electrons of the cation A (column < IV), associated with a positive charge ($+Ze$), is compensated by an excess of electrons of the anion B (column > IV), credited with the opposite charge ($-Ze$). This balance results in an asymmetric bond position toward B, which confers a partially ionic character to the bond. The crystal structure, called zinc blende, consists of the A- and B-like intercalated fcc (face-centered cubic) sublattices, each atom being at the center of a regular tetrahedron formed with atoms of the other species. The dependence of the elastic properties on the ionicity of the chemical bond in zinc blende semiconductors, that has a direct impact on the phonon properties, is detailed in Ref. [1].

Ideally, the atom displacement $\vec{u}$ associated with a phonon in a semiconductor can be described as a plane wave, which is written as $\vec{u} = \vec{u}_0 \exp[j(\omega t - \vec{q} \cdot \vec{r})]$, where $\vec{u}_0$, t, and $\vec{r}$ refer to the amplitude, to time, and to the atom position, respectively. As such, a phonon is best represented in the reciprocal space where the dispersion, that is, the relation between the pulsation (ω) and the wavenumber (q), is explicited. In practice, an insight along the high-symmetry directions of the first BZ, that covers the whole q-domain from the BZ-center ($q = 0$) to the Brillouin zone boundary ($q_{BZB} = \pi/a \sim 10^8\ cm^{-1}$, where a is the lattice constant), will do. Experimentally, full insight into such dispersion curves can be achieved by using INS or IXS, as already mentioned. On the theoretical side, the reference is the adiabatic bond charge model as originally worked out by Weber [2] for the purely covalent semiconductors from column IV. Precisely, this is based on a description of the chemical bond in terms of a regular electronic charge, as indicated above. A successful variant for the partially ionic III–V's was developed by Rustagi and Weber [3] later.

2.1.2 $q \sim 0$ Optical Modes: Concept of Polaritons [4, 5]

Optical spectroscopies provide a selected insight at $q \sim 0$ only (refer to Section 2.1.3). At this limit, two broad families of phonons can be distinguished: optical modes that have a propagating character upon approaching the zone center, and acoustical modes, whose dispersion converges to zero at $q = 0$. These correspond to out-of-phase and in-phase vibrations of the anion and cation sublattices, respectively. In particular, due to the cubic symmetry of the zinc blende structure, the transverse (TO) and longitudinal (LO) optical modes, corresponding to atom vibrations perpendicular to and along the direction of propagation, respectively, should be degenerate exactly at $q = 0$.[1)] For a deeper insight, let us formalize that, in polar materials, optical modes do create in principle a polarization $\vec{P}$, with temporal (ω) and spatial (q) dependencies similar to those of the underlying atom displacements. Such elastic waves, with mechanical-/electrical-mixed character are referred to as phonon polaritons. The dispersion relation $\omega(q)$ is derived by combining the relevant sets of equations that carry the two characters. For each set, the restriction on $q \sim 0$ provides much simplification.

2.1.2.1 Maxwell Equations

At $q \sim 0$, the wavelength is large enough – quasi-infinite, in fact – that the atom displacements are similar over several lattice cells (introducing the notion of quasirigid sublattices) with the consequence that the accompanying polarization field $\vec{P}$ has a macroscopic character (spatially averaged over one lattice cell). As such, its propagation can be formalized via the Maxwell equations that bear upon field quantities having meaning only on a macroscopic basis, precisely. The crystal is viewed as an effective medium. In the case of a nonmagnetic semiconductor, an elimination of the induction from Ampere equation via Lenz equation leads to $\vec{q}(\vec{q}\cdot\vec{E}) - q^2\vec{E} = -\mu_0\omega^2(\varepsilon_0\vec{E} + \vec{P})$, where μ_0 and ε_0 refer to the magnetic permeability and dielectric permittivity of vacuum, respectively. A combination with Gauss equation, that is, $\vec{q}\cdot\vec{E} = -\varepsilon_0^{-1}\vec{q}\cdot\vec{P}$, leads to

$$\vec{E} = \left[\frac{\omega^2}{c^2}\vec{P} - \vec{q}(\vec{q}\cdot\vec{P})\right] \times \left[\varepsilon_0\left(q^2 - \frac{\omega^2}{c^2}\right)\right]^{-1} \tag{2.1}$$

1) To figure this out, represent an optical mode at $q = 0$, that is, a vibration of the perfectly rigid cation sublattice against the perfectly rigid anion one, and realize that the wave vector – nul, in fact – could as well be oriented perpendicular to (TO) or along (LO) the sublattice displacements.

where $\sqrt{\mu_0 \varepsilon_0}$ is the velocity of light in vacuum. The electric fields carried by a TO ($\vec{q} \perp \vec{P}$) and a LO ($\vec{q} // \vec{P}$) mode in a zinc blende system (the cubic symmetry implies isotropy) follow directly as

$$\vec{E}_{\mathrm{L}} = -\frac{\vec{P}}{\varepsilon_0}; \quad \vec{E}_{\mathrm{T}} = \omega^2 [\varepsilon_0 (q^2 c^2 - \omega^2)]^{-1} \vec{P} \tag{2.2}$$

If the so-called retardation effects are neglected, that is, by taking $c \rightarrow \infty$ (which becomes valid for finite/nonnegligible q values), $E_{\mathrm{T}} = 0$. So, reasonably far from the center of the BZ – an estimate is given below, a TO mode basically reduces to a purely mechanical vibration. For a LO mode, a Coulombic interaction exists that adds to the mechanical restoring force.[2)] This is at the origin of the TO–LO splitting at $q \sim 0$ in a polar crystal. Now, exactly at $q = 0$, Equation 2.2 leads to $\vec{E}_{\mathrm{T}} = \vec{E}_{\mathrm{L}}$, providing the required TO–LO degeneracy. What happens at intermediate q value is the object of the development hereafter. Note that the TO and LO modes may also be characterized by a condition on the relative dielectric function of the crystal – with respect to vacuum, denoted by ε_r, as introduced via the Gauss equation

$$\varepsilon_0 \varepsilon_r \vec{E} = \varepsilon_0 \vec{E} + \vec{P} \tag{2.3}$$

The compatibility with Equation 2.2 implies the respective conditions for the TO and LO modes as

$$\varepsilon_r = q^2 c^2 \omega^{-2}; \quad \varepsilon_r = 0 \tag{2.4}$$

2.1.2.2 **Mechanical Equations**

The lattice dynamics takes place in the IR spectral range, and involves two oscillators (at least), that is, the nuclei of (A, B) atoms (including the core electrons, strongly bound) – also referred to as ions hereafter – plus the peripheral electrons (from the chemical bonding, weakly bound). The latter vibrate naturally at high frequency (visible), and are therefore able *a fortiori* to follow the comparatively slow dynamics (IR) of nuclei (much heavier corpuscles). The two oscillators are characterized by the displacements with respect to their equilibrium positions, denoted by u_{A}, u_{B}, and y, respectively. Now, placing the analysis at $q \sim 0$ brings in a major advantage that the space-related phase term of the plane wave that ideally describes each oscillator (nuclei, electrons) just disappears, and with it an obligation to position vectorially each individual oscillator in the crystal. Therefore, a model based on a scalar representation of the crystal (linear chain approximation) – a phenomenological one – should do in principle. At this limit, the relevant IR oscillator for an optical mode naturally appears to be the A–B bond itself, and the associated displacement is

2) Consider a $q \sim 0$ propagation along the [1 1 1] crystal direction, where A (polarized +) and B (polarized −) planes alternate. For a TO mode, the A and B planes slide on themselves, so that no polarization is created with respect to the static situation, that is, $E_{\mathrm{T}} = 0$. For a LO mode, the average distance between the A and B planes over a time period has become shorter than in the static case, resulting in the creation of some polarization, that is, $E_{\mathrm{L}} \neq 0$.

the bond stretching $u = u_A - u_B$ representing, in fact, the relative displacement of the quasirigid A- and B-sublattices. By considering Hooke-like mechanical restoring forces between first neighbors only for A and B (in first approximation), that is, proportional to the bond stretching (harmonic approximation), plus the Coulombic forces, that imply the local field, termed E_e, the equations of motion per bond and per peripheral electron, respectively, are formulated as

$$\mu\ddot{u} = -k_0 u + (Ze)E_e \tag{2.5}$$

and

$$m\ddot{y} = -k'_0 y + eE_e \tag{2.6}$$

where $\mu = (m_A^{-1} + m_B^{-1})^{-1}$ is the reduced mass of the A–B bond, and (m, e) are the mass and elementary charge of an electron, respectively. Care must be taken that E_e and E are not equivalent, that is, one has a microscopic local character (E_e) and the other a macroscopic-average character (E), so that Equations 2.1 and 2.2 cannot be directly combined with Equations 2.5 and 2.6. A prerequisite is to express E_e via the macroscopic parameters of the system (u, P, E).

2.1.2.3 **Lorentz Approach**

An elegant method developed by Lorentz, as schematically represented in Figure 2.2, assimilates each bond to a discrete electric dipole $\vec{p}$ (resulting from the added contributions of the ions and peripheral electrons) being immersed into the dipolar field created by the other – identical – bonds from the whole crystal. A so-called Lorentz sphere $S(O, R)$ is introduced, centered on the dipole under consideration (O)

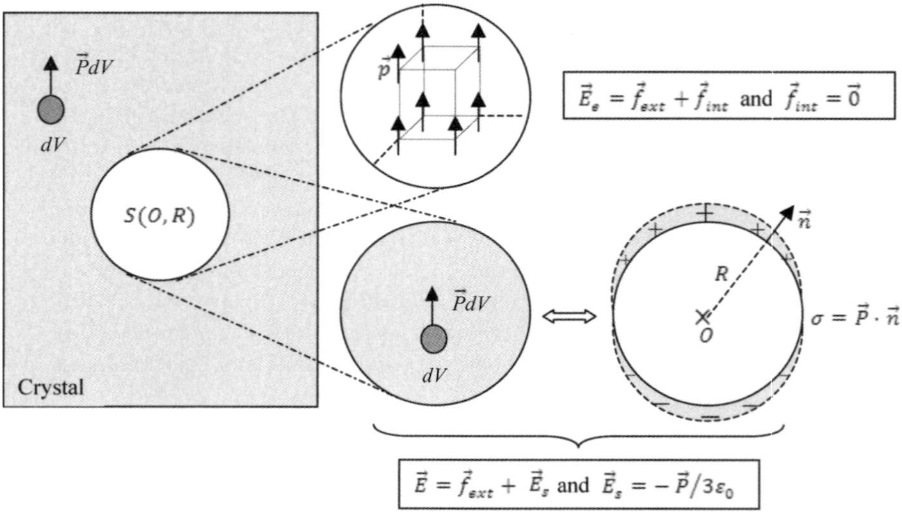

Figure 2.2 Schematic view of the Lorentz approach for calculation of the local field $\vec{E}_e$.

and with arbitrary radius (R), such that the inside and the outside of the sphere (extending to the infinite) are, respectively, perceived from O as a discrete collection of dipoles $\vec{p}$ arranged on a regular array – a cubic one in the present case, and a uniform dipolar continuum – characterized by polarization $\vec{P}$ (an average dipolar moment per crystal unit volume) as compatible with the Maxwell equations. The two regions create their own electric fields in O, denoted by $\vec{f}_{\text{int}}$ and $\vec{f}_{\text{ext}}$, respectively, that add to form $\vec{E}_e$. The macroscopic-average electric field $\vec{E}$, as for it, is obtained by adding to $\vec{f}_{\text{ext}}$ the electric field $\vec{E}_s$ created in O by $S(O, R)$ where a uniform dipolar continuum with polarization $\vec{P}$ substitutes for the discrete collection of dipoles $\vec{p}$. This comes to a vision of the crystal in terms of a uniform dipolar continuum, that is, an effective medium, which is in the spirit of the Maxwell equations. Then, $\vec{E}_e = \vec{E} - \vec{E}_s + \vec{f}_{\text{int}}$. The cubic symmetry implies $\vec{f}_{\text{int}} = 0$ (a full treatment for a simple cubic lattice is detailed in Refs. [6, 7]). $\vec{E}_s$ can be calculated in O by substituting a surfacial distribution of charge (σ), for example, for the volumic polarization $\vec{P}$ inside S (O, R), that is, a nonhomogeneous one, as the equivalence is written as $\sigma = \vec{P} \cdot \vec{n}$, where $\vec{n}$ is a unit vector normal to $S(O, R)$ pointing outside. This leads to $\vec{E}_s = -\vec{P}/(3\varepsilon_0)$.

Thus,

$$\vec{E}_e = \vec{E} + \frac{\vec{P}}{3\varepsilon_0} \tag{2.7}$$

with

$$P = N[(Ze)u + (\alpha_+ + \alpha_-)E_e] \tag{2.8}$$

where $(Ze)u$ and $\alpha_i E_e$ (subscript i stands for $+$ or $-$) are the dipolar moments per bond and per atom, (α_+, α_-) denote the individual polarizabilities of the (A, B) ions (due to electrons), and N is the number of bonds per crystal unit volume.

2.1.2.4 **Effective Charge/Force Constant**

Equations of motion (2.5) and (2.6) reformulate via Equations 2.7 and 2.8 on the basis of the macroscopic observables u and E, which makes them compatible with the Maxwell equations. The general form remains unchanged, but the force constants and charges are "renormalized to the macroscopic scale," coming to a terminology of "effective" force constants/charges, as identified by a star hereafter. Equations 2.5 and 2.6 lead to

$$\mu(-\omega^2 + \omega_T^2)u - (Ze)^* E = 0 \tag{2.9}$$

and

$$m\omega_e^2 y - e^* E = 0 \tag{2.10}$$

where $\omega_T = \sqrt{k_0^*/\mu}$ and $\omega_e = \sqrt{k_0'^*/m}$ are resonance frequencies as obtained by taking $E = 0$ – of a transverse type for phonons. Note that in Equation 2.10, ω (IR range) is neglected in front of ω_e (visible range).

2.1.2.5 **Combined Electrical/Mechanical Equations: Dispersion of Polaritons Modes**

Equations 2.3 and 2.7–2.9 allow to explicit ε_r as

$$\varepsilon_r = \varepsilon_\infty + (\varepsilon_S - \varepsilon_\infty)\left(1 - \frac{\omega^2}{\omega_T^2}\right)^{-1} \tag{2.11}$$

where ε_S and ε_∞ refer to the relative static and high-frequency dielectric constants, as obtained by taking extreme ω values in Equation 2.9, that is, $\omega \to 0$ and $\omega \to \infty$ ($\omega \gg \omega_T$), respectively. In Equation (2.11) the first term refers to electrons and the second term refers to ions. Equations 2.4 lead to

$$\omega_L^2 = \omega_T^2 \frac{\varepsilon_S}{\varepsilon_\infty}; \quad \frac{q^2c^2}{\omega^2} = \varepsilon_\infty\left(1 + \frac{\omega_L^2 - \omega_T^2}{\omega_T^2 - \omega^2}\right) \tag{2.12}$$

for the LO (Lyddane–Sachs–Teller relation) and TO modes, respectively. A dimensionless *oscillator strength* is introduced to estimate the "strength" of the ionic resonance, that is, $S = \varepsilon_s - \varepsilon_\infty = \varepsilon_\infty(\omega_L^2 - \omega_T^2)/\omega_T^2$. Naturally, S scales as the fraction of bonds in the crystal (see a generalization to multioscillator systems in Section 2.1.4.1). In fact, the same expression of ε_r can be obtained just by writing P as $N(Ze)^* u + ne^* y$, that is, directly in terms of the effective charges (macroscopic scale), which allows to skip a treatment of the local field E_e, but $(Ze)^*$ should satisfy

$$\omega_L^2 - \omega_T^2 = \frac{N(Ze)^{*2}}{\mu\varepsilon_0\varepsilon_\infty} \quad (\varepsilon_r = 0) \tag{2.13}$$

and also, the electronic susceptibility χ_∞, as given by $ne^* y = \varepsilon_0\chi_\infty E$, should be formulated as

$$\varepsilon_\infty = 1 + \chi_\infty. \tag{2.14}$$

The latter relation simply expresses that at pulsations well beyond those of optical modes, the polarization of the crystal is all due to electrons. What emerges from Equation 2.12 is that the LO mode has no dispersion, thus simply referred to as the LO phonon. In contrast, the TO mode is highly dispersive but the dispersion is confined close to the BZ-center. In fact, the asymptotes are reached for $q \sim 5\omega_T/c$, corresponding to $\sim 10^{-5} q_{BZB}$. They are defined by those limit situations where the polarization wave (E_T) is no more coupled to the nuclei (u). This may occur either for $E_T = 0$, in which case the TO polariton is like the TO phonon of a diamond-type (nonpolar) crystal, with "purely mechanical" pulsation ω_T; or for $u = 0$, the typical situation at $\omega \to \infty$ ($\omega \gg \omega_T$) where the propagation of the polarization wave is all governed by electrons – the dispersion takes the linear form $\omega^2 = q^2c^2/\varepsilon_\infty$. TO modes are referred to as phonon polaritons in the strong electrical/mechanical-coupling regime. At low q values, the low- and high-frequency polariton branches are of electromagnetic (EM) and elastic types, respectively. The situation is reversed at large q values. For example, the dispersion of the phonon polaritons in gallium phosphide (GaP) – reproduced from Ref. 8 – is shown in Figure 2.3, left panel.

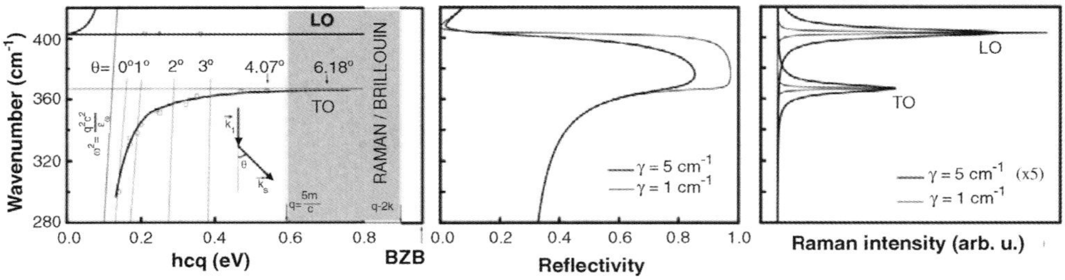

Figure 2.3 Dispersion of polariton modes of GaP (left panel, adapted from Ref. [8]). The corresponding IR reflectivity and (TO, LO) Raman spectra are shown in the central and right panels, respectively, for two values of the phonon damping γ, as indicated. The spectra were calculated from Equations 2.15 and 2.18, respectively, by using the set of input parameter ($\varepsilon_{\infty} = 9.1$ [9], TO–LO $= 365$–$405\ \mathrm{cm}^{-1}$, $C_{\mathrm{F-H}} = -0.47$ [10]). The shaded zone in the left panel describes the specific q-domain as explored by using the standard Raman/Brillouin scattering geometries, that is, more backscattering-like. The low and high q-limits are $\sim 10^{-5} q_{\mathrm{BZB}}$ and $\sim 10^{-3} q_{\mathrm{BZB}}$, typically. The low-frequency polariton dispersion curve is accessible by Raman scattering by using forward-like scattering geometries, that is, small θ angles as indicated (valid for the excitation line 632.8 nm).

2.1.3
Vibration Spectra

The vibrational techniques differ by the light/matter interaction mechanisms brought into play.

2.1.3.1 IR Spectroscopies: A Direct Light/Optical-Mode Interaction [4, 5, 11]

As their name implies, optical phonons may be revealed via a direct interaction with an EM radiation of similar pulsation, which opens the field of IR spectroscopies. Now, EM radiations have a transverse character, and as such may couple to TO polaritons only, not to the LO phonon. Basically, when an IR beam penetrates the crystal, it propagates under the form of a TO polariton. Maxwell equations indicate that the phase velocity reduces from c (in vacuum) to c/n (in the crystal), where $n = \sqrt{\varepsilon_r}$ is the refractive index of the crystal. The *TO* polariton is – obviously – forced at the same pulsation as the exciting radiation, so that the modification of the phase velocity is all due to a renormalization of the wavenumber q_0 (in vacuum) to $q = nq_0$ (in the crystal). From Equations 2.11 and 2.12, n is real outside the (TO, LO) band and imaginary inside. Thus, radiations with pulsation $\omega_T < \omega < \omega_L$ cannot propagate into the crystal in principle, and thereby should be totally reflected. In fact, total reflection never occurs as part of the incident energy is absorbed by the crystal, which is accounted for by adding a dissipative force into the phonon-related equation of motion. This is naturally taken as antiproportional to the displacement velocity (Coulomb-like), that is, as $-\mu\gamma\dot{u}$, γ being the so-called phonon damping – a simple rule states *the smaller* γ, *the better the crystal quality*. This comes *in fine* to add $-j\gamma\omega$ to the denominator of ε_r in Equations 2.11 and 2.12, which thus takes an imaginary character. This is enough to confer on n, a partially real character within the optical band, thus allowing some propagation. If we refer to the standard setup for IR reflectivity/transmissivity pressure and temperature measurements, corresponding to quasinormal incidence on a unique vacuum/crystal interface, the energy reflection and transmission coefficients, that give the IR reflectivity and transmissivity spectra – as derived on the basis of a pure normal incidence – take the well-known forms [12]

$$R(\omega) = \left|\frac{1-\sqrt{\varepsilon_r}}{1+\sqrt{\varepsilon_r}}\right|^2; \quad T_r(\omega) = \frac{4\sqrt{\varepsilon_r}}{\left(1+\sqrt{\varepsilon_r}\right)^2}. \tag{2.15}$$

For example, the normal incidence IR reflectivity spectrum of GaP is shown in Figure 2.3, central panel. Note that it peaks close to the asymptotic TO frequency, and goes through a minimum close to the LO frequency. The light energy that is neither reflected nor transmitted is absorbed by the medium, and, at thermal equilibrium, reemitted as thermal radiation at the same pulsation (Kirchoff's law, $A(\omega) = E_m(\omega)$ where $A(\omega)$ denotes absorbance). The spectral emissivity E_m is defined as the ratio of the sample-to-blackbody intensities at the same temperature and angular conditions. A basic law of energy conservation is fulfilled, that is, $E_m(\omega) = 1 - [R(\omega) + T_r(\omega)]$, that simply gives the energy balance upon incident light beam. If one considers opaque materials, spectral emissivity is nothing but $[1 - R(\omega)]$. In fact, the spectral emissivity gives access to the same information on vibrational properties as reflectivity and transmissivity do. It is implicit that all four coefficients (R, T_r, E_m, and A) are temperature dependent.

2.1.3.2 **Raman Scattering: An Indirect Light/Optical-Mode Interaction [13, 14]**

By penetrating the crystal, a visible excitation interacts resonantly with electrons (peripheral ones). In a classical description, these behave as oscillating dipoles P_e, and as such scatter light in all directions – but the oscillating one – according to the Hertz mechanism. Most of the light is scattered quasielastically, which is referred to as Rayleigh scattering. Less probably (10^6–10^{12} times less) part of the incident energy is taken to set one optical phonon in motion, or on the contrary one optical phonon is consumed during the light/matter interaction. The pulsation of the scattered light (ω_s) is accordingly shifted with respect to the incident one (ω_i) – downward for a Stokes process and upward for an anti-Stokes one – of the pulsation of the optical phonon (Ω) brought into play. Such inelastic scattering is referred to as the first-order Raman scattering (a single optical phonon is involved). Another formulation is that the optical phonon modulates the electronic susceptibility χ, that is, the ability of electrons to polarize under the electric field E_v carried by the visible excitation ($P_e = \varepsilon_0 \chi E_v$), the Raman scattering being due, then, to the change in polarization $dP_e = \varepsilon_0\, d\chi E_v$.

2.1.3.2.1 **Raman Geometry** The light/phonon interaction is governed by energy and wavevector conservation laws similar in every respect to those implied in a collision process. For Stokes scattering, these come to $\omega_i - \omega_s = \Omega$ and $\vec{k}_i - \vec{k}_s = \vec{q}$, respectively, where subscripts "i" and "s" refer to the incident and scattered radiations (visible range), the right term being related to the phonon (IR range). Depending on the scattering geometry, a continuous domain of q values can be explored. Extreme values correspond to twice the incident wavenumber ($\vec{k}_i$ and $\vec{k}_s$ antiparallel, that is, backward scattering, or backscattering) and to the close neighborhood of the center of the BZ ($\vec{k}_i$ and $\vec{k}_s$ parallel, that is, forward scattering), as schematically shown in Figure 2.3, left panel. Now, the dispersion of visible light in the crystal is quasivertical at the scale of the phonon dispersion, so that, in fact, $2k_i \sim 10^{-3} q_{BZB}$ still refers to the BZ-center. The former classical formalism used for polaritons and IR spectroscopies – based on the Maxwell equations and a linear chain model for mechanical equations – thus remains basically valid for the Raman scattering. Another aspect is that the limit q value corresponding to the asymptotic behavior of the polariton, that is, $q_{asymp} \sim 5\omega_T/c$, is reached for an angle between $\vec{k}_i$ and $\vec{k}_s$ as small as $\theta \sim 5°$ typically. Thus, the dispersion of the polaritons is accessible via the Raman scattering only by using small θ values, that is, in a quasiperfect forward scattering geometry, provided samples with a thickness smaller than the penetration depth of the exciting beam are used ($\sim$1 μm typically out of resonance conditions). With usual scattering geometries, that is, backscattering-like ones, only the LO mode and the asymptotic TO mode – a purely mechanical vibration – are accessible.

2.1.3.2.2 **TO Frequency** This relates directly to the effective bond force constant (k_0^*, cf. Section 2.1.2.4). Now, a basic rule is that *the bond force constant scales as the inverse bond length* (strictly, the harmonic approximation – cf. Section 2.1.2.2 – is not valid). By applying a hydrostatic pressure (resp. increasing the temperature), the lattice of AB compound shrinks (resp. expands), and therefore the TO frequency is increased (resp. decreased). The rule can also be used to probe the local environment of an impurity, say, C in substitution for B in AB-like matrix. Basically different local distortions of the impurity A–C bond, depending on the local environment, that is, more or less C-rich, should provide as many AC-like TO frequencies (more detail is given in Section 2.3.2.1). There exists a limit. however. Indeed, such so-called phonon localization is observed only if the fluctuation in the TO frequency is larger than the dispersion of the TO mode – the TO dispersion in the pure AC crystal providing a natural reference. This is known as the Anderson's criterion [15].

2.1.3.2.3 **(TO, LO) Raman Lineshape** According to the Hertz-dipole formalism, the intensity scattered by a unit dipole p at pulsation ω scales as $\omega^4\langle p^2\rangle$, where the brackets refer to a so-called fluctuation spectral density, that is, to averaging over a time period. As the Raman process between the light (visible) and the optical (IR) modes is indirect, mediated by peripheral electrons (visible), the Raman polarization is essentially nonlinear. In fact, $P_{NL} = dP_e$ is taken as the general bilinear form of the phonons (u) and electrons (y) displacements at the two different pulsations, that is, as a linear combination of products (uy_v, yy_v, yu_v, and uu_v) where subscript "v" refers to "visible" and the absence of subscript refers to "IR." Since the lattice response is entirely negligible in the visible (quasistatic/Born–Oppenheimer approximation), that is, $u_v = 0$, the last two terms are just omitted. Now, we consider Equation 2.10 that establishes a generic proportionality between the electron displacement and a macroscopic electric field in the crystal. This is valid both in the IR (y) and in the visible (y_v) ranges. Only, in the second case, the macroscopic field E_v of the visible radiation should substitute for the macroscopic field E of the phonon (the LO one). On this basis, P_{NL} is expressed as $d\chi E_v = (d_u u + d_E E)E_v$, where d_u and d_E are constants (when using standard excitations, i.e., nonresonant ones) that identify with the partial derivatives of χ with respect to u and E, that is, $d_u \approx \frac{\partial\chi}{\partial u}$ and $dE = \frac{\partial\chi}{\partial E}$ (we recall that u and E are differential parameters in essence, in that they refer to a relative displacement of two sublattices and to the macroscopic field that proceeds from such relative displacement). It is explicit then that the Raman scattering by a LO mode is due to modulation of χ via both the atom displacement u ("deformation potential" mechanism) and the related macroscopic field E ("electro-optic" effect). For a TO mode, E is turned off, that is, the Raman scattering occurs via the "deformation potential" mechanism only. Here, we treat the general case of a LO mode, and indicate how the resulting Raman cross section modifies for a TO mode. For a Stokes process, the generic LO Raman signal as derived via $\langle P_{NL}^2\rangle$ takes the form

$$I_{\omega_s=\omega_i-\Omega} \sim \omega_s^4 \cdot \left\{ \langle E^2\rangle + 2\frac{d_u}{d_E}\langle uE\rangle + \left(\frac{d_u}{d_E}\right)^2 \langle u^2\rangle \right\}. \tag{2.16}$$

2.1.3.2.4 **Fluctuation–Dissipation Theorem** $\langle E^2\rangle$, $\langle u.E\rangle$, and $\langle u^2\rangle$ are evaluated via the fluctuation–dissipation theorem. This relates $\langle R^2\rangle$ at a pulsation ω, referring to the "response" (R) of the crystal to an external "stimulus" (S_t), as the dissipative part of the linear response function L ($L = R/S_t$) according to $\langle R^2\rangle = \mathrm{Im}(L)(\hbar/2\pi)[n(\omega)+1]$, where $n(\omega)$, the Boltzmann factor, accounts for the temperature dependence. In the case of multiple responses/stimuli, a general formulation is $\langle R_i R_j\rangle = \mathrm{Im}(L_{ij})(\hbar/2\pi)[n(\omega)+1]$, where $L_{ij} = R_i/S_{tj}$ is the crossed linear response function. Our set of (mechanical and electrical) equations being linear, the formalism applies. The L-matrix is referred to as the generalized Nyquist susceptibility.

For doing so, a proper external stimulus vector $S_t = (P_{ext}, F, f)$ as defined per crystal unit volume is associated with the overall response vector $R = (E, u, y)$. Note that S_t and R may consist of concrete as well as abstract physical quantities, putting forward, in the second case, that one cannot go against the principle of causality. In fact, (F, f) and P_{ext} represent pseudoexternal mechanical forces and a pseudolongitudinal (E exists only for LO modes) plane wave of polarization density to be added in the original mechanical and electrical equations, respectively. It is just a matter to add F/N and f/n as rightmost terms of Equations 2.5 and 2.6, where n and N are the number of electronic and ionic oscillators per crystal unit volume, respectively, and to reformulate Equation 2.3 by adding P_{ext} to the polarization term (by definition, P_{ext} is already defined per crystal unit volume).

In fact, $\langle E^2\rangle$ and $\langle u^2\rangle$ basically relate to the reactualized forms of Equations 2.3 and 2.9, respectively, and $\langle uE\rangle$ to a mixture of both. While Equation 2.3 gives ε_r, that expresses via the natural observables (ε_∞, ω_L, and ω_T), Equation 2.9 in its present form is not that easy to handle, due to N and $(Ze)^*$ that are not directly accessible. So, a novel parameter Z that relates to the oscillator strength S via $S = Z^2\omega_T^{-2}\varepsilon_0^{-1}$ is introduced in substitution to $(Ze)^*$ by writing the compatibility with Equation 2.13, which comes to $Z = \sqrt{N/\mu}(Ze)^*$. The subsequent change of variable $v = \sqrt{N\mu}u$ allows to generate a counterpart to Equation 2.9 with fully determined parameters. Care must be taken that Z involves N and as such is defined per crystal unit volume (not per oscillator, as in Equation 2.9). Thus, the generalized force constant F is not divided by N when added to the (v, Z)-based version of Equation 2.9. Eventually, the reactualized set of electrical/mechanical equations for a LO mode ($\varepsilon_r = 0$) is written as

$$\begin{pmatrix} P_{ext} \\ f \\ F \end{pmatrix} = \begin{pmatrix} -\varepsilon_0 & -ne^* & -Z \\ -ne^* & n^2 e^{*2}\varepsilon_0^{-1}\chi_\infty^{-1} & 0 \\ -Z & 0 & G(\omega) \end{pmatrix} \begin{pmatrix} E \\ y \\ v \end{pmatrix} \tag{2.17}$$

where $G(\omega) = (-\omega^2 + \omega_T^2 - j\gamma\omega)$ describes the damped resonance due to the ions in ε_r. Here, γ accounts for the finite linewidth of the Raman features. The L_{ij} terms are eventually obtained by taking the inverse of the above 3×3 matrix. By using the $u \rightarrow v$ change of variable, we can arrive at

$$\begin{aligned}(I\omega_s)_{LO,TO} \sim \omega_s^4 \cdot [n(\omega)+1] \cdot Im\{-(\varepsilon_0\varepsilon_r)^{-1}[1 + C_{F\text{-}H}\omega_T^2 G(\omega)^{-1}]^2 \\ + C_{F\text{-}H}^2 \omega_T^4 Z^{-2} G(\omega)^{-1}\}\end{aligned} \tag{2.18}$$

where $C_{F\text{-}H} = (N\mu)^{-1/2} Z\omega_T^{-2}(d_u/d_E)$ is the so-called Faust–Henry coefficient, that measures the relative efficiencies of the "deformation potential" and "electro-optic" scattering mechanisms [16]. Equation 2.18 provides directly the LO Raman lineshape of a polar semiconductor compound on the basis of the restricted set of intangible observables (ε_∞, ω_T, ω_L, and C_{F-H}), plus the phonon damping γ. For the LO mode ($\varepsilon_r = 0$), the first term clearly plays a major role – in fact, preliminary insight into the LO mode can be obtained via $Im(\varepsilon_r^{-1})$. In contrast, it turns negligible for the TO mode ($\varepsilon_r = q^2c^2/\omega^2$) as for usual backscattering-like geometries $q \gg \sqrt{\omega_T/c}$, as already mentioned. For practical use the Raman cross section of the *TO* mode thus reduces to the imaginary part of the second term in Equation 2.18. Alternative insight into the TO mode can be obtained by realizing that the TO resonance is fixed by $\langle v^2 \rangle$ only, which is expressed as $G^{-1} - \varepsilon_0\varepsilon_r{}^{-1}Z^2G^{-2}$, and again the second term vanishes due to the above considerations on ε_r. Thus, in fact, a TO insight is given by $Im(G^{-1})$, that scales as $Im(\varepsilon_r)$, corresponding *in fine* to the following Lorentzian with linewidth at half height ($\gamma/2$) centered at the pole of ε_r

$$(I_{\omega_s})_{TO} \sim \left[(\omega - \omega_T)^2 + \left(\frac{\gamma}{2}\right)^2\right]^{-1}. \tag{2.19}$$

For example, the TO and LO Raman lineshapes of GaP – as calculated from Equation 2.18 – are shown in Figure 2.3, right panel.

2.1.3.3 **Brillouin Scattering: An Indirect Light/Acoustical-Mode Interaction**

From a qualitative point of view (see a more formal classical description in Section 2.1.5.3), propagation of a sound (acoustical) wave in a medium produces a periodical density fluctuation that diffracts visible light in the same way X-rays are diffracted by the atomic planes in a crystal. A difference, here, is that the modulation propagates with the sound velocity, hence producing a Doppler effect on the diffracted light. With this, the energy difference between the incident and the diffracted lights is the energy of the acoustic mode. In fact, the basic principles of light/matter interaction are similar for the Brillouin and the Raman scatterings. Only in the former case, the visible radiation is scattered by an acoustical phonon. So, the Brillouin signal takes a Lorentzian lineshape as typically accounted for by Equation 2.19, but where ω_T is replaced by the pulsation of the acoustical phonon, denoted by ω_{ac}. In particular, the energy and momentum conservations are identical for the Raman and the Brillouin scatterings, so that the latter takes place at $q \sim 0$ also. At this limit, the dispersion of an acoustical mode is quasilinear, that is, of the type $\omega_{ac} = q v_{ac}$, where v_{ac} refers to the velocity of the acoustical wave. A substitution into the energy and momentum conservations, making the reasonable approximations $k_s \sim k_i$ and $n_s \sim n_i$ where n is the refractive index of the crystal and subscripts s and i refer to the scattered and incident visible radiations, leads to [17]

$$\omega_{ac} \sim 2 n_i \omega_i v_{ac} c^{-1} \sin \frac{\theta}{2} \tag{2.20}$$

where θ is the angle between $\vec{k}_s$ and $\vec{k}_i$ inside the crystal. Dealing with different θ values allows exploring part of the acoustical dispersion curve: the Brillouin shift (ω_{ac}) is maximum in the backscattering geometry ($\theta \sim 180°$) and reduces to zero in the forward geometry ($\theta \sim 0°$). Also, Equation 2.20 may be used to determine n_i or v_{ac}, where v_{ac} relates directly to the elastic constants of the crystal, via the density ϱ, as

$$c_{11} = \varrho v^2_{ac,[100,LA]}; \quad C_{44} = \varrho v^2_{ac,[100,TA]}; \quad C_{12} = C_{11} + C_{44} - \varrho v^2_{ac,[100,TA]}. \tag{2.21}$$

From an experimental point of view, because of the proximity of the Brillouin lines with the elastically scattered Rayleigh line, high-resolution apparatus should be used. Such a high-resolution apparatus is the Fabry–Pérot (FP) interferometer. Modern FP interferometers were introduced mostly by J.R. Sandercock [18] with the tandem triple pass interferometer (see Section 2.2.2.4, for detail).

2.1.4
Some Particular Cases

2.1.4.1 Multioscillator System

Equations 2.15 and 2.18 remain basically valid for multioscillator systems, such as alloys (detail is given in Section 2.3.2.1). Then, ε_r builds up by summing over the whole series of p ionic oscillators in the crystal,

$$\varepsilon_r = \varepsilon_\infty + \sum\nolimits_p S_p \omega^2_{T,p} G(\omega)^{-1}_p \tag{2.22}$$

where ε_∞ is the sum of the related values in the parent crystals weighted by the corresponding fractions of oscillators. A similar summation applies to the C_{F-H} terms in Equation 2.18, the constituent parameters being all p-dependent and thereby rewritten with subscript p. The full expression as derived on the general basis of mechanically coupled oscillators is [19]

$$(I_{\omega_s})_{\mathrm{LO,TO}} \sim Im\Big\{-(\varepsilon_0\varepsilon_r)^{-1}\Big[1+\sum\nolimits_p C_{\mathrm{F\text{-}H},p}\omega_{\mathrm{T},p}^2 G(\omega)_p^{-1}\Big]^2 + \sum\nolimits_p C_{\mathrm{F\text{-}H},p}^2\omega_{\mathrm{T},p}^2 Z_p^{-2} G(\omega)_p^{-1}\Big\} \tag{2.23}$$

if we omit the Boltzmann and ω_s^4 factors. A similar expression was earlier derived by Mintairov *et al.* for mechanically independent oscillators [20]. A basic rule remains valid that both S_p – that relates to Z_p^2, and $C_{\mathrm{F-H},p}$, scale as the fraction of oscillator p in the crystal. By writing that ε_r turns equal to zero at each LO frequency $\omega_{\mathrm{L},p}$ [cf. Equation 2.4], it becomes

$$S_p = \varepsilon_\infty \frac{\omega_{\mathrm{L},p}^2-\omega_{\mathrm{T},p}^2}{\omega_{\mathrm{T},p}^2}\prod\nolimits_{m\neq p}\frac{\omega_{\mathrm{L},m}^2-\omega_{\mathrm{T},p}^2}{\omega_{\mathrm{T},m}^2-\omega_{\mathrm{T},p}^2}. \tag{2.24}$$

In principle, N_p should follow from direct comparison with the oscillator strength in the p-type pure crystal. In fact, the procedure is rather hazardous as close LO modes couple via their macroscopic field E, resulting in a strong distortion of the original/uncoupled *LO* signals. Therefore, the observed $\omega_{\mathrm{L},p}$ frequencies may just be not reliable [19].

2.1.4.2 **Multilayer System [12]**

Now, we treat briefly multilayer systems, focusing on the case most commonly found in practice of a thin layer (thickness a) deposited on a semi-infinite substrate analyzed by optical spectroscopies at quasinormal incidence. The layer and the substrate are assumed to be homogenous and isotropic dielectrics.

The layer and the substrate provide separate Raman (Brillouin) signals as given by $I_{\mathrm{L}}^0(\mathrm{e}^{-2\alpha a}-1)$ and $I_{\mathrm{S}}^0\mathrm{e}^{-2\alpha a}$, where I_{L}^0 and I_{S}^0 refer to the signals from the raw layer taken as semi-infinite ($a \rightarrow \infty$) and the raw semi-infinite substrate, respectively. These expressions are obtained simply by taking into account that the signal from an arbitrary depth z of a given medium with Raman (resp. Brillouin) cross section I and absorption coefficient α at the considered wavelength (out of resonance conditions, no distinction is made in practice between the α values of the incident and scattered radiations) scales as $I\mathrm{e}^{-2\alpha z}$. The exponential term represents attenuation of the incident beam on its way forth down to depth z and of the scattered beam on its way back to the surface.

Regarding IR reflectivity, the question is how the light propagates in a multilayer stack? A basic law, as inferred from the Maxwell equations, is that the tangential components of the macroscopic $\vec{E}$ and $\vec{B}$ fields are preserved across the vacuum/layer and layer/substrate interfaces. This is expressed in terms of the individual fields related to the incident, reflected and transmitted waves, the coexisting fields on one side of a surface/interface just adding vectorially (linearity of the Maxwell equations). Each equation separates in two, one corresponding to phase conservation, and the other being related to amplitudes. The former involves the incident/reflection/refraction angles, and comes to Descartes laws. The latter reformulates by considering that the layer-related amplitudes at the vacuum/layer and layer/substrate interfaces just differ because of a phase difference $\delta = k_0 n_L a$ that develops at the crossing of the layer, where n_L is the refractive index of the layer and k_0 the wavenumber in vacuum. Then, a correlation can be established between the net amplitudes of the macroscopic fields $\vec{E}$ and $\vec{B}$ at the two interfaces via a 2×2 so-called *transfer matrix*, with components $m_{11} = \cos\delta$, $m_{12} = m_{21} = j(c/n_L)\sin\delta$, and $m_{44} = \cos\delta$. In a next stage, the macroscopic fields are expressed in terms of the incident (E_0), reflected (E_r), and transmitted (E_t) fields related to the whole system. The reflection coefficient related to the amplitude, that is, $r = E_r/E_0$, is then written as [21]

$$r = \frac{n_L(n_0 - n_S)\cos\delta + j(n_0 n_S - n_L^2)\sin\delta}{n_L(n_0 + n_S)\cos\delta + j(n_0 n_S + n_L^2)\sin\delta} \tag{2.25}$$

where n_S is the refractive index of the substrate, that is, the square root of the related relative dielectric function – including a damping term. The IR reflectivity coefficient is eventually inferred via $R = |r|^2$. A similar expression can be derived for the transmission coefficient. The procedure generalizes to multilayers, the transfer matrix of the entire multilayer stack being the product of the individual transfer matrices from each layer.

2.1.4.3 **Multicomponent System (Composite) [22]**

The simplest composite consists of k monoatomic components randomly dispersed as spherical inclusions with uniform size (radius R_k). The crucial issue, here, is how to calculate an effective dielectric function for the whole system, for subsequent injection into the IR reflectivity coefficient, and also, in principle, into the Raman cross section. The concept of effective dielectric function is valid provided macroscopic fields can be defined – in the sense of the Maxwell equations, which requires that both R_k and the correlation length ξ_k per component (average spacing between type-k inclusions) are much smaller than the optical wavelength λ. An additional constraint, for practical use, is that R_k is large enough that electronic confinement effects can be neglected, in which case the individual dielectric constant ε_k associated with each component can be approximated to the bulk value.

In each pure crystal, the dielectric function $\varepsilon_k = (\varepsilon_0\varepsilon_r)_k$ (a macroscopic parameter) may relate to the atom polarizability α_k (a microscopic parameter) by combining Equation 2.8 – a simplified form for monoatomic crystals, that is, $P_k = N_k \alpha_k E_e$ – with Equations 2.7 and 2.3, leading to the so-called Clausius–Mossotti formula

$$\frac{\varepsilon_k - \varepsilon_0}{\varepsilon_k + 2\varepsilon_0} = \frac{1}{3\varepsilon_0} N_k \alpha_k. \tag{2.26}$$

This establishes a relation between the *microscopic* (α_k) and the *mesoscopic* (ε_k) scales in the composite. By analogy, the effective dielectric function ε_e of the whole composite (*macroscopic* scale) is written as

$$\frac{\varepsilon_e - \varepsilon_0}{\varepsilon_e + 2\varepsilon_0} = \frac{1}{3\varepsilon_0} \sum_k X_k n_k \alpha_k \tag{2.27}$$

where X_k is the number of type-k inclusions per crystal unit volume, and n_k is the number of type-k atoms per inclusion. By using Equation 2.26 and realizing that

$X_k n_k / N_k$ corresponds to the fraction p_k of component k in the crystal – assuming that the density of atoms is the same in all k pure materials, it becomes

$$\frac{\varepsilon_e - \varepsilon_0}{\varepsilon_e - 2\varepsilon_0} = \frac{1}{3\varepsilon_0} \sum_k \frac{\varepsilon_k - \varepsilon_0}{\varepsilon_k + 2\varepsilon_0} \tag{2.28}$$

that provides the searched relation between the *mesoscopic* (ε_k) and the *macroscopic* (ε_e) scales. If we consider a two-component (a, b) system, two limit cases can be distinguished depending on the relative fractions (p_a, $p_b = 1 - p_a$) of the constituents. When one component (a) is so diluted that each type-a inclusion can be viewed as isolated in the host matrix formed by the other component (b), that is, when $p_a \ll p_b$, Maxwell Garnett [23] proposes to derive a univocal expression of ε_e just by replacing ε_0 (where subscript "0" refers to vacuum, i.e., the "host" medium within the microscopic-to-mesosocopic description of the a- and b-type pure crystals) by ε_b (the host medium is crystal "b" within the mesoscopic-to-macroscopic description of the composite) in Equation 2.28. In contrast, when the two components are in similar proportion in the crystal, that is, when $p_a \sim p_b$, Bruggeman [24] comes to a vision where all inclusions of any type, that is, a- or b-type, are immersed into the same (a,b)-like effective medium. In this case, the latter plays the role of the host medium which comes to replace ε_0 by ε_e in Equation 2.28, ending up in a fully symmetrical expression with respect to species a and b.

2.1.5
Selection Rules [5, 17]

The phonons involved in the IR reflectivity, Raman, and Brillouin processes are identified from symmetry considerations. The restriction to $q \sim 0$ brings in a simplification that the atom displacements are invariant under a lattice translation, so that only the point group should be considered (T_d for zinc blende crystals). The phonon symmetry is derived by relating the species of the point group to the species of the site group per atom that have the symmetry of an atom displacement, that is, of a translation vector (T_x, T_y, T_z). These are expressed in terms of the irreducible representations of the point group as determined from group theory, which comes eventually to symmetry F_2, with three dimension, for both acoustical and optical modes in zinc blende structure. Now, as phonon properties are invariant under symmetry operations of a crystal, the light–phonon interaction should remain unchanged also. Thus, the first step is to symmetrize the relevant tensor for the light-matter process under consideration in terms of the irreducible representations of the crystal. The active phonons for a given process are those with symmetries that belong to the development of the related tensor.

2.1.5.1 **Raman Scattering**

Only optical modes are involved in the Raman scattering. Now, on top of the basic selection rules imposed by the symmetry, one should consider setup-dependent ones related to the polarizations of the incident ($\vec{e}_\mathrm{i}$) and scattered radiations ($\vec{e}_\mathrm{s}$). In fact, in a classical approach, the Raman cross section of a *TO* mode is expressed as

$(I_{\omega_s})_{TO} \sim \left| \vec{e}_i \; \vec{\vec{R}}_\nu \; \vec{e}_s \right|^2$, where $\vec{\vec{R}}_\nu$, the so-called Raman tensor, refers to the polarization ν of the phonon mode. $\vec{\vec{R}}_\nu$ is a contraction of u (first-rank tensor) and $d_u \sim \frac{\partial \chi}{\partial u}$ (third-rank tensor) in the Raman polarization, and therefore a second-rank tensor as represented by a 3×3 matrix. The group theory indicates that a third-rank tensor has only one nonzero component in the zinc blende symmetry, denoted by d hereafter, corresponding to indices xyz and cyclic permutations. Thus for a phonon polarized along $x = [0\,0\,1]$, d appears in the yz and zy positions of the related Raman tensor, denoted by $\vec{\vec{R}}_x$, the remaining terms being nul. Similar considerations apply to $\vec{\vec{R}}_y$ and $\vec{\vec{R}}_z$. So, in the system of crystallographic axes x, y, and z, the Raman tensors take the form

$$\vec{\vec{R}}_x = \begin{pmatrix} 0 & 0 & 0 \\ 0 & 0 & d \\ 0 & d & 0 \end{pmatrix}; \quad \vec{\vec{R}}_y \begin{pmatrix} 0 & 0 & d \\ 0 & 0 & 0 \\ d & 0 & 0 \end{pmatrix}; \quad \vec{\vec{R}}_z \begin{pmatrix} 0 & d & 0 \\ d & 0 & 0 \\ 0 & 0 & 0 \end{pmatrix} \tag{2.29}$$

for the zinc blende structure. Regarding the symmetry, $\frac{\partial \chi}{\partial u}$ and $\frac{\partial \chi}{\partial E}$ are equivalent since both u and E are vectors (first-rank tensors). So, the Raman tensors are the same for TO and LO modes. Only, the d value is different. For example, in a zinc blende crystal, backscattering onto the (0 0 1), (1 1 0), and (1 1 1) crystal faces correspond to (TO forbidden, LO allowed), (TO allowed, LO forbidden) and (TO allowed, LO allowed) unpolarized configurations, respectively.

2.1.5.2 IR Absorption

Again, this is concerned with optical modes only. The relevant tensor, here, refers to the electric dipolar moment (in a classical/molecular approach, the incident light induces oscillating electric dipoles in the crystal, which radiate the reflected and transmitted beams), that is, a vector (first-rank tensor). The irreducible representation of such a tensor is of the F_2 type, so that the optical mode of the zinc blende structure is IR active. A general rule is that *in crystals with a center of inversion, the Raman active modes are IR inactive, and reciprocally.*

2.1.5.3 Brillouin Scattering

This is concerned with acoustical modes. To introduce the Brillouin tensor, $\vec{\vec{B}}$, we should be more explicit than that discussed in Section 2.1.3.3 on the nature of the interaction between the incident light (visible) and the phonon (IR). Again, the interaction is mediated by peripheral electrons. Basically, the fluctuations in strain caused by the acoustical wave generate local χ -fluctuations ($d\chi$), which in turn are responsible for light scattering (according to the Hertz mechanism). This is at the point that the difference between the Raman and Brillouin scatterings becomes clear. Optical modes as detected by Raman scattering are not specially q-sensitive (cf. Figure 2.1), so that a basic formalism at the limit $q \to 0$, as summarized in Sections 2.1.2 and 2.1.3, is sufficient. In contrast, some $\vec{q}$-dependence should be there for Brillouin scattering because a rigid translation of the whole crystal (an acoustical mode at $q = 0$) cannot produce any local χ -fluctuation. With this respect, a general expression of $\vec{\vec{B}}$ as a contraction of e, the strain induced by the acoustical phonon (second-rank tensor), and $\frac{\partial\chi}{\partial e}$ (fourth-rank tensor) does not seem sufficiently explicit. The tradition is to express $\vec{\vec{B}}$ in relation to the strain-induced variation in ε (where $\varepsilon = \varepsilon_0\varepsilon_r$) – rather than in χ, as given by the photoelastic effect at $q \sim 0$. The distortion of the electronic dielectric constant – a tensor – is then written as $\overrightarrow{\overrightarrow{d\varepsilon}} = qu\ \vec{\vec{T}}$, where u is the local atom displacement and the elements of tensor $\vec{\vec{T}}$ are of the form $\varepsilon_{\alpha\alpha}\varepsilon_{\beta\beta}$ time $\frac{B_{\alpha\beta}}{qu}$, $\vec{\vec{B}}$ being the product of ε^{-1} by the Pockel's tensor. Eventually, the Brillouin selection rules are fixed by the product $\left|\vec{e}_i\ \vec{\vec{T}}_\nu\ \vec{\vec{e}}_s\right|^2$ – with conventional notation. The calculations were performed in several high-symmetry directions for all the symmetry groups as given in Ref. [25].

2.1.6 When Departing from Pure Crystals . . .

The formalism developed above is rigorous for crystals. A large class of materials, such as glasses, remain out of this formalism. While a rigorous description of the dynamics and the vibrational properties of glassy materials is still lacking today, nevertheless, it is possible to extend the previous description in an approximate way.

The dynamics of a perfect crystal is represented by the phonon dispersion curves. The disorder inherent to glasses will act in two manners on the phonon dispersion curves. First, it induces distributions of bond lengths and angles, which results in broadening of the dispersion curves. The second consequence is the loss of translational symmetry, and hence of the notion of Brillouin zone. Basically, all dispersion curves collapse in their common projection on the frequency axis, that is, they merge into some density of states. This is true in particular for optical modes that address directly a local physical property, that is, the bond force constant, and as such are highly sensitive to the local atom arrangement. In fact, the loss of translational symmetry shows itself via partial release of the Raman and infrared $q \sim 0$ selection rules. Another effect is that the IR and Raman linewidths are generally broader in glasses than in crystals. An example is given in Figure 2.4. Brillouin modes are much less affected. As they correspond to vibrations with long characteristic wavelength, that is, of the crystal taken as a whole, they are much less sensitive to disorder and most often the Brillouin lines are comparable in a crystal and in the glass of same composition.

2.2 Optical Vibrational Spectroscopies Under Extreme Conditions

2.2.1 A Specific Impact/Identity in the Field

The field of high pressure is extremely multidisciplinary, as may be noticed on the Web site of both the international organizations for high pressure (http://www.ct.

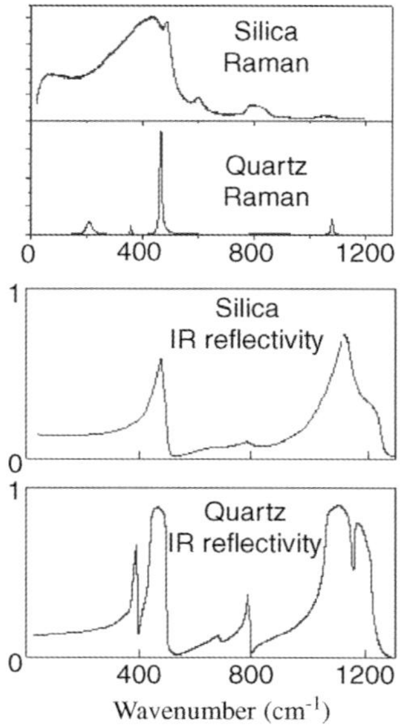

Figure 2.4 Comparison of the Raman responses (top) and the IR reflectivity (bottom) of glassy SiO_2 (silica) and crystalline SiO_2 (α-quartz).

infn.it/airapt/, http://www.ehprg.org/). All the scientific areas of "hard" sciences are covered, but Mathematics, and the industrial applications is an ever-developing field, from the synthesis of new materials, the hot pressing, and the food processing. In order to make progress, scientists of many fields have to collaborate into any project. The high-temperature field may be opened to less fundamental domains, but deals also with numerous industrial problematics such as glasses, ceramics, refractory materials, thermal barriers, cements, nuclear industry, which are some of the main examples. And thermodynamics is clearly connected to both parameters, temperature and pressure.

In the following, we take specific examples in various scientific fields to illustrate the power of optical spectroscopies under high pressure and temperature.

2.2.1.1 **Solid-State Physics**

The main effect of high pressure is to modify the interatomic distance and, thereby the intensity and the hierarchy of the different interactions involved in interatomic bonding. The modification of interatomic distances and of crystals lattice parameters directly affects the Brillouin zone dimensions and the electronic properties. The electronic bandwidth increases under pressure and the relative position of the various bands or localized levels changes. This can produce insulator–metal transition through a gap closing or a Mott–Hubbard process. The magnetic properties can be modified through electronic transfer from one band to another one (spin transition, dismutation). Dimerization processes inducing a gap opening (Peierls transition) leads to the existence of two different bond lengths with different compressibilities. The pressure acts against this distortion, the largest distance being more compressible than the shortest one, and when the distances are almost identical, the dimerized state (insulating) becomes unstable with respect to the symmetrical one (metallic).

We present here the example of the discovery of a nonmolecular phase of nitrogen. Nitrogen is known to be a very stable molecular diatomic, because of the extremely strong triple bond. An advantage can be taken from the strength of the molecular bond: due to the difference of energy between 1/3 triple and the single bond, it would enable to stock a large energy density by transforming the molecular materials into a polymeric network – that could be utilized at the back-transformation to the molecular state. Calculations [26] predicted a pressure-induced molecular solid to single-bonded atomic crystal, whose structure was predicted to be cubic *gauche* (cg-N) [27]. M.I. Eremets *et al.* [28, 29] and M.J. Lipp *et al.* [30] have studied nitrogen as a function of pressure at high temperature. The characterization techniques were the Raman scattering and X-ray diffraction. They proved that at 110 GPa and ~2000 K, solid N_2 transforms indeed to a new crystalline form. The Raman scattering spectra at 110 GPa and various temperatures are shown in Figure 2.5. The spectrum at 1990 K is the first one ever measured in pure N_2 where the vibron of the molecule is not observed. It is replaced by a phonon approximately 840 cm^{-1}. X-ray diffraction spectra were obtained in the new polymeric structure, and Rietveld refinements were performed. This procedure led to a cg-N structure (space group $I2_13$) with $a_0 = 3.4542 \pm 0.0009$ Å.

In this structure, shown in Figure 2.6, all nitrogen atoms have a coordination number of three, and all the bonds are identical. The cg-N structure was stable in the downstroke down to 42 GPa. This enabled to fit the volume in the 42–134 GPa range with a Birch–Murnaghan equation of states with the following parameters: $B_0 = 298$ GPa $\pm$ 0.6 GPa (or $B_0 = 301.0 \pm 0.9$ GPa for Vinet equation of state), with $B' = 4$, fixed, and $V_0 = 6.592$ Å [29]. cg-N is hence stiff solid with a bulk modulus value characteristic of strong covalent solids.

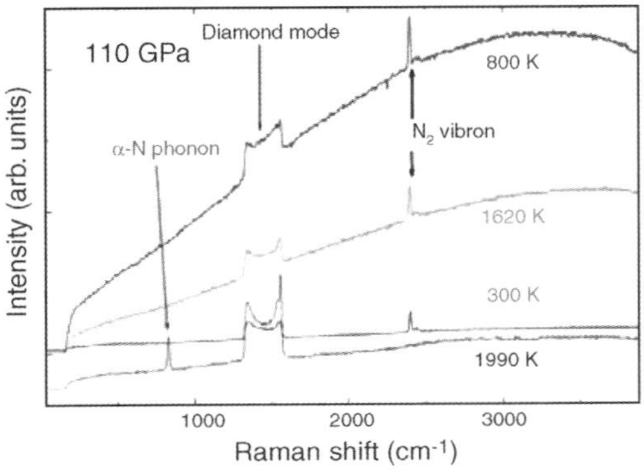

Figure 2.5 The Raman spectra of nitrogen at 110 GPa at several temperatures. Reprinted with permission from Ref. [29]. Copyright 2010, American Institute of Physics.

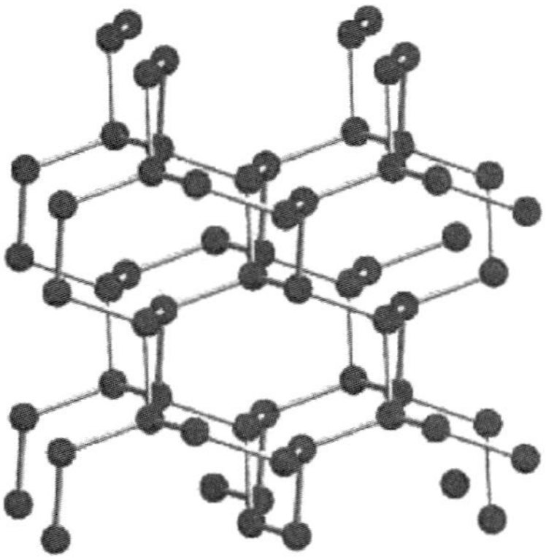

Figure 2.6 Cubic gauche structure of the polymeric nitrogen. Reprinted by permission from Macmillan Publishers Ltd: Nature Materials, Ref. [28], copyright 2010.

2.2.1.2 **Earth Sciences**

It is not necessary to demonstrate the importance of SiO_2 not only in everyday life, but also in geosciences. It is a major constituent of the Earth mantle and its high-pressure/-temperature behavior is therefore fundamental in many aspects. It exists under many different crystalline and amorphous phases.

Here, we present results on an isostructural crystal, the berlinite $AlPO_4$, whose phase diagram is similar to that of SiO_2. This compound is obtained by replacing the silicon atoms in the quartz structure alternatively by phosphorus and aluminium, resulting in the doubling of the crystallographic unit cell. The number of studies increased a lot after the publication of a "memory glass" property [31], that is, at 14 GPa at ambient temperature, this compound is subjected to a structural transformation to an amorphous phase; at the downstroke, the sample recovered its initial crystalline structure, with the same orientation. Amorphization was detected by energy dispersive X-ray diffraction on a powder, whereas the reversibility and recovering of the orientations was detected by birefringence measurements on a single crystalline sample, with measurements at ambient conditions, before and after a pressure run to 15 GPa. It was proved by Brillouin scattering under pressure [32] that the transition was indeed reversible with the same crystalline orientation before and after pressurization, and moreover that the high-pressure phase was elastically anisotropic. The first proof that the high-pressure phase was not amorphous was given by a Raman study [33] where it was shown that above 14 GPa, there were narrow Raman lines, footprint of a crystallinity. Representative spectra are shown in Figure 2.7. The issue was definitely solved by X-ray diffraction measurements [34, 35] when the structure of the first and a second high-pressure phases were identified. The first high-pressure phase corresponds to a coordination number of 6 around the aluminium atoms, and four around the phosphorus, whereas the second one, starting around 47 GPa corresponds to the increase of coordination number from 4 to 6 around the phosphorus.

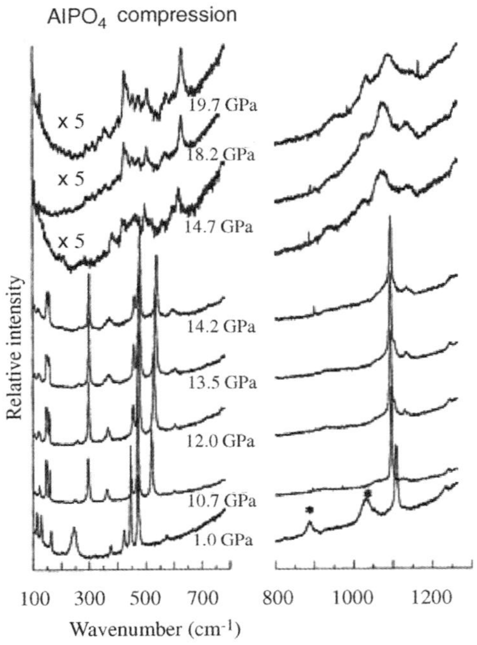

Figure 2.7 The Raman spectra taken during compression. Reprinted figure with permission from Ref. [33]. Copyright 2010 by the American Physical Society.

2.2.2 Specificities and Instrumentation for High-Temperature and High-Pressure Investigations

2.2.2.1 Temperature and Emissivity

Performing optical spectroscopy (infrared, Raman, and Brillouin) in the high-temperature range is dominated by the following main characteristic: the thermal emission of the sample and of its close environment (sample holder, heating device, and so on). These spectroscopies act in the visible or infrared spectral range, which is just the domain where thermal emission stands, for temperatures between room temperature and 2000–3000 K. To give some idea of the importance of thermal emission in our problem, one can compare the typical amplitude of Raman or Brillouin scatterings, which is too weak to be detected by the eye, and the thermal emission of a sample heated at 2000 °C, which needs specific precautions to avoid dazzling, as dark goggles or welder masks. For infrared transmission or reflection, even if this simple comparison does not stand, the ratio of intensities of the required signal versus thermal emission is comparable. In fact, as we will see hereafter, this ratio of intensities can easily reach several orders of magnitude. Therefore, to have a better understanding of the specificities of high temperature for vibrational optical spectroscopy, it is essential in a first step to recall the basic features of thermal emission. A better understanding of its physical concepts will allow to find solutions for performing IR, Raman, and Brillouin spectroscopies up to high temperatures.

The thermal radiation characteristics of a material are given by its spectral emissivity, as introduced in Section 2.1.3.1. The luminance $L(\omega, T)$ emitted by a material as thermal radiation is in fact the product of two terms: luminance of blackbody radiation, that is, Planck's law, which fixes dependence on frequency and temperature; multiplied by spectral emissivity, which reflects the properties of the material itself.

$$L(\omega, T) = E_m(\omega, T)L_0(\omega, T) = E_m(\omega, T)C_1\omega^3\left[e^{\frac{C_2\omega}{T}}-1\right]^{-1} \quad (2.30)$$

with $C_1 = 2h.c^2 = 1.191 \times 10^{-16}\,\mathrm{W\,m^2sr^{-1}}$ and $C_2 = 1.439 \times 10^{-2}\,\mathrm{mK}$. The maximum of Planck's law L_0 stands in the mid-infrared at room temperature and progressively shifts to high frequencies ($4000\,\mathrm{cm^{-1}}$ at $2000\,^\circ$C), to arrive close to the visible range at $6000\,^\circ$C.

This equation, together with Kirchoff's law, reflects all the consequences of temperature on optical vibrational spectroscopy. To give some quantitative idea of the effect of Planck's law in the temperature range of interest, typically 1000–2000 °C, the luminance close to 500 nm (wavelength of the current Raman and Brillouin lasers) increases by about one order of magnitude every 300 °C. The optical conditions for a sample reaching 2000 °C are much more severe than at 1000 °C.

One can consider the consequences of thermal radiation on each of the three spectroscopies, IR, Raman, and Brillouin scatterings. For the Raman and Brillouin scatterings, thermal radiation acts as a background term superimposed on the scattered light. It is the same situation for infrared reflectivity and transmittivity, and all derivatives (diffuse or specular reflection). The experimental routes to extract the vibrational information from thermal background will be rather comparable and mainly defined according to the specificities of each method. Brillouin scattering is the favorable case (contrary to room temperature where Brillouin scattering is a much less routine technique than the Raman or the infrared techniques) due to the high contrast between the narrow sharp Brillouin lines and the thermal background. Besides this, the Raman and the infrared need specific noticeable adaptations of the optical spectrometer to be performed up to 2000 °C. Infrared spectroscopy is a peculiar case: above some temperature threshold (300–1000 °C, depending of the spectral range in infrared: the lowest the frequency, the lowest the temperature), an elegant way to solve the problem of thermal emission is to measure it directly, without lightening by a source as for reflection or transmission. As previously exposed, emissivity is directly connected to transmittivity or reflectivity, and the vibrational information can be extracted from emissivity measurements. Thermal emission is not a parasitic signal but the signal to be recorded, and, as it is stronger and stronger upon heating, one can think of easier and easier measurements! But reality is not that simple. All these aspects will be detailed hereafter in the corresponding subsections.

First, let us emphasize that the first way to limit difficulties due to thermal emission on optical vibrational spectroscopies (this does not concern the emissivity spectroscopy, except at low frequencies – far-IR – due to room temperature thermal emission) are to limit the effect of the first term $E_m(\omega, T)$ in Equation 2.30 no real material acts as a blackbody, the spectral emissivity is always smaller than one. Then, all means to limit effective absorption in the sample can be used: chemical purity, homogeneity, and quality of the polishing. If there is no source of absorption in the visible range, no emission will occur in this range! For example, silica of high purity heated near its T_g (1200 °C) emits negligible intensity in the visible range, whereas another one containing 3d ion impurities will emit a strong orange–yellow light. Improving optical quality of the surface (polishing) and of volume (minimizing texture effects) is important too: it is a way to limit effective absorption cross section through multiple reflections, and, for materials heated in a furnace, it is also a way to limit diffuse scattering of light emitted by the furnace itself. Of course, these ideas cannot be employed in many cases. However, when realistic, they must be tried, due to their efficiency, as they directly eliminate a large part of the problem at its source.

2.2.2.2 **High-Pressure Optical Cells, Diamond–Anvil Cells**

To perform optical measurements under high pressure, there are typically two kinds of apparatuses, the "hydrostatic pressure" cells [36], and the diamond–anvil cells [36, 37]. The cells of first kind are normally large volume cells, where a fluid or a gas under high pressure is led into the experimental volume through a capillary. The windows, sealed under several techniques, are in sapphire. The maximum pressure reached in such a cell is approximately 1.5 GPa. However, the versatile setup is the diamond– anvil cell (DAC), because of many advantages. First of all, diamond is transparent in a broad wavelength range, it is the hardest natural material, and it has a very high thermal conductivity. With a DAC, pressure equivalent to that of the center of the Earth (360 GPa) can be reached!

The principle of the DAC is extremely simple (Figure 2.8): two anvils squeeze a metallic gasket. A hole drilled at the position of the center of the anvils forms the experimental volume. It contains the sample, a pressure-transmitting medium and a pressure gauge, generally a piece of ruby. The pressure-transmitting medium should maintain hydrostatic – or at least "quasihydrostatic" conditions in the broadest possible pressure range. 4 : 1 methanol–ethanol, or 16 : 3 : 1 methanol–ethanol–water mixture are well adapted up to 10–15 GPa. Above this pressure, neon or better helium are better suited, but present the disadvantage to need special high-pressure equipment to be loaded in the DAC.

The general shape of the anvils is that of jewellery brilliant where the tip is cut to form the culet. Standard −50 GPa- anvils are shown in Figure 2.8. Recently [38], a new type of anvil was proposed. Progress in diamond machining enabled to polish the sides of the stones as a cone, instead as a faceted polygone (Figure 2.9). This enables to have a large lateral support, leaving the complete table for the optical access.

Naturally, because by definition pressure is a force divided by a surface, the larger the pressure one wants to reach, the smaller the surface. To reach the 100 GPa range, beveled anvils have to be used, with a typical dimension of 20–100 μm for the culet, 100–300 μm for the bevel, and a bevel angle of 7–8° [39].

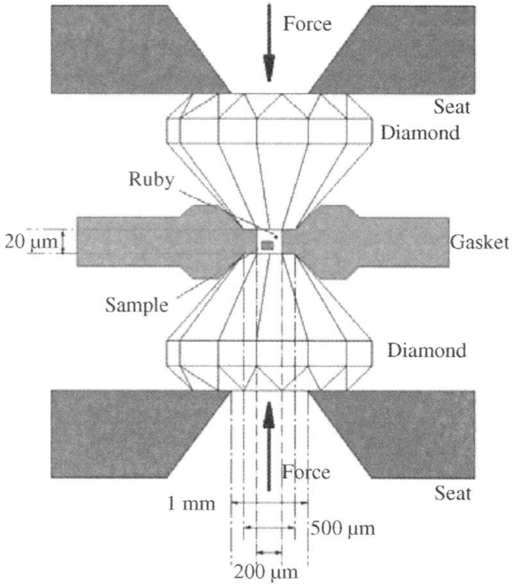

Figure 2.8 Typical size of the diamond anvil setup for "standard" anvils.

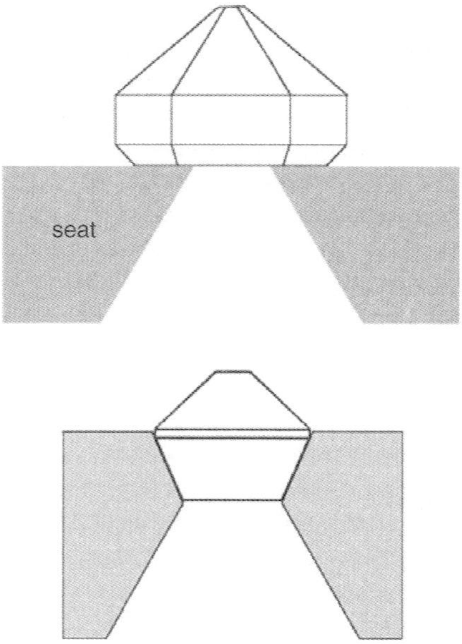

Figure 2.9 Schematic view of conventional (top) and conical "Boehler–Almax" (bottom) high-pressure anvils. Reprinted from Ref. [38] entitled 'New anvil designs in diamond cells', by permission of the publisher (Taylor & Francis Group, http://www.informaworld.com).

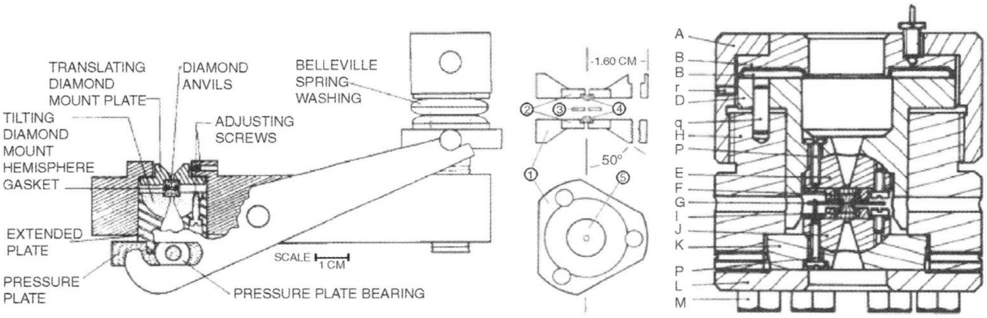

Figure 2.10 Various types of diamond–anvil cells. Left: Piermarini–Block lever arms DAC (Reprinted with permission from Ref. [40]. Copyright 2010, American Institute of Physics); Center: Merrill–Bassett miniature DAC (Reprinted with permission from Ref. [41]. Copyright 2010, American Institute of Physics); Right: LeToullec membrane DAC (5 cm in diameter) (Reprinted from Ref. [42] entitled 'The membrane diamond anvil cell: A new device for generating continuous pressure and temperature variations', by permission of the publisher (Taylor & Francis Group, http://www.informaworld.com).

The classification of the diamonds is important in view of the application one needs. Pure diamonds, that is, without impurities – type IIa, are requested for infrared measurements. For the Raman and Brillouin scatterings, the main problem comes from luminescence of the anvils. Therefore, they have to be selected, by comparing the intensity of the first-order (1330 cm^{-1}) and the second-order Raman scatterings with the luminescence around 800 cm^{-1}.

There are many types of DAC [36, 37], and some of them are presented in Figure 2.10, with the lever-arms DAC of Piermarini and Block [40], used for optical measurements, the miniature Merrill–Bassett [41] with beryllium seats mostly used for X-ray diffraction measurements, and the LeToullec DAC [42], the first driven by a deformable membrane. Figure 2.11 presents one of the first "pneumatic" DAC [43] in which the pressure is varied with nitrogen gas pressure acting on the piston. As can be seen from the various designs, the differences between them are mostly in the way the force is transmitted to the anvils, lever arms, and Belleville spring washers, screws, gas pressure. Depending on the utilization they are made for, the optical

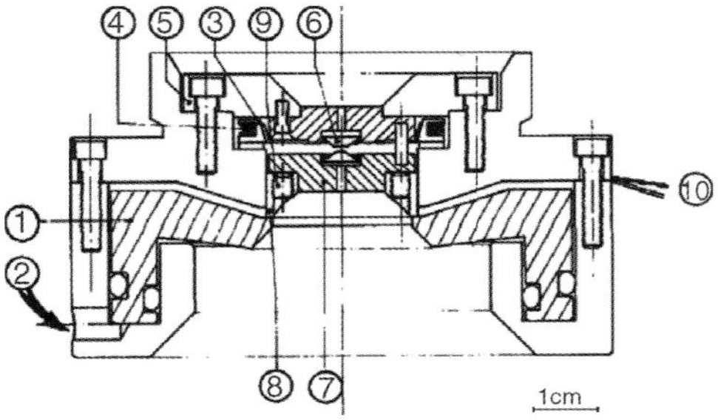

Figure 2.11 One of the early DACs (reprinted from Ref. [43] with IUCr's copyright permission, http://journals.iucr.org/), where the pressure is varied by applying a gas pressure.

aperture may vary in the visible. As an example, in the DACs presented here, the aperture is quite small for the Merrill–Basset and the Fourme DACs conceived for X-ray diffraction, because the diamond anvil seats are made in beryllium. On the contrary, using the Boehler–Almax design for the anvils, apertures as large as 50° (cone angle) may be reached with many DACs.

2.2.2.3 High-Temperature Instrumentation

Instrumentation about high temperature can be divided in two parts: how to apply temperature and how to measure it. For applying temperatures, two large routes are employed: furnaces (with heating by Joule effect) or "optical heating." This second term concerns all devices based on heating by light absorption (lasers, image furnaces, and so on.).

Furnace is the tool universally used for moderated temperatures (below 1000 °C). They can be used for higher temperatures, but some limits must be kept in mind. First, in a furnace, whatever the device, a large volume of matter is heated: sample, of course, but also sample holder, and mainly the internal walls of the furnace, or the heating device itself. These hot elements give a strong thermal radiation on the sample – radiation is the efficient way of energy transfer at high temperatures. This transfer is the mechanism that brings thermal energy to the sample by radiation, but this light can be reflected or scattered by the sample to the spectrometer and detector. The apparent thermal background will be the sum of thermal emission of the sample, and of the contribution of the sample surrounding. A furnace devoted to high-temperature optical measurements must have its output window adapted to the following optical devices to limit this effect, through fairly low numerical apertures.

Second limitation of furnaces is their use with specific atmospheres: metallic heating devices can be used in air up to approximately 1600–1700 °C (platinum); above which one must generally use systems under vacuum or reducing atmosphere (with tungsten heating elements for instance). The device must include specific windows compatible with the spectral range under consideration. If this is not a real difficulty for the Raman and Brillouin, it can be more complicated to solve for infrared, where no material apart from diamond is transparent on the whole vibrational spectral range. Besides these limitations, the problem of temperature measurement in a furnace is somewhat easy to solve, at least up to temperatures in the vicinity of 1700 °C: a well-suited thermocouple inside the furnace, in good contact with the sample holder or the sample itself, gives acceptable values of temperature. Above, pyrometric measurements are more adapted.

A specific system of furnace often used in micro-Raman spectrometry is the "hot-wire" device. It consists of a thin metallic wire, heated by Joule effect, where a small hole (typically some tens of micrometers) is made, and the sample is put into this hole. The device is easily compatible with the geometry of a Raman microscope, even with short front distances. It is an easy mean to perform measurements on liquid samples (the hole in the wire forms a crucible where the sample is molten). Other advantages are the fact that the hot zone is small compared to a furnace, limiting the thermal background. Temperature measurement can be done on the wire, the temperature of which is close to that of the sample.

The second class of heating devices deals with the "optical" ones. Some efforts were carried out through image furnaces, where the light of an intense source (as arc lamps) was focused onto the sample. But currently, high-power lasers are used (mainly CO_2 ones – $\lambda = 10.6\,\mu m$, with typical powers of some tens to hundreds Watt). Induction furnaces used for conducting samples enter also this category, as heating with the Raman excitation laser itself [44]. Several advantages appear compared to a "Joule" furnace: (i) Only the sample is heated, not its neighborhood. This limits the thermal emission to the sample alone. (ii) As the mass of matter is limited, the systems allow rapid changes of temperature (some seconds from room temperature to 2000 °C). This is often essential, especially for liquid samples, where interaction between them, the surrounding atmosphere, and the sample holder can lead rapidly to noticeable changes of chemical composition. One can probe some metastable states, of limited lifetimes. A way to limit the effect of the sample holder is to suppress it, by putting the sample (somewhat spherical) in equilibrium on a gas flow (Figure 2.12). This so-called "aerodynamic levitation" device was already instrumented on different spectroscopic tools [45, 46]. It works well on solids as on liquids (a droplet in levitation), allows to control atmosphere via the choice of the sustaining gas, and is particularly interesting for glass-forming liquids as heterogeneous nucleation due to contact with the crucible is strongly limited.

Another advantage of heating by optical beams as CO_2 lasers is their compatibility with diamond–anvil cells, as the heating power can be focused on very small samples (and also owing to the diamond transparency at $10.6\,\mu m$).

But these optical devices display also some disadvantages. First, the sample must absorb in the spectral range of the light source. It is not possible to heat neither pure silicas with an arc lamp for instance (light is transmitted) nor metals with a CO_2 laser (light is mainly reflected). However, in the latter case, a possible solution would be to use YAG: Nd^{3+} lasers at $1.06\,\mu m$. This illustrates well that it is always essential to think emissivity when performing high-temperature experiments.

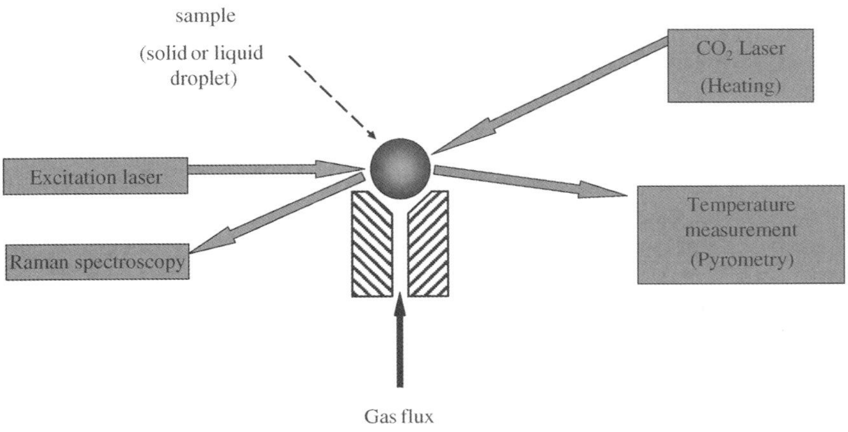

Figure 2.12 Typical aerodynamic levitation device, used here for the Raman scattering experiments.

The second disadvantage of optical heating devices versus Joule furnaces concerns measurement of the temperature, which drives us to present the following topic. As only the sample is heated, it is generally hazardous to measure temperature by devices as thermocouples (apart if the sample is big compared to the thermocouple, and this one in good contact with the sample, ideally inside it). But this is difficult to realize, in fact. Temperature monitoring is most often done by pyrometry in "optical heating" devices. And once more, spectral emissivity of the sample must be kept in mind. Pyrometry consists of measuring luminance of the sample and of an internal temperature-calibrated blackbody source, at some well-chosen wavenumber where the sample spectral emissivity is known. This allows precise ($\Delta T < 10\,^\circ C$) temperature measurements up to more than 2000 °C. But it must be underlined that such accuracy is completely dependent on the knowledge of the spectral emissivity of the material under consideration (the main part of the uncertainty originates from emissivity knowledge). Precise temperature measurements in the high temperature are never easy, and this topic would be for itself the subject of a review article. One can give some ideas about an elegant way, consisting of using the specific properties of the Christiansen point [47], which is generally observed in polar dielectric materials; for this point, standing just above the highest frequency LO mode, the refraction index is equal to 1 and the extinction one is very low. The reflectivity and transmittivity are near to zero, so emissivity is equal to 1 and the material acts as a blackbody for this wavenumber. Performing pyrometry at this wavenumber will limit uncertainty on temperature to that of the luminance alone, and obviously this Christiansen point pyrometry constitutes the best way to measure temperatures. On the other hand, these measurements are rather uneasy to handle, as they necessitate an infrared spectrometer as pyrometer (see Section 2.2.2.6 below). Christiansen point generally is most often situated in the mid-infrared, close to $1000\,cm^{-1}$ for oxides that constitute a large part of the materials of interest for high temperatures. Besides this, pyrometric measurements also need a direct access, or through windows transparent at the wavenumber of pyrometric measurement: this is obvious to say, but it needs generally two sets of windows (one for Christiansen – temperature measurements, one for vibrational spectroscopy) apart from infrared methods where the same spectral range is used.

A particular method of temperature measurement for the Raman scattering must be cited here: the ratio of Stokes over anti-Stokes spectra gives access to temperature. It is one of the rare direct methods of measurement of temperature, other methods, in general, are indirect as they are based on some calibration through a well-known temperature reference: blackbody for pyrometry, cold junction for thermocouples, and so on. Moreover, the temperature is determined at the exact point, which is the source of the Raman light. The Stokes-to-anti-Stokes ratio gives access directly to temperature (provided the ω^4 scattering correction and the spectrometer calibration response are done). Accuracy is never so good as at very low temperatures, as Stokes and anti-Stokes components tend to be equal at infinite temperatures. To be precise, this necessitates to have a good knowledge of the spectrometer response, and must be done on a wide spectral range, not only on one point. In spite of these limitations, this remains an efficient way to determine temperature [48].

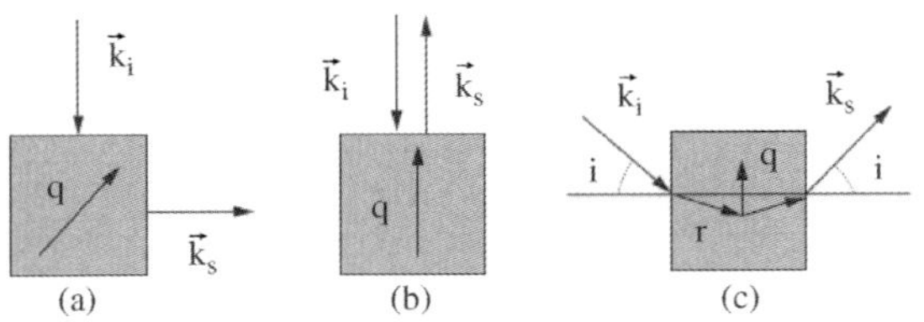

Figure 2.13 Most commonly used geometries in Brillouin scattering, (a) 90° scattering, $|q| = 2^{1/2}nk$; (b) backscattering geometry, $|q| = 2nk$; (c) platelet geometry, $|q| = 2nk \sin r = 2k \sin i$.

2.2.2.4 **Brillouin Devices**

The scattering geometries commonly used in order to determine the elastic constants are shown in Figure 2.13a. In Section 2.1, the modulus of the phonon wavevector is equal to $\sqrt{2}nk$, where $k \sim k_i \sim k_s$ and $n \sim n_i \sim n_s$. In Figure 2.13b, $|q| = 2nk$. Such backscattering geometry produces the largest wavenumber shift. Figure 2.13c shows an interesting geometry – referred to as the platelet geometry – for Brillouin scattering under pressure (provided that an *ad hoc* cell is available), because measurement of the refractive index n is not required. Indeed, in the crystal the modulus of the light wavevector is equal to nk. Owing to the incidence angle, $|q| = 2nk\sin r$, where r is the angle of the refracted photon inside the crystal with respect to the normal to the surface. Using the Descartes law of refraction, one obtains $|q| = 2k \sin i$, where i is the known angle of incidence. In particular, the knowledge of the relative orientation of $\vec{k}_i$ and $\vec{k}_s$ with respect to the crystallographic axes allows the determination of that of $\vec{q}$. The selection rules [25] allow to deduce the related combination of elastic constants.

In any case, the frequency difference between the elastically scattered light (Rayleigh) and the scattering by the acoustic mode (Brillouin) is small [less than $6\,\text{cm}^{-1}$ (180 GHz)]. Moreover, the intensity ratio between Rayleigh and Brillouin signals is typically 10^6–10^9 in solids. Therefore, a high-resolution apparatus is mandatory. The grating Raman spectrometer is not sufficient for that purpose, even the triple grating ones. The best-adapted setup is the Fabry–Pérot interferometer. This interferometer is composed of two plane mirrors, maintained extremely parallel to each other. The transmission of such a system is given by the well-known expression:

$$T_{FB} = [1 - A_M(1 - R_M)^{-1}]^2[1 + 4F^2\pi^{-2}\sin^2 p\pi]^{-1} \tag{2.31}$$

where $A_M(R_M)$ is the absorption (reflection) by the mirrors, the integer p the order of interference equal to 1/2e where e is the distance between the mirrors, and F the finesse, defined by $\Delta\omega/\delta\omega$, with $\Delta\omega$ the distance between two adjacent interference orders, also called free spectral range (FSR), is equal to 1/2e (in cm^{-1}), and $\delta\omega$ is the width at half maximum of a transmission peak (see Figure 2.14). In fact, the finesse is due to both the reflectivity and the planeity of the mirrors, and one can write $F^{-2} = F_R^{-2} + F_p^{-2}$, where $F_R = \pi\sqrt{2}(1 - R_M)^{-1}$ and $F_p = m/2$ for a mirror polished at λ/m.

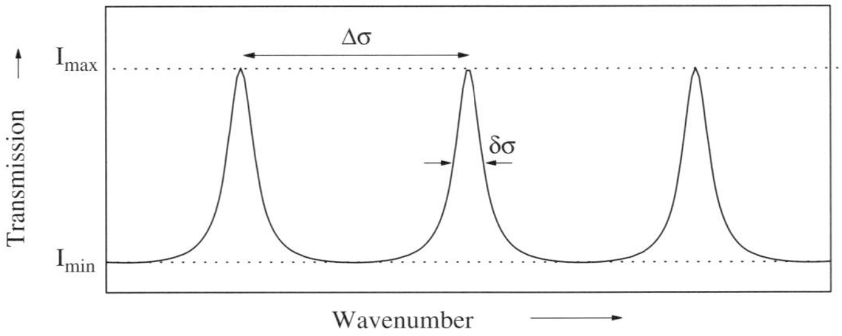

Figure 2.14 Transmission of a single pass Fabry–Pérot. Here, an alternative notation for ω is σ.

Another important quality of the Fabry–Pérot interferometer is the contrast, defined as the ratio of the maximum-to-minimum transmitted light, that is, $C = I_{\max}/I_{\min} = [4R/(1-R)^2] + 1$. The resolving power is given by $\mathfrak{R} = \omega/\delta\omega = F\omega/\Delta\omega$. Taking realistic values, that is, $R = 0.92$, $\omega = 2.\ 10^4\ \text{cm}^{-1}$ (500 nm), and $e = 2$ mm gives $F = 37.7$, $C = 576$, free spectral range $= 2.5\ \text{cm}^{-1}$, and $\mathfrak{R} = 3 10^5$.

Such an instrument is already quite efficent, but is often not sufficient, specially when working with solids in the backscattering geometry where the Brillouin efficiency is not high enough to enable a sufficient signal-to-noise ratio. Two improvements have been made to the system, the multipass and the introduction of the tandem Fabry–Pérot.

Suppose the beam passes N times through the interferometer. In that case, the transmission will be $T_{FB,N} = T_{FB,1}^{N}$. Using Equation 2.31, one can easily compute the finesse and the contrast for the N pass system, coming to $F_N = F_1\left[1/\sqrt{2^1/N-1}\right]$ and $C_N = C_1^N$

From this, it is clear that the multipassing of an interferometer is extremely advantageous, especially in terms of contrast.

In the classical multipass setups, there are still some problems, that is, it is difficult to change the free spectral range, and with one measurement it is impossible to decide whether one line is the Stokes component coming from the interference order, N, or the anti-Stokes one, of the order $N + 1$. In that case, one has to make two measurements with different free spectral ranges. A solution to increase the FSR is to use two interferometers in series, with slightly different FSR. In that case, the transmission through the system is shown in Figure 2.15. The two lateral peaks of the individual FP are attenuated by the second FP. A simple and elegant way to build a tandem FP was proposed by J.R. Sandercock [18], as shown in Figure 2.16. The two FP in tandem are scanned at the same time by a single piezoelectric stack thanks to deformable parallelogram geometry. The optical path is such that the parallel beam goes through the system three times, providing a high contrast and a large FSR. Moreover, the parallelism of the mirrors is independent of the scanning of the setup, giving a good stability to the system.

2.2.2.5 **Raman Devices**

For high pressure, there is no specificity of the spectrometer. The only adaptation is that, because of the small size of the samples and of the size of the diamond–anvil

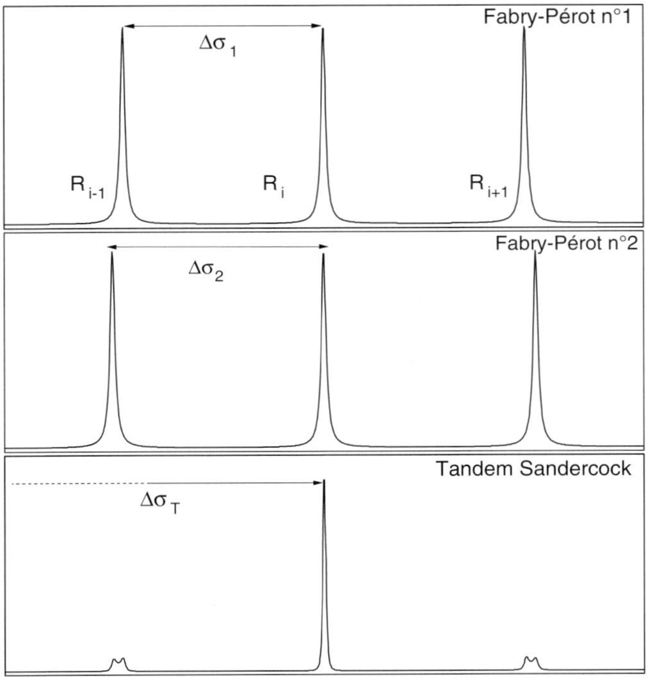

Figure 2.15 Transmission through two individual interferometers, and through the tandem. Here, an alternative notation for ω is σ.

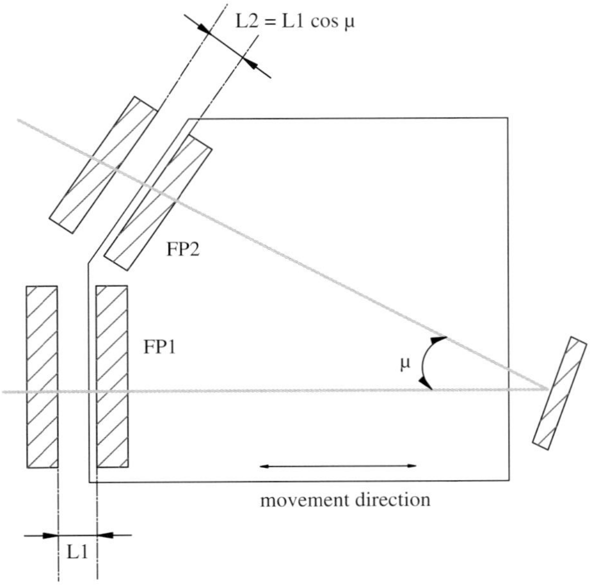

Figure 2.16 Scheme of the tandem interferometer proposed by Sandercock.

cell, one should use micro-Raman with long working distance microscope objectives.

Contrary to Brillouin devices and to high pressures, the Raman scattering at high temperatures needs specific improvements to limit thermal background. If not, the Raman measurements in classical configuration (green excitation laser) will be limited to approximately 1000–1200 °C, depending on the Raman scattering efficiency of the considered material. Above, the thermal background is too intense and precludes detecting the Raman information. Several ways are possible to push up this limit, and addition of some of these ways is needed to reach 2000 °C. First, let us recall that limiting emissivity itself (see above) is one of the most efficient ways, when realistic. A second way is to remove thermal background, that is, to record the spectrum of the sample with excitation laser, followed by a second record without laser (which consists of the thermal flux emitted by the sample and its environment), and to subtract the second from the first. This makes it possible to increase the temperature limit by 100–200°C. Another way is to record anti-Stokes spectrum rather than Stokes one [49], which results in a small gain in thermal background rejection. All these solutions allow pushing up the saturation limit of the detector. But, it is clearly insufficient to increase the limit by 1000°C. Some solutions are possible for this goal; they will be rapidly reviewed hereafter.

2.2.2.5.1 **Spatial Filtering** The idea here is to limit the source volume of thermal emission to the only source volume of the Raman scattering. The hot zone of the sample and its immediate environment is always larger than that of the Raman source volume. Besides this, as thermal background can be much larger than the Raman signal, it contributes to the recorded signal even if not in ideal optical conditions. It is of interest to limit the size of the sample surface that is imaged at the spectrometer entrance slit to the only zone where the excitation laser is focused. This is done by an afocal system with two convergent lenses and a pinhole positioned exactly at their common focal point. It is the basic idea of a confocal microscope. Such spatial filtering was used by Gillet *et al.* [50] and McMillan *et al.* [51] in a confocal microscope, and then by Yashima *et al.* [52] (with UV excitation for the latter) in the macro-Raman configuration, today it is often used in the high-temperature Raman devices.

2.2.2.5.2 **UV Shifting of the Laser Line** As the thermal radiation maximum lies in the near infrared, its influence progressively decreases upon increasing the excitation laser frequency. One can position the laser excitation at higher frequencies, in the UV range, to help removing the thermal background. This was pointed out first by Farrow and Nagelberg [53], and then by different groups, with good results: Both Yashima *et al.* [52] and Fujimori *et al.* [54] were using an argon line (363.8 nm), and Zouboulis *et al.* [55] a quadrupled YAG: Nd^{3+} laser (266 nm). Today, the availability of UV lasers (tripled, quadrupled, and even quintupled YAG: Nd^{3+}) and that of UV-dedicated spectrometers from the different Raman manufacturers opens this way to many groups, and clearly this constitutes one of the promising directions for the future. However, even if UV excitation displays advantage of an increased Raman efficiency through the ω^4 scattering factor, specificities of UV excitation must be kept in mind: more frequent absorption and luminescence processes, smaller penetration depth inducing preferential analysis of the surface. Besides, IXS (cf. Chapter 4) provides a huge high-frequency shift, which is not affected by thermal radiation problems.

2.2.2.5.3 **Temporal Filtering** The driving idea is to take benefit of the fact that the Raman scattering is a rapid effect (characteristic times beyond the picosecond scale). So if taking a pulsed laser, with some detection device synchronized to it (opened during the laser pulses; closed in-between), one can reject the thermal background by a factor equal to the ratio $r = t_2/t_1$ between the period t_2 separating two pulses and the pulsewidth t_1. If compared to a continuous laser of the same averaged power, this allows to access to very high temperatures: typical pulsed lasers (532 nm doubled YAG: Nd^{3+}) have $t_2 = 1$ ms and $t_1 = 30$ ns, then a rejection factor of 3.10^4. Let us recall that between 1000 and 2000 °C and for the green spectral range, thermal emission increases by about one order of magnitude for every 300 °C, which means that temporal filtering is a way to increase the high-temperature limit from 1000 to 2200 °C. Temporal filtering appears as the efficient way to access to very high temperatures. It was recently extensively discussed [56], we will limit here to instrumental aspects and to a short comparison of pulsed 532 nm laser with continuous 355 nm: UV continuous device is more efficient for "moderately" high temperatures (up to about 1700 °C), the pulsed one is better above (see [56] for details).

2.2.2.5.4 **Specific Instrumentation** The pulsed Raman spectroscopy mainly requires the laser source and the detection device. However, all the optical path must be compatible with the high peak powers inherent to a pulsed laser (this generally precludes high-magnification objectives in micro-Raman configuration). Lasers are not a problem since they are commercially available. For a long time, detectors were a serious problem. The first paper about temporal filtering for the high-temperature Raman scattering was the pioneering work by Exharos *et al.* [57, 58] on silicas and boron nitride; they used a photodiode array detector (PDA). They were followed by Bernadez *et al.* [59] and by the Grenoble group [60–63] that monitored plasma-assisted growth of diamond films, with similar PDA devices. During these years, the development and the impressive performances of CCD detectors have made them inescapable in a Raman spectrometer. However, they are intrinsically incompatible with nanosecond timescales, whereas PDA devices were. Devices based on photomultipliers and gated electronics were also used for detection [44, 64–70]. Some attempts were also made through positioning, between sample and spectrometer, some rapid optical door, based for instance on a Pockels device [56]. But today, a more efficient solution is given by intensified CCDs (ICCD). These are now easy to find commercially and offer satisfactory performances and lifetimes. The light intensifier can be driven by a high voltage (opaque or transparent depending on the voltage) and constitutes an optical door integrated to the detector. These ICCDs today appear as the best solution for the time-resolved Raman measurements, and they are used by several groups by the world [56, 71–78]. Goncharov *et al.* [71], who have used them for simultaneous high P and high T measurements feel that ICCDs could allow the Raman scattering up to huge temperatures (5000 K).

This concept of temporal filtering through pulsed spectroscopy could also thought to be applicable for Brillouin scattering. In fact, it is not as simple as that a specific difficulty comes from the extremely narrow linewidth needed for a Brillouin excitation laser, as Brillouin lines stand close to the Rayleigh lines. The short pulse durations (some tens of nanoseconds) of a pulsed laser induce rather large frequency linewidths, typically of 1 cm^{-1}, incompatible with Brillouin scattering.

2.2.2.6 **Infrared Devices: Emissivity Measurements (Temperature and Pressure)**

As already underlined, the temperature and the infrared spectral range are intimately connected since radiative thermal exchanges essentially stand in this range. About 30 years ago, first attempts of infrared spectroscopy at high temperature were carried out by reflection and transmission techniques [11]. The problem was, as for scattering techniques, to extract the reflection or transmission information from the thermal background emitted by the sample and its environment. But in this case, the problem was more complex, when using Fourier-transform interferometer, which is the current tool in infrared spectroscopy. The sample and its environment emit a thermal background that goes directly to the detector, where it constitutes a continuous background superimposed to the interferogram; but, the thermal background goes back to the Michelson interferometer, where it is modulated by the scanner, a part of it is reflected to the detector through the sample, which part constitutes an interferogram in phase opposition with the main one, resulting in an underestimated final spectrum. This difficulty has a solution in the particular case of the Genzel-type interferometer (with light focused on the beamsplitter), where it is possible to spatially filter the second interferogram through a judicious misalignment of the interferometer [11]. As Genzel-type interferometers are no more commercially available today, more details will not be given here, but can be found in Ref. [11]. A consequence is that today no solution for performing reflection or transmission infrared spectroscopy can be built based on commercial spectrometers. In fact, the high-temperature domain is much less investigated in infrared spectroscopy than in Raman scattering.

As already rapidly noted in Section 2.2.2.1, the best solution for opening high temperatures to infrared techniques consists of emissivity measurements. The thermal beam emitted by the heated sample is directly recorded by the spectrometer, and the reference beam (obtained by removing the sample in transmittivity or replacing it by a mirror in reflectivity) is given by a blackbody reference furnace, ideally heated at the same temperature, and with same directional conditions. Spectral emissivity is the ratio of these two spectra.

There are very few reports about spectral emissivity measurements, with the aim to use them for vibrational spectra determination and condensed matter application. Often, they are devoted to determine thermal balances, without focusing on vibration properties. While dealing with lattice dynamics, one can cite Sova *et al.* [79] and Rozenbaum *et al.* [80], who presented reports based on CO_2 laser heating. The latter authors have developed a device devoted to emissivity measurements (Figure 2.17),

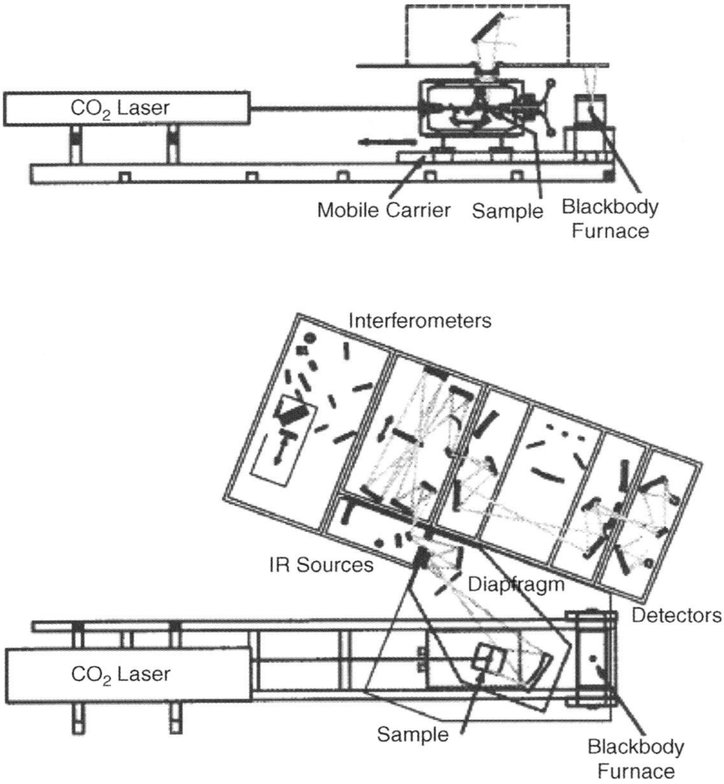

Figure 2.17 Experimental setup of infrared emissivity measurements. Reprinted with permission from Ref. [80]. Copyright 2010, American Institute of Physics.

operational in a large spectral range, from far-infrared to the visible range. Again, one of the main advantages concerns the way to solve the problem of thermal fluxes emitted by the sample environment. The solution developed by Rozenbaum *et al.* [80] is optical heating by using a power CO_2 laser. The sample is a flat disk, heated by a CO_2 laser. The flux emitted by the sample perpendicular to the surface is collected, focused, and injected as source in the FT-IR spectrometer. The reference spectrum is given by a blackbody furnace, which is translated at the exact place of the sample. All the emission parts (laser, sample, and blackbody furnace) are firmly attached as a whole, and can be moved without any temperature perturbation. For practical reasons (good working point for the furnace, better stability, no need for a temperature change, and no time to wait for equilibrium), the temperature of the blackbody furnace is kept constant ($\theta_{BB} \sim 1400\,^{\circ}C$) and the blackbody reference spectrum at the sample temperature is deduced from the sample one at 1400 °C via Planck's law. By the way the system is operational on a much larger temperature range than the blackbody furnace: up to very high temperature ($\rightarrow 3000\,^{\circ}C$), down to low temperatures, and close to room temperature. By using this system, one limits the parasitic thermal emission fluxes, even though these are not completely removed.

A subsequent analysis of the same group [81] has shown that recording accurate results needs to take into account the remaining parasitic fluxes emitted by all optical components situated in the immediate neighborhood of the beam, that is, typically, a diaphragm or an optical mount limits the diameter of the optical beam, but its matter constitutes a source of thermal emission at the spectrometer temperature. This correction (refer to Equation 2.32) is more important for intermediate temperatures, and the low-/intermediate-frequency range.

$$E(\theta_{\mathrm{Obj}}) = \frac{\mathrm{FT}(I - I_{\mathrm{S}})}{\mathrm{FT}(I_{\mathrm{BB}} - I_{\mathrm{S}})} \cdot \frac{L[\theta_{\mathrm{BB}} - L(\theta_{\mathrm{S}})]}{L[\theta_{\mathrm{Obj}} - L(\theta_{\mathrm{S}})]} \tag{2.32}$$

where I, I_{BB}, and I_{S} denote, respectively, the interferograms of the sample at temperature θ_{Obj}, of the blackbody at temperature θ_{BB}, and of the sample at the temperature θ_{S} of its surrounding (close to room temperature). FT means Fourier transform. In that way, all effects due to parasitic thermal emission are removed.

This device works well up to very high temperature (2500 °C and even more is available), on solid materials as on liquids. By CO_2 heating, it is possible to heat the center part of a solid disk sample above its melting point, in that way the liquid is sustained in a self-crucible, of identical chemical nature (at least for congruent melting). The sample must be rather large (typically $\sim$1 cm^2) and only the hot molten zone acts as a source for the measurement.

Compared to the Raman and Brillouin scatterings, infrared measurements (specular reflectivity as emissivity) need more data processing to access the relevant vibrational information. A fitting process with specific dielectric function models is necessary. Often, the three-parameter classical Born–Huang dispersion model (cf. Section 2.1), or the four-parameter factorized form, is used (see Ref. [11] for a review). These models describe quite correctly the lattice mode spectral range, but are quite less efficient to reconstruct higher frequencies, where transmission becomes progressively semitransparent. Besides this, for high temperatures, anharmonic effects become more important. More sophisticated semiquantum models of dielectric functions are necessary, which take into account the frequency-dependence of damping [82].

2.2.3 Acoustical Modes

2.2.3.1 General Presentation

As discussed in the introductory part of this chapter, the methods to measure the elastic properties are ultrasonics (US), inelastic neutrons scattering (INS), IXS, and Brillouin scattering. In conjunction with DAC, INS is not usable because its needs large samples. In principle, US also requires large samples, but two developments enabled measurements in the DAC. In "classical" US, the tranducers vibrate around 10–50 MHz, which corresponds to wavelength in the millimeter range, that is, much too large to use in a DAC. Some years ago, a gigahertz ultrasonic delay line, working only with longitudinal waves was developed [83]. This was a big improvement,

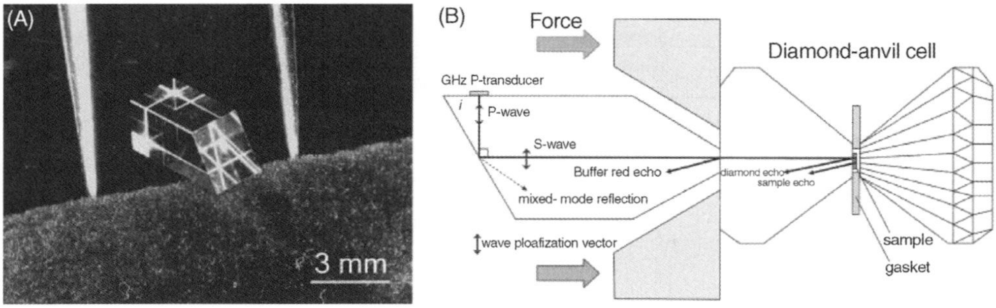

Figure 2.18 Ultrasonics in a DAC. (A) Converter from longitudnal to transverse wave. It is made of a YAG crystal that produces pure mode with well-defined polarization direction. (B) Scheme of the whole setup. Reprinted from Ref. [84]. Copyright 2010 National Academy of Sciences, U.S.A.

because with such a setup, the wavelengths are now compatible with the typical dimensions of the sample in a DAC. About a decade later, the same group invented a longitudinal to transverse converter enabling the study of the elastic properties in crystals, by the US method ([84]; see Figure 2.18).

This technique presents the great advantage with respect to Brillouin scattering to permit measurements whatever the electronic state of the sample is. However, for the moment, it seems to be limited to the 10 GPa range, because of the necessity that the diamond culet and the sample have to remain strictly parallel, which is not the case when the pressure is increased.

The second technique is the picosecond US [85], which has been applied very recently to the studies under high pressure. This technique is based on the pump-probe technique: a Ti:sapphire laser produces ultrashort pulses of 100 fs corresponding to a wavelength of ~800 nm. The laser beam is split into pump (~20 mW) and probe (~3 mW) beams (Figure 2.19) by a beamsplitter. The pump is focused on one surface of the sample, whereas the probe is focused on the opposite one. Both focus points are of the order of 3 μm. The probe beam is scanned on the sample surface using a *X–Y* piezoelectric stage to locate the pump beam. The effect of the pump laser pulse is to create a sudden and small temperature rise (1 K), which produces a longitudinal acoustic strain field. The thermal and acoustic effects alter the optical reflectivity of the sample in two ways: the photoelastic effect and the surface displacement as the acoustic echo reaches the surface. The first modification contributes to the change of both real and imaginary parts of the reflectance, and the second one to the change of the imaginary contribution only. The variation of the optical path length of the probe enables to detect the reflectivity modification as a function of time. The detection is carried out by a stabilized Michelson interferometer that allows the determination of the reflectivity imaginary part change (at high pressure, interferometric measurements are mandatory to efficiently detect acoustic pulses with a low-frequency spectrum due to sound attenuation). The possibilities of this setup were tested by performing measurements of the elastic properties of the quasicrystal AlPdMn [86]. Two echoes corresponding to longitudinal acoustic waves were recorded at each pressure, through the relative

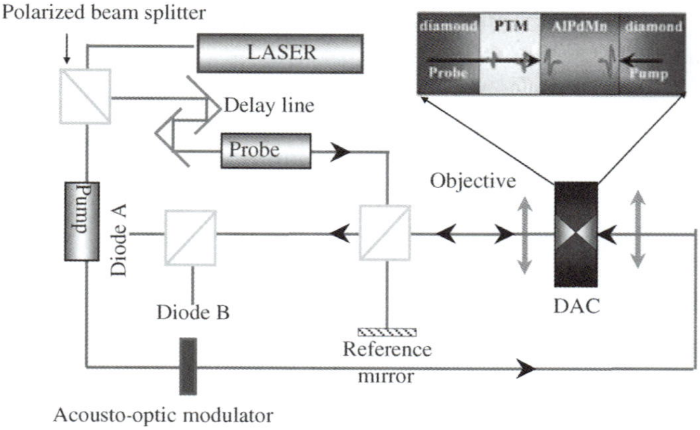

Figure 2.19 Experimental setup used to perform picosecond-laser acoustics studies at high pressure (Reprinted figure with permission from Ref. [86]. Copyright 2010 by the American Physical Society). The arrow indicates the schematic illustration of generation and detection process for AlPdMn in DAC. Inset: Enlargement of the ultrasounds in the DAC. *PTM*: pressure-transmitting medium. There is a thin layer (~1 μm) between the sample and the diamond (pump side).

variation of the reflectance imaginary part. The first echo corresponds to a single way between the two surfaces of the sample, the second one being reflected twice more at the AlPdMn/PTM interfaces.

In order to deduce the sample thickness, the experiment was first performed at ambient conditions. The knowledge of the sound velocity, and the measurement of

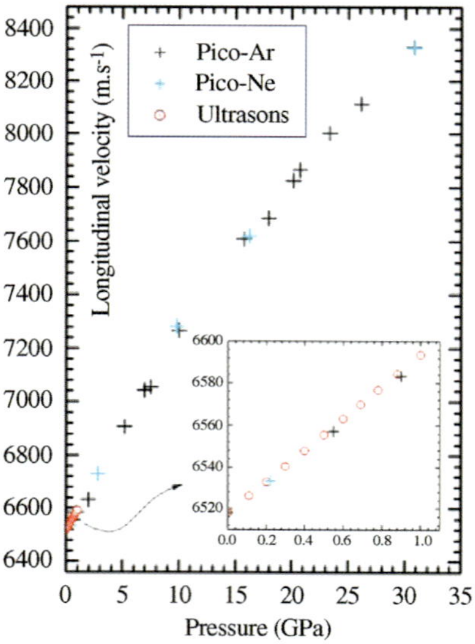

Figure 2.20 Pressure dependence of the longitudinal sound velocity of AlPdMn. The red circles were obtained from classical ultrasonic measurements, and the black and cyan crosses from the picosecond measurements presented here. Reprinted figure with permission from Ref. [86]. Copyright 2010 by the American Physical Society.

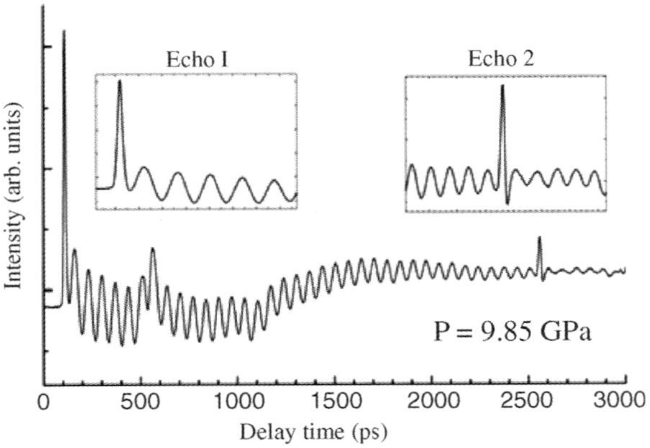

Figure 2.21 Spectrum obtained of AlPdMn at 9.85 GPa. The two insets present enlarged parts of the first and second echoes. Reprinted figure with permission from Ref. [86]. Copyright 2010 by the American Physical Society.

the transit time gives the thickness through $v = 2d/t$ with a high accuracy. The transit time is measured under pressure with the same accuracy. The pressure dependence of the thickness is obtained from the equation of state [87], and the sound velocity finally deduced as a function of pressure.

This technique is particularly promising for several reasons. First, it may be utilized with any kind of sample, transparent or opaque, and of dimensions perfectly adapted to the DAC. Second, it is not perturbed by the deviatoric stresses induced by the pressure-transmitting medium, as shown in Figure 2.20, where the black crosses were obtained using argon as a pressure-transmitting medium, and the cyan ones with neon (neon is much softer as argon), contrary to gigahertz method. It also does not suffer from the curvature of diamonds imposed by very high pressure. Another great advantage of this technique is that it enables the measurement of the attenuation: subtracting the thermal background from the oscillations shown in Figure 2.21 permits the calculation of the attenuation in the pressure-transmitting medium. The principal disadvantage of this technique is that for the moment, the authors did not found the possibility to measure shear waves. The case of the quasicrystal, AlPdMn is specially favorable because it is elastically isotropic, that is, $(C_{11} - C_{12})/2 = C_{44}$. Therefore, the precise determination of the equation of state using X-ray diffraction gives the bulk modulus $B = (C_{11} + 2C_{12})/3 = C_{11} + 4C_{44}/3$. Hence, with both measurements, the two elastic moduli are determined.

2.2.3.2 **Examples**

H_2O is a chemical species with one of the richest known phase diagrams with more than 12 different crystallographic phases, 3 amorphous structures (low-density amorphous, high-density amorphous, and very high-density amorphous) with first-order transition between them, and two liquid phases (low-density liquid, high-density liquid). This property, together with the importance of water for life, and the particular physical properties of this compound, led to an extremely rich literature. The elastic properties of H_2O were studied as a function of pressure and temperature. At ambient temperature, the succession of phases is water → ice VI at approximately 1.05 GPa and ice VI → ice VII at 2.2 GPa. The structure of ice VI is tetragonal (space group $P4_2/nmc$), and that of ice VII, cubic (space group Pn3m). The first Brillouin scattering studies of H_2O at ambient temperature under high pressure [88] were performed in the backscattering geometry (see Section 2.2.2.4). In this geometry, the knowledge of the refractive index is mandatory. Fortunately, it was known in the liquid, and measured at the liquid – ice VI equilibrium. It was extrapolated to high pressure with the hypothesis that n is a linear function of the Eulerian strains, using the measured equation of state of ices [89]. In that way, the sound velocity was computed from the results presented in Figure 2.22. The square of the sound velocity is equal to the ratio of a combination of elastic moduli by the density. In this chapter, the results have been analyzed in terms of Murnaghan's equation of state by plotting the logarithm of the elastic modulus C versus log (ϱ/ϱ_0) (see Figure 2.23). The slope in the liquid range is approximately 6.8, consistent with that obtained in liquids, and that obtained both in ice VI and in ice VII is approximately 4.3, in good agreement with the B′ value deduced from the X-ray equation of state [89], which is 4.1.

These measurements performed up to 30 GPa have been extended up to 67 GPa [90] and performed on D_2O up to 34 GPa [91] with results similar to those obtained on H_2O.

The results on ice VII have been improved by using the platelet geometry [92]. We recall that such geometry presents the advantage of not implying the measurement of the refractive index (cf. Section 2.2.2.4), and further enables to measure the transverse acoustical phonons, that are often not seen in the backscattering geometry. The authors not only used the platelet geometry but also developed a new method of measurements for single crystals grown under pressure, like ice in the present case. The crystallographic orientation of the sample was not determined by X-rays. In order to overcome this lack, the authors did rotate the DAC around the anvils axis, measuring the Brillouin spectra every 10 degrees. Now, the square of the velocities of a cubic crystal may be expressed as a function of six variables:

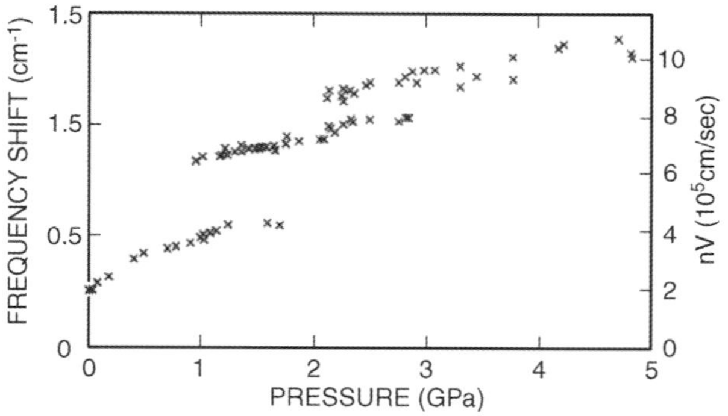

Figure 2.22 Brillouin frequency shift and product nV of water, ice VI, and ice VII. Reprinted figure with permission from Ref. [88]. Copyright 2010 by the American Physical Society.

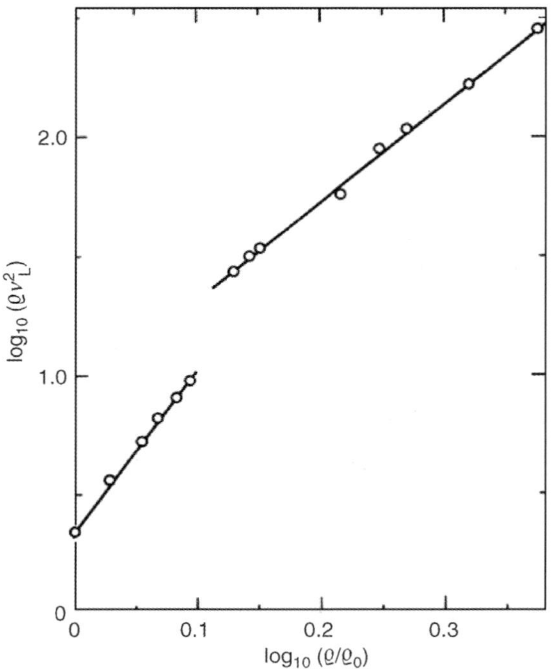

Figure 2.23 Plot of $\log_{10}(\varrho v_L^2)$ as a function of $\log_{10}(\varrho/\varrho_0)$.

$v_i^2 = f(C_{11}/\varrho, C_{12}/\varrho, C_{44}/\varrho, \theta, \varphi, \chi)$, where the subscript i stand for the three acoustic modes; ϱ is the density; and θ, φ, and χ are the Euler angles relating the laboratory and the crystal frames. A least-square fit procedure enabled the determination of the C_{ij}/ϱ, and subsequent use of the X-ray determined equation of state [89] did provide the value of the individual elastic moduli as a function of pressure (Figures 2.24 and 2.25).

The same technique was applied [93] to the measurement of the elastic moduli of the more complex ice VI, stable at ambient temperature between 1.05 and 2.1 GPa

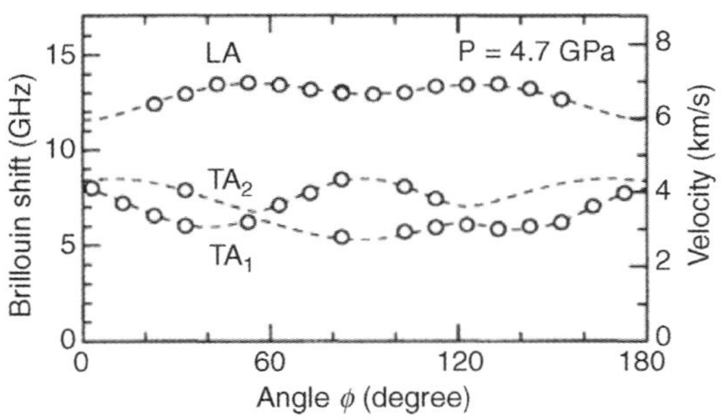

Figure 2.24 Brillouin shift and sound velocities in ice VII at 4.7 GPa. Reprinted figure with permission from Ref. [92]. Copyright 2010 by the American Physical Society.

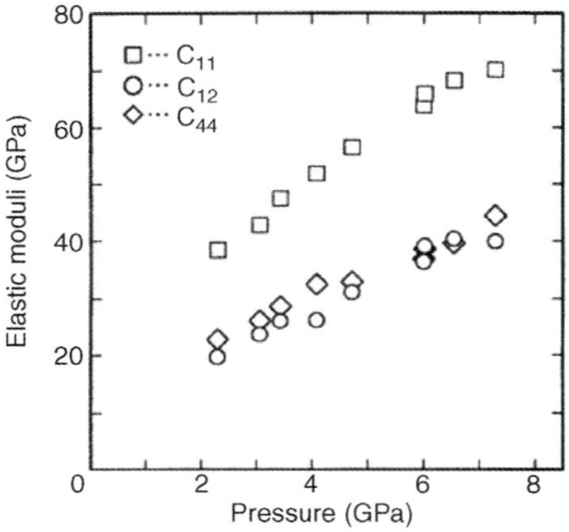

Figure 2.25 Pressure dependence of thee elastic moduli of the cubic ice VII at 300 K. Reprinted figure with permission from Ref. [92]. Copyright 2010 by the American Physical Society.

(see Figure 2.26). This phase is tetragonal and has therefore six elastic moduli, C_{11}, C_{12}, C_{13}, C_{33}, C_{44}, and C_{66}.

At the same time, the elastic moduli of ice VI were determined also by Brillouin scattering [94], but on large (125 mm^3) single crystals at $-2\,^\circ$C and 0.72 GPa, using

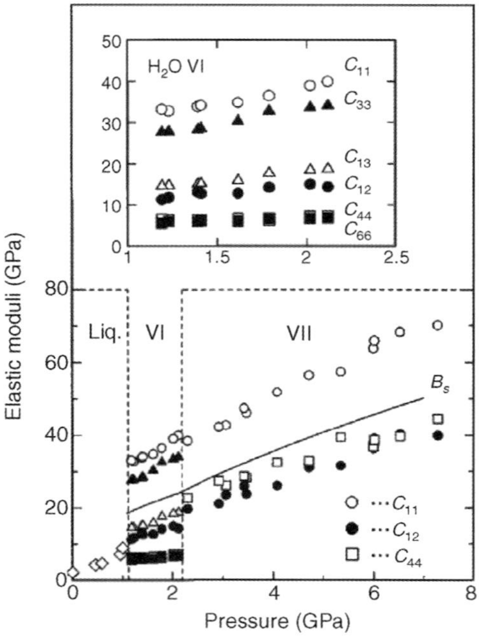

Figure 2.26 Pressure dependence of elastic moduli in a liquid, tetragonal ice VI, and cubic ice VII at 300 K. The solid line indicates the adiabatic bulk modulus. Reprinted figure with permission from Ref. [93]. Copyright 2010 by the American Physical Society.

Table 2.1 Comparison of the elastic moduli (in GPa) obtained in the DAC and on large single crystals.

	[93] (300 K, 1.23 GPa)	[94] (271 K, 0.72 GPa)	[94] With rotation 45° [96]
C_{11}	32.8	26.8	31.1
C_{12}	11.8	14.5	10.3
C_{13}	14.7	12.8	12.8
C_{33}	27.8	26.2	26.2
C_{44}	6.3	6.3	6.3
C_{66}	5.9	10.4	6.1

90° geometry. The accuracy of the Brillouin frequency shift is certainly much better under these conditions, but the determination of the refractive index is mandatory.

The results on large crystals were confirmed in a work on ice III and ice VI [95], where the Brillouin scattering was measured at 271 K and between 0.62 and 0.82 GPa. The results obtained by both techniques are compared in Table 2.1. There are three important differences between the two sets of results: (i) 50% difference in C_{66}, (ii) 20% difference in C_{11}, and (iii) inversion between C_{12} and C_{13}. The origin of these discrepancies is not clear. Again, in many of the experiments, the orientation of the crystals was not made by X-rays. In the DAC, it was deduced from the fitting procedure, while for the large crystal it was made by measuring the birefringence. This is the explanation proposed by H. Shimizu [96] who calculated that with a 45° rotation of the crystalline axes around the z-direction, the values are close to those obtained in the DAC.

The adiabatic bulk modulus can be deduced from the elastic moduli via the relation

$$B_S = \frac{C_{33}C_{11} - 2C_{13}^2 + C_{33}C_{12}}{C_{11} + 2C_{33} - 4C_{13} + C_{12}}. \tag{2.33}$$

By using the values of Table 2.1, one can find $B_S = 17.8$ GPa for the large crystal and 19.5 GPa in the DAC.

Quartz is almost as present in our everyday life as water, although it is not vital for human life. It is the major rock-forming mineral, and it plays a fundamental role in the Earth's crust. Moreover, both cristobalite and tridymite exist in two varieties, with several amorphous modifications. In the simplified phase diagram shown in Figure 2.27, the silicon atoms are tetrahedrally coordinated with oxygen, but in the stishovite where the coordination number is 6. At ambient conditions, the stable structure is the α-quartz, which is trigonal (space group $P3_121$). When compressed at ambient temperature, the α-quartz remains stable up to approximately 20–25 GPa (there is no direct transformation to coesite or stishovite structures). At that pressure, there is a progressive destabilization of the structure, leading to an amorphous phase [97, 98]. This transition in such an important material provoked an abundant literature, both on the experimental and the theoretical sides. We will restrict ourselves here to the elastic properties. It should be emphasized at this point that

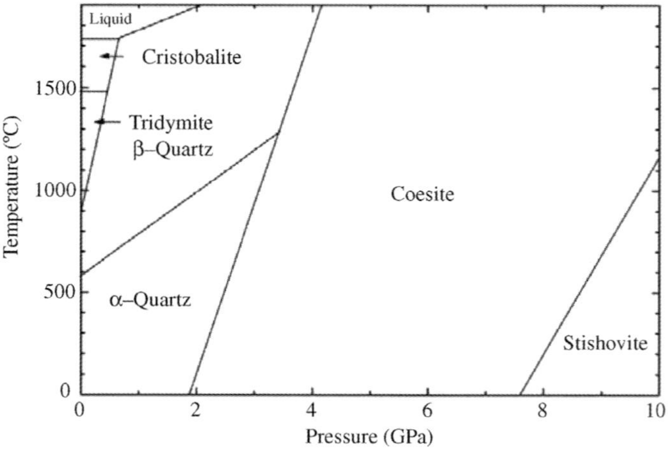

Figure 2.27 Schematic phase diagram of SiO_2.

sound waves propagate in all materials, that is, none of the concerned technique is able to detect whether a sample is in a crystalline or an amorphous phase.

The transition pressure (~20 GPa) led to the use of Brillouin scattering as the only possible technique. The first result was obtained in the backscattering geometry [99]. The experiment used either argon or 1 : 4 ethanol–methanol mixture as a pressure-transmitting medium, with no difference between the results. The longitudinal modes were obtained for phonons propagating parallel and perpendicular to the [0 0 1] axis. The experiment was performed up to 25 GPa, during both pressure increase and pressure decrease. During upstroke, the Brillouin peaks began to broaden above 15 GPa, but sharpen again at the downstroke below 10 GPa. The astonishing result presented in this chapter is that the recovered sample, supposed to be amorphous, is elastically anisotropic; indeed, as shown in Figure 2.28, the LA modes measured at ambient after an upstroke to 20 GPa are at 1.3 cm^{-1} along the *c*-axis, and at 1.45 cm^{-1} in the perpendicular direction.

It has been later observed by transmission electron microscopy [100] that the first step of amorphization occurs by the formation of planar defects, followed by the formation of the amorphous phase on the defect sites. Indeed, these planar defects are due to the formation of one crystalline phase at 21 GPa [101].

The most recent results obtained by Brillouin scattering on quartz ([102], see Figure 2.29) were obtained in the platelet geometry (cf. Section 2.2.2.2), by using the technique developed by H. Shimizu, and presented in the preceding paragraph for water. The experiment was performed using neon or helium as a pressure-transmitting medium (with no measurable difference). It should be noted here that this is the crystal with the lowest symmetry on which the Shimizu's technique has been used. One of the goals of the experiment was to determine which of the stability criteria (Born criterion) is involved in the α-quartz high-pressure destabilization. It seems to be necessary to remind once more that in trigonal crystals (32, 3*m*, and 3*m* classes), the Born criterion gives rise to *only two* conditions [103, 104], that is,

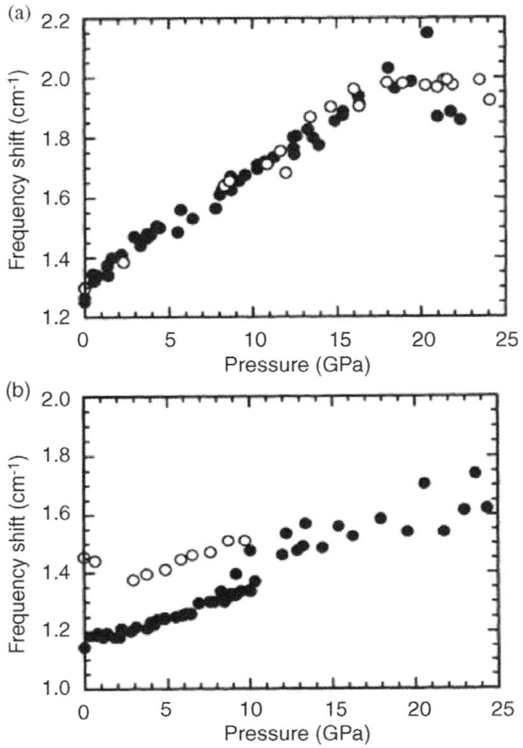

Figure 2.28 LA modes in quartz versus pressure. Solid (open) symbols correspond to upstroke (downstroke). (a) [0 0 1] direction and (b) perpendicular direction. Reprinted figure with permission from Ref. [99]. Copyright 2010 by the American Physical Society.

$B_1 = (c_{11} + c_{12})c_{33} - 2c_{13}^2 = 2((c_{11} - c_{66})c_{33} - c_{13}^2) > 0$ and $B_2 = (c_{11} - c_{12})c_{44} - 2c_{14}^2 = 2(c_{66}c_{44} - c_{14}^2) > 0$. These conditions are labeled B_2 and B_3 in [102]. The main finding of this chapter is that the condition labeled here B_2 decreases above 15 GPa, and reaches 0 around 45 GPa.

Unfortunately, it has been later shown by US measurements [104] that there are some inconsistencies in [102], and that the attribution of the modes is not correct. The Born criterion both increase with pressure. c_{44} increases with pressure, while c_{66} decreases.

It might be emphasized that this powerful technique should be handled with care, because the orientation of the studied crystal is not checked with X-rays.

SiO_2 at very high temperature

Silica is an important material for its technological application, as well as for its unusual properties. For instance, in the glassy state its compressibility increases with pressure, as do its refractive index and elastic moduli with temperature.

Glassy silica is made up of a three-dimensional open network of SiO_4 tetrahedra linked by bridging oxygens. The very strong Si–O bonds have an almost constant

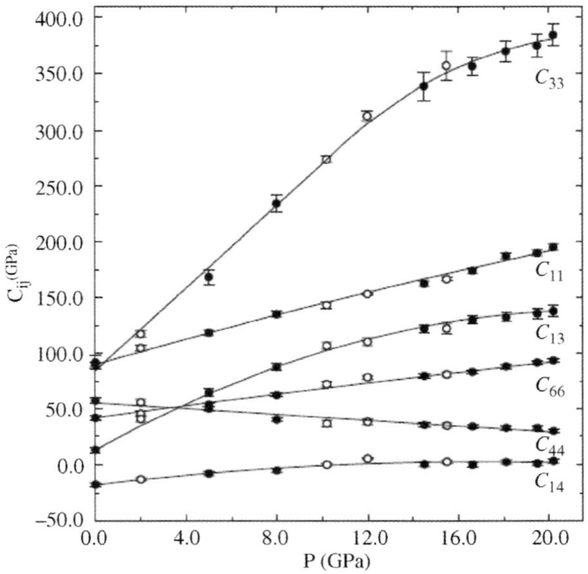

Figure 2.29 Elastic moduli of quartz as a function of pressure (Reprinted figure with permission from Ref. [102]. Copyright 2010 by the American Physical Society). Solid (empty) symbols represents results obtained during (de)pressurization.

length, and the disorder of the structure is given by the wide range of values adopted by Si–O–Si angles. As a result, the thermal-expansion coefficient of the glass is small, like that of the liquid.

The physical properties of liquid SiO_2 remain not well known, however, due to the experimental difficulties related to a very high glass transition temperature

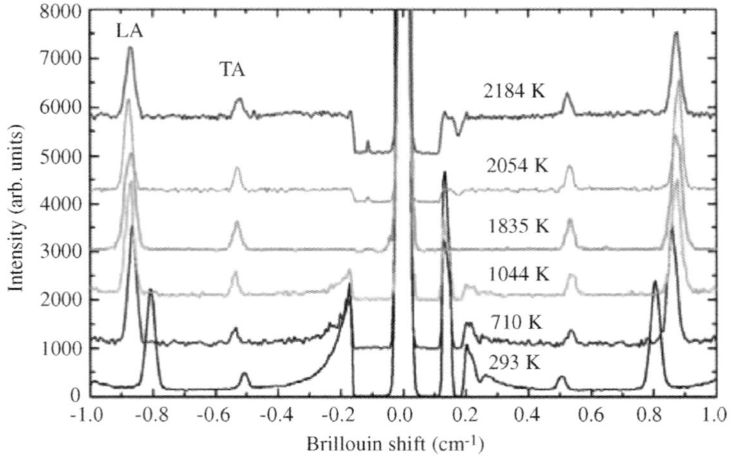

Figure 2.30 Brillouin scattering spectra of SiO_2 obtained at various temperatures in the 90° geometry. The central peak is the Rayleigh line, and the peaks to the left and to the right are the anti-Stokes and Stokes Brillouin lines, respectively.

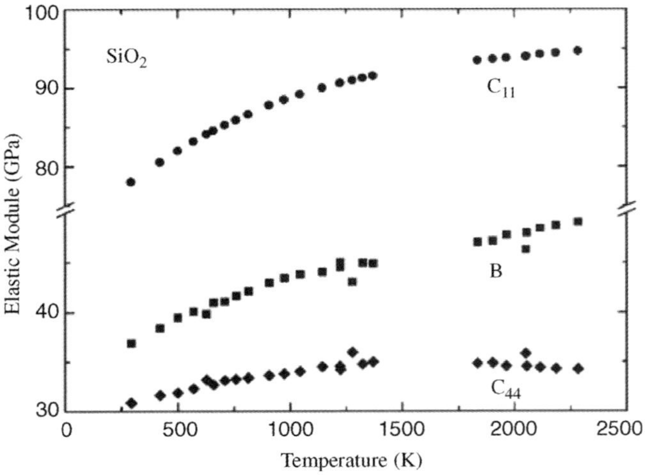

Figure 2.31 Longitudinal (C_{11}, circles), transverse (C_{44}, diamonds), and bulk (squares) moduli against temperature. Reprinted figure from Ref. [105] by permission from EDP Sciences.

(1450 K) and extremely high viscosity, that is, above 2000 K, the melting point of cristobalite.

Brillouin scattering studies have been performed up to 2300 K in a 90° geometry, which enabled to measure the velocities of both the longitudinal and the transverse acoustic modes [105]. The sample was first melted inside a heating wire made of platinum–iridium alloy and cooled down before the start of the experiment. Typical Brillouin spectra obtained using this setup and a tandem Fabry–Pérot interferometer are shown in Figure 2.30 for various temperatures. From the frequency shift, the sound velocity has been obtained through a modeling of the refractive index as a function of temperature. From the longitudinal and transverse sound velocities, the elastic moduli could be deduced from $C_{11} = \varrho . V_L^2$ and $C_{44} = \varrho . V_T^2$, respectively. Here, ϱ is the density calculated with the assumption of a constant thermal expansion coefficient and V_L (V_T) is the longitudinal (transverse) sound velocity. From the two elastic moduli of an isotropic medium, the bulk modulus is given by $B = C_{11} - 4C_{44}/3$. The three moduli are shown in Figure 2.31 as a function of temperature. There, the temperature range without experimental points is due to the recrystallization of the sample, which prevents any measurement. One striking result of this experiment is the continuous increase of the bulk modulus, that is, the decrease of the compressibility, up to the highest temperature reached.

2.2.4 Optical Modes

2.2.4.1 Pressure Aspect

Carbon dioxide presents a good example of the power of optical spectroscopies for the determination of the structural and vibrational properties of matter. At ambient

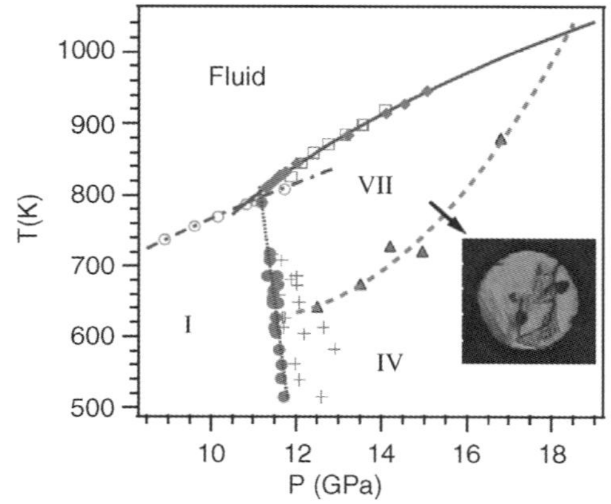

Figure 2.32 Phase transition points of CO_2 from the Raman scattering (Reprinted figure from Ref. [107] by permission from EDP Sciences). The curves are fit to the data. The melting points of phase IV and phase VII are well fitted by the Simon–Glatzel law: $T = T_0[1 + (P - P_0)/a]1/c$ with $T_0 = 805$ K, $P_0 = 11.15$ GPa, $a = 4.6(8)$ GPa and $c = 3.8(6)$. Photograph: sample of phase VII (with pressure sensors) at 770 K and 11.8 GPa.

conditions, CO_2 is a simple molecular gas, and an important atmospheric constituent in the sense that it is one of the powerful global warming gases. Subjected to large compression, the tendency of the molecular gases is that the intramolecular bonding weakens with respect to the intermolecular one, leading finally to nonmolecular solids. Such an example was given in Section 2.2.1.2 in the case of nitrogen. CO_2 was found also to lose its molecular character at high pressure and high temperature [106].

The phase diagram was explored by different techniques such as X-ray diffraction, Raman, and infrared spectroscopies. However, the full complexity of this phase diagram is certainly far from being known, and the recent Raman measurements led to the discovery of a new structure along the melting curve ([107], see Figure 2.32), CO_2-VII. The symmetry of this phase was determined by X-ray diffraction, and is orthorhombic Cmcb (identical to Cmca with an inversion of *a*- and *b*-axes).

Like many phases of carbon dioxide, this could be maintained down to 300 K (Figure 2.33) in a metastable state. In the cited study, the authors prove that phase VII is strictly molecular, with a C=O bond length not far from that in phase I. Moreover, they could determine the internal energy difference between phases IV and VII, that is, 2.5 meV/molecule at 726 K. This strongly supports that phase IV, and most likely phase II, are molecular phases too, in contradiction with the phase diagram presented in Figure 2.34.

Amorphous materials, amorphous-amorphous transition, pressure-induced amorphisation. The vibrational spectrum of a disordered material, glass or amorphous, is characterized by a low-frequency excitation, commonly referred to as the "Boson peak." It is observed in the Raman scattering and in inelastic neutron scattering. The origin of this boson peak is still controversial, but is in most cases ascribed to an

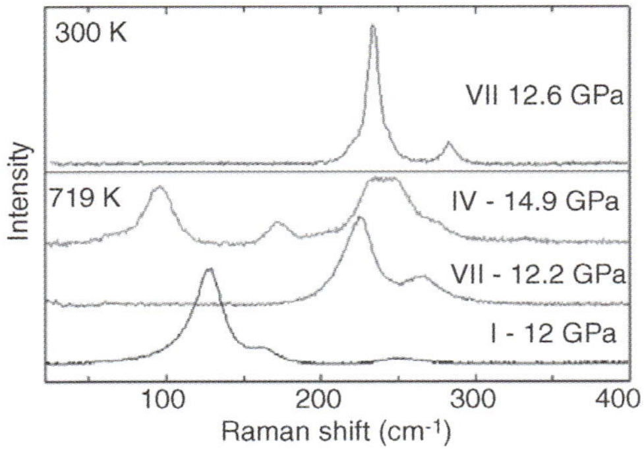

Figure 2.33 The Raman spectra of CO_2 collected on quenched phase VII at 300 K (top) and along an isotherm at 719K going from phase I to phase IV through phase VII (bottom). Reprinted figure from Ref. [107] by permission from EDP Sciences.

excess density of vibrational states whose origin is in transverse vibrational modes [108, 109]. The pressure dependence of the boson peak could be used to clarify his properties, but it clearly depends on the studied materials. For example, in glassy As_2S_3, "pressure effects on the Boson peak are manifested as an appreciable shift of its frequency to higher values, a suppression of its intensity, as well as a noticeable change of its asymmetry leading to a more symmetric shape at high pressures" [110]. On the contrary, in poly(isobutylene) densified by 20% at 1.4 GPa, if the intensity decreases and there is a shift to higher frequencies, "surprisingly, the shape of the boson peak remains unchanged even at such high compression" [111]. The temperature dependence of the Boson peak in glasses will be discussed in the following Section 2.2.4.3, through the example of silica.

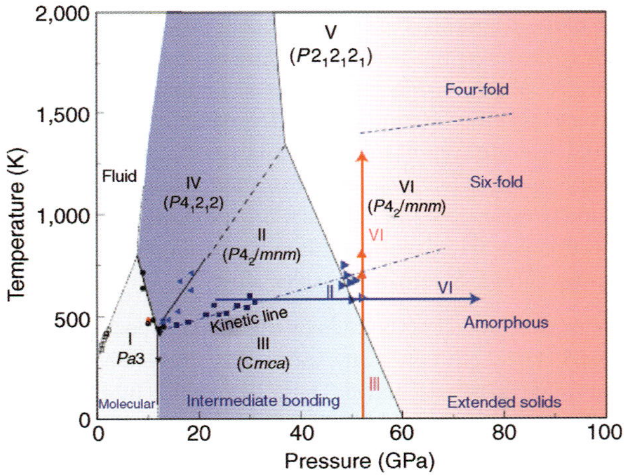

Figure 2.34 Phase diagram of CO_2 showing the molecular (white), intermediate (blue), and nonmolecular (red) phases. The arrows represent typical experimental paths. Reprinted by permission from Macmillan Publishers Ltd: Nature Materials, Ref. [106], copyright 2010.

The story of first-order amorphous–amorphous transition was open with the case of the low-density amorphous (LDA) to high-density amorphous (HDA) transition in water ice [112]. The order of the transition was deduced from ΔV measured by the displacement of the piston in a piston-cylinder apparatus. It was later confirmed directly by neutron diffraction in a Paris–Edinburgh Press and by the Raman scattering in a diamond–anvil cell [113]. In this work, the authors show *in situ* the coexistence of the two forms of amorphous ices, and the transformation from one form to the other one at constant pressure.

The compression of crystals leads to denser structures, crystalline, or amorphous. This pressure-induced amorphization may be due to potential barrier, and may happen either suddenly like in ice at low temperature and high pressure [114] or in several steps. The case of $Co(OH)_2$ illustrates this last process [115]. At ambient temperature, the Raman and infrared absorption spectra show a decrease of the intensity and a large broadening of the modes around 11 GPa, characteristic of a pressure-induced amorphization. The transition is shown to be reversible. However, X-ray diffraction show that the compound retains crystalline order at least up to 30 GPa, that is, much higher than the "amorphization" pressure. Because the X-rays intensities are insensitive to the positions of the hydrogen atoms, the authors conclude that the Co–O sublattice remains ordered, while the hydrogen sublattice becomes disordered. In $Sc_2(MoO_4)_3$ the story presents similarities, in the sense that in a first step the MoO_4 tetrahedra become disordered around 4 GPa, as shown by the large broadening of the internal modes of the molybdate ions. Finally, a complete amorphization of the crystal does occur only above 12 GPa [116]. In that case, the amorphization is an irreversible transformation.

2.2.4.2 **Temperature Aspect**

Searching to access optical modes in the high-temperature range deals most often with large fields of interest. First of them concerns condensed matter physics and physical chemistry: phase transitions, chemical species, anharmonicity, metastability, and mechanisms of conductivity. A specific case is the physics and chemistry of glasses and their related liquids. The goals proceed from a structural point of view, chemical composition, behavior around T_g, and effects of annealing. Glasses at high temperature constitute a multidisciplinary field in which scientists from the fields of condensed matter physics, chemistry, and geoscience act and interact. In order to display the main features of optical mode spectroscopy at high temperatures, typical applications will be reviewed hereafter concerning diamond, refractory oxides (spinels), and glasses.

Diamond was one of the most studied compounds by the Raman spectroscopy at high temperatures [48, 59–65, 117–121]. Obviously, infrared methods here are not relevant since no first-order mode is infrared active. The reason of this interest is the need for monitoring crystal growth of diamond films by CVD methods. Figure 2.35 displays the dependence of the diamond Raman line during plasma-assisted CVD deposition. The increase of the line intensity evidences an increasing thickness of the film; the line position is temperature dependent and stress dependent. It can be used as a thermometer; during growth, the film surface is supposed to be stress-free and

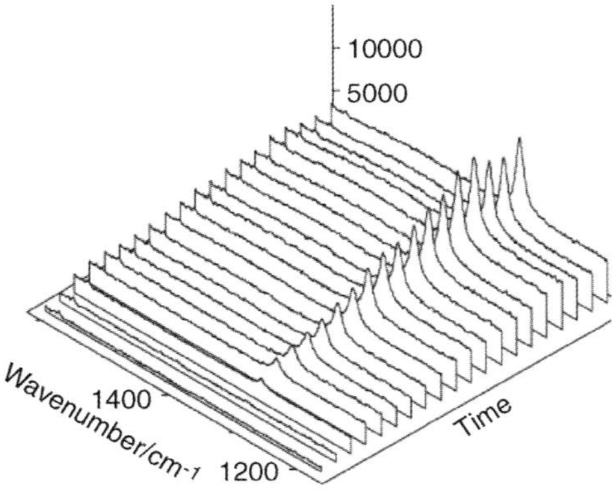

Figure 2.35 The Raman spectra recorded during the growth of a diamond film on a silicon substrate, inside a plasma-assisted CVD reactor ($T = 850\,^\circ C$). The figure covers 24 h of deposition. Reprinted figure from Ref. [120] by permission from Wiley.

the line position can be used as a thermometer. If temperature is known by another way, one can monitor the level of stresses inside the film, during annealing treatments, for instance. Note that even if the temperature in this example is not so high (850 °C), the light emitted by the plasma gives a background signal comparable to much higher temperatures.

Cubic boron nitride is rather analogous to diamond and was the object of several studies at high temperatures [44, 58], with a particularly spectacular *in situ* evidence of the cubic-to-hexagonal transformation [44] at 1570 °C (see Figure 2.36). Aluminum nitride of related structure was also investigated [122].

Refractory oxides obviously take a large place in high temperature *in situ* characterization. Among them, the spinel $MgAl_2O_4$ takes a peculiar place, for geoscientists

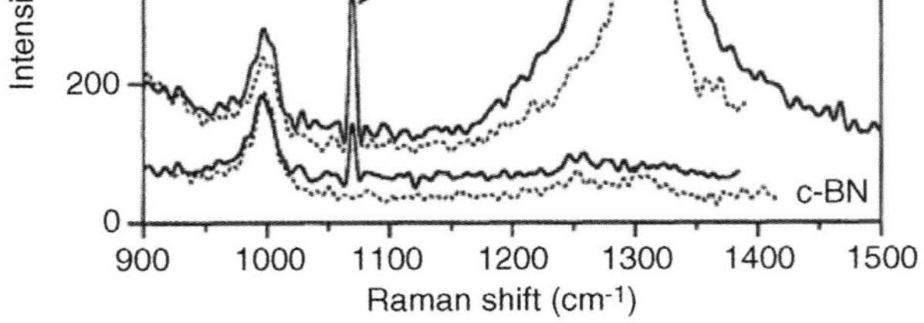

Figure 2.36 The Raman spectra exhibiting the cubic to hexagonal form of BN at 1840 K (cal. line denote a frequency–calibration line). Reprinted figure with permission from Ref. [44]. Copyright 2010 by the American Physical Society.

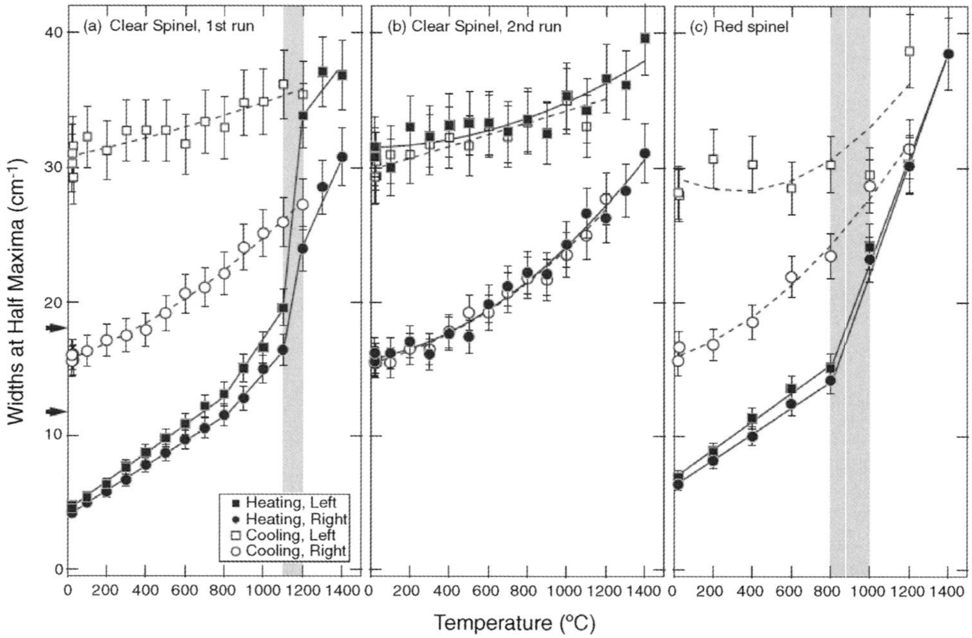

Figure 2.37 The left and right widths at half maximum for the Raman *Eg* mode of $MgAl_2O_4$ at high temperatures, on two spinels of different origin. The gray zone indicate the irreversible order–disorder transition. The widths in the nonstoichiometric synthetic spinel at room temperature are indicated by arrows. Reprinted figure from Ref. [74] by permission from the Mineralogical Society of America.

as for condensed matter physicists. One of the reasons of this interest is the irreversible order–disorder transition occurring in this compound above 1000 °C: the cations are ordered below this range (Al on octahedral sites and Mg in tetrahedral ones), and become progressively disordered above. This compound was the object of several studies dealing with this transition, through the infrared emittance and the Raman experiments [74, 81, 123]. Figure 2.37 displays the temperature dependence of linewidths obtained from the Raman measurements [74] (as disorder induces appearance of new lines of the low-frequency wing of the main lines, it results a strong asymmetry of the resulting lines). This figure evidences the increasing of disorder in a temperature range slightly dependent of the thermal history of the sample. Disorder is characterized by an inversion parameter x (the real formula being $(Mg_{1-x}Al_x)[Mg_xAl_{2-x}]O_4$, where parentheses stand for tetrahedral sites and square brackets stand for octahedral ones. De Sousa Meneses *et al.* [81] have succeeded in determining accurately this inversion parameter from intensity of infrared lines. The same authors have extended their infrared emittance measurements up to the liquid state ($T_{melting}$ ~2125 °C), with evidence of a noticeable increase of the emittance in the mid-infrared range (3000–5000 cm^{-1}) beginning more than 200 °C below melting (Figure 2.38). At the same time, vibration mode frequencies are not modified, whereas they undergo a clear jump at melting temperature (Figure 2.39). The increase in emittance is explained by a Debye-like relaxation thermally activated,

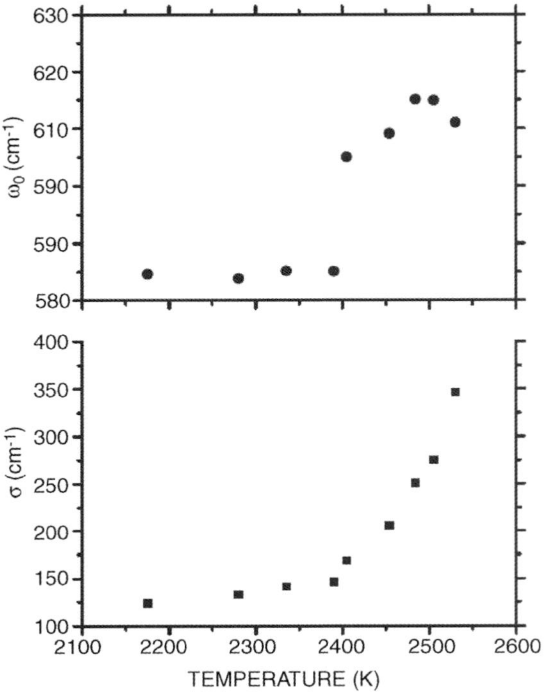

Figure 2.38 Effect of the solid to liquid phase transition on frequency and full width at half maximum of a typical vibration mode of $MgAl_2O_4$. Reprinted figure from Ref. [81] by permission from the Institute of Physics.

with activation energy unaffected at the melting. Similar behaviors of increase of absorption in the mid-IR were also observed in other oxides (MgO and Al_2O_3), but not at all in silica [80]. The short characteristic time of the relaxation leads to think that it is some *electronic* contribution. A deeper understanding of this phenomenon is still lacking.

Another class of refractory materials was also object of several reports: zirconia and yttria-stabilized zirconia [56, 67, 124, 125]. The sequence of phase transitions, the effect of size particle (nanostructured ZrO_2), and the need for a better knowledge of stabilized zirconia (a material often used in high-temperature devices) were the aim of these studies.

Another class of refractory materials was also object of several reports: zirconia and yttria-stabilized zirconia [56, 67, 124, 125]. The sequence of phase transitions, the effect of size particle (nanostructured ZrO_2), and the need for a better knowledge of stabilized zirconia (a material often used in high-temperature devices) were the aim of these studies.

REPETITIVE:

Glasses, as a generic topic, are the object of numerous studies. The high-temperature range correspond to many interesting physical subjects: physics of the glass tran-

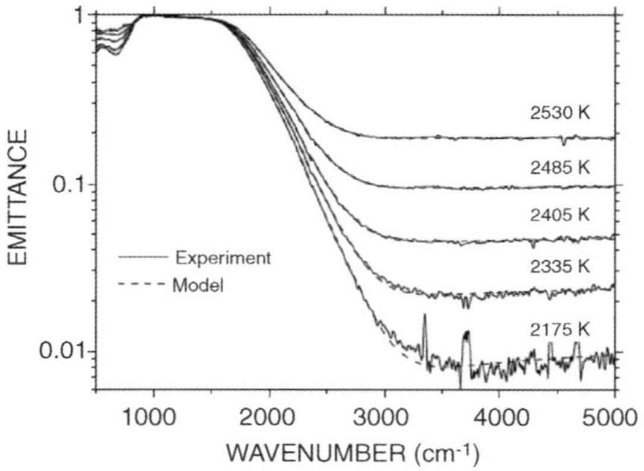

Figure 2.39 Set of emittance spectra of $MgAl_2O_4$ in both solid and liquid states. Melting is around 2400 K. Results show appearance of a relaxation process inducing an emittance increase in the mid-IR. Reprinted figure from Ref. [81] by permission from the Institute of Physic.

sition, structure of the liquid and of the undercooled liquid, coordination number in the liquid and the glass, local changes on recrystallization, and monitoring of stresses and of structural effects of annealing for the main ones, which are mainly not only silicates but also aluminates [51, 57, 69, 77, 80, 126–143].

Silica takes an important place in this list with important contributions, starting by Exharos's one that can be considered as a pioneering work [57] of the high-temperature Raman spectroscopy. The glass transition of silica is approximately 1200 °C, meaning that measurements of the liquid above this temperature and of the glass upon annealing in this range are of the highest interest for a better knowledge of this model glass. Many works were undertaken, some of them were undertaken recently (Raman: ([144–146] (Figure 2.40), [51, 139, 147]) and infrared: [79, 80, 148]). The main results concern the nature of the structural relaxation close to T_g (decrease of S–O–Si average bond, increase of Si–O bond stretching force, temperature dependence of the "ring " modes D_1 and D_2), and structure of the liquid at very high temperature, with breaking of some intertetrahedra bonds and formation of terminal Si=O bonds. It was also shown that the boson peak persists not only in the supercooled liquid but also in the stable liquid state at higher temperature, with three distinct regimes (glass, supercooled, and molten). The existence of the boson peak in the stable liquid could be supported by a specific topology of the potential energy landscape [145].

Molten salts. Besides all these very high temperatures, sometimes the word "extreme" applies only to the temperature but also to the chemical nature of the sample under study. Some liquids are much aggressive at intermediate temperatures and constitute serious problems to imagine a sample environment compatible with temperature, corrosive conditions, and optical measurements. It is typically the case of molten salts, mainly fluorides, for which specific devices were conceived operating at somewhat intermediate temperature (typically 1000 °C) [149–151].

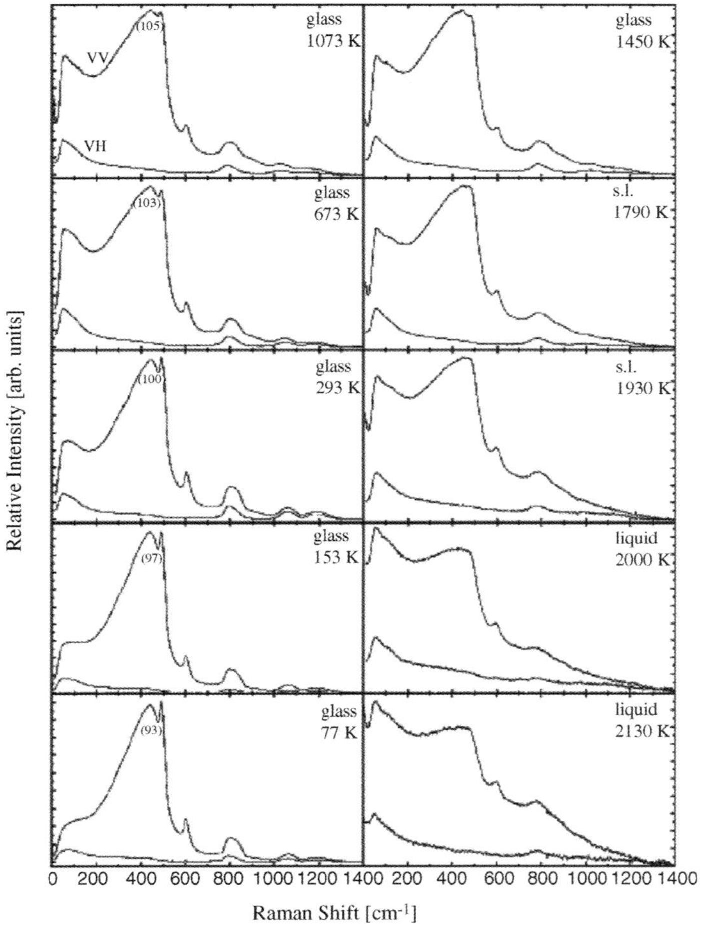

Figure 2.40 The Raman spectra of silica in glassy, supercooled (s.l.), and liquid states. Reprinted with permission from Ref. [146]. Copyright 2010, American Institute of Physics.

2.3 Perspectives

2.3.1 Instrumentation

2.3.1.1 Natural Development of Existing Setups

The past years were marked by an significant increase of activity in the high-temperature Raman scattering, mainly due to important technological improvements, with the availability of ICCD devices. This impulse will presumably continue on the next years due to the interest for high *T* in many domains (see hereafter) One can think in a realistic future to a new breakthrough with ICCDs operating in the UV range, which would mix temporal filtering and UV-shift advantage.

Infrared methodology is up to date largely less developed than the Raman methodology for high temperatures. The emissivity measurements systems remain rare, at least aimed by an aim in solid (or molten) state and material sciences. However, it undoubtedly constitute the most efficient way to attain very high temperatures, up to 2500–3000 °C – the highest temperature results reported in this chapter were obtained through IR emissivity measurements.

2.3.1.2 Innovative Combinations of X-ray and Vibrational Spectroscopies

It is clear that the data obtained by vibrational spectroscopies alone are not sufficient to understand the properties of the compounds, especially when pressure or temperature are used as external thermodynamic parameters. In any case, one needs the structure of the sample under examination, the symmetry, the interatomic distances, the density, and the dependence of these quantities with the pressure/temperature. In other words, X-ray diffraction and/or X-ray absorption give prerequisite information in order to interpret any result given by the other techniques. Another one problem is that under extreme conditions, especially when the measurements are performed at both high pressure and high temperature, the reproducibility of a given thermodynamic condition is extremely difficult. When a phase transformation is generated at high pressure, measurement of the maximum of physical properties is highly desirable because it may happen that the new phase is (meta)stable in a small part of the *P*, *T* phase diagram. It is the reason why in several synchrotron radiation facilities, Raman, infrared, or Brillouin setups are build in beamline stations to allow the simultaneous measurement of structural and vibrational/elastic properties. One good example is given by the 13-BM-D station (GSE-CARS, see Figure 2.41) of the Advanced Photon Source (Argonne, USA). This X-ray

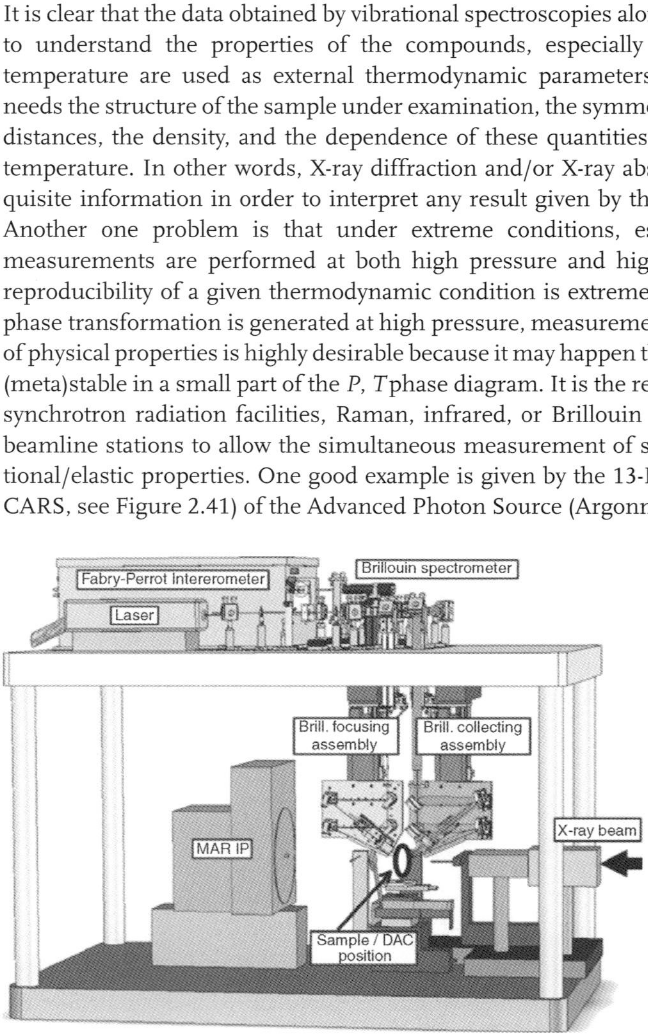

Figure 2.41 Schematic view of the setup installed on the 13-BM-D beamline of the APS. Reprinted with permission from Ref. [152]. Copyright 2010, American Institute of Physics.

diffraction beamline is installed on a bending magnet and equipped with a MAR image plate detector for the X-rays. A Brillouin interferometer has been installed on an optical table above the X-ray beam. An optical system enables to make the measurements in a diamond–anvil cell without any movement, simultaneously with the X-ray diffraction [152]. With this setup, the authors could measure the elastic properties of MgO together with the X-ray diffraction and of NaCl at high pressure and high temperature, with an oven around the diamond anvils.

The number of experimental stations developing capability of the Raman spectroscopy together with the X-ray measurements (both diffraction and absorption) is growing fast. For example, the ID13, as well as the SNBL beamline of the ESRF (Grenoble, France) and the SAMBA beamline at SOLEIL (Paris, France) offer the possibility of micro-Raman measurement both in-line and off-line, but for the time being not simultaneously with the X-ray diffraction measurement.

2.3.2 Physical Phenomena

2.3.2.1 Phonons (Zone-Center): A Natural "Mesoscope" into the Alloy Disorder

Since the pioneer extended X-ray absorption fine structure (EXAFS) measurements performed by Mikkelsen and Boyce with $In_{1-x}Ga_xAs$ [153] it is well known that each bond tends to retain its natural bond length in an alloy, corresponding to a 1-bond $\rightarrow$ 1-length picture of the lattice relaxation of zinc blende alloys. The naive relation quoted in italics in Section 2.1.3.2.2 between the bond length and the bond force constant suggests a similar 1-bond $\rightarrow$ 1-TO generic scheme for the lattice dynamics. Such behavior is well accounted for by the modified random element isodisplacement (MREI) model [154], where each mode shifts smoothly between the frequencies of the parent and the impurity (ω_{imp}) modes when x changes, its intensity scaling as the related bond fraction. No singularity is expected. Only one input parameter is needed per bond, that is ω_{imp}. The whole trend can be figured out by imagining that the bond of like species is equivalent, immersed into the same VCA-type continuum (Figure 2.42a). In this sense the MREI model is VCA-like.

When confronted with the Raman/IR data the 1-bond $\rightarrow$ 1-*TO* generic scheme comes into two main types: (i) pure 1-bond $\rightarrow$ 1-mode, that is, well separated A–B and A–C TO modes that shift smoothly between the parent and the impurity modes with antagonist intensity variations when x changes and (ii) 2-bond $\rightarrow$ 1-mode, that is, a unique (A–B, A–C)-mixed TO mode that shifts regularly between the parent TO modes with quasistable intensity. There exists also a sort of intermediary type (exceptional), referred to as (iii) modified two-mode, with two (A–B, A–C)-mixed TO modes, that is, a dominant one of type (ii) that joins the parent TO modes, plus a minor one connecting the impurity modes, that does not allow to fully discard type (i). The leading systems in each case are InGaAs [type (i)], ZnTeSe [type (ii)], and InGaP [type (iii)]. A more detailed classification was proposed by Genzel *et al.* [155]. The basic MREI-VCA classification in two main types – (i) and (ii) – is supported by a theoretical criterion worked out by Elliott *et al.* [156] within the coherent potential approximation (CPA). This predicts type (i) if the optical bands of the A–B and A–C bonds do

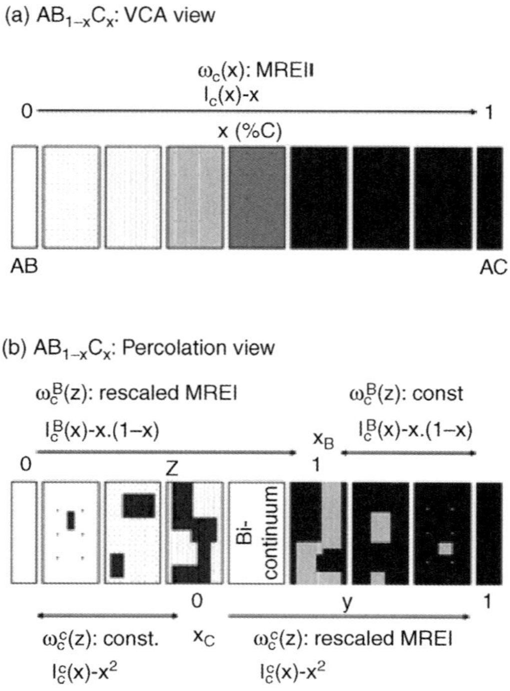

Figure 2.42 Schematic views of a random $AB_{1\text{-}x}C_x$ alloy according to the VCA (a) and percolation (b) schemes. The gray scales reinforces when the local coposition becomes more like that of the corresponding pure crystal. A MREI-like correlation between the AC-like TO frequencies (ω) and the actual (x) or rescaled (y, z) alloy compositions is emphasized. The intensity aspect (I) is indicated within brackets. For each (ω,I) couple, subscript and superscript B (C) refer to the A–B (A–C) bond and to the pale AB-like (dark AC-like) regions, respectively. Similar descriptions apply to the A–B and A–C bonds. Reprinted figure with permission from Ref. [159]. Copyright 2010 by the American Physical Society.

overlap in the alloy; type (ii) otherwise. The optical bands here are simply built up by linear extrapolation of the parent TO and LO frequencies onto the related ω_{imp} values. Generally, MREI-VCA (experimental) and Elliott-CPA (theoretical) classifications were found to be consistent [157]. However, in the past few years, a careful reexamination of the available vibrational data in the literature has revealed that even the InGaAs, ZnTeSe, and InGaP do not obey the phonon mode behaviors that they are supposed to represent [19].

Pagès *et al.* proposed that, owing to the local character of the bond force constant, even basic understanding of the phonon (TO) behavior in alloys should rely on detailed insight into the local topologies of the (B, C) substituting species, guiding to the percolation site theory. In fact, all the apparent anomalies in the vibration spectra of the "leading" alloys above could be explained within a so-called percolation model, introducing a vision of a random zinc blende $AB_{1\text{-}x}C_x$ alloy at the *mesoscopic* scale in terms of a composite of the AB-like and AC-like interpenetrated host regions, as originating from natural fluctuations in the x-value at the local scale (Figure 2.42b).

For a given bond, each region brings a specific TO mode, corresponding to a 1-bond → 2-mode (TO) behavior. In a ω_T versus x plot, one should consider that each individual MREI-like TO branch is split into a symmetrical double-branch attached at its two ends to the parent and impurity modes, as characterized by the splitting Δ in the dilute limits. The microscopic mechanism behind is a slight difference in bond length for a given bond species in the AB-like and AC-like regions. The difference is of the order of ~1% as estimated from *ab initio* calculations [158], which seems to be enough to create phonon localization – cf. Anderson's criterion in Section 2.1.3.2. The intensity of each TO submode scales as the fraction of like bonds in the crystals, that is, as (1-x) for A–B bonds and as x for A–C ones, weighted by the scattering volumes of the corresponding AB-like (1-x) or AC-like host region. Further, each subbranch is characterized by two regimes on each side of the related bond percolation threshold (PC), that is, $x_C \sim 0.19$ for the AC-like region and $x_B \sim 0.81$ for the AB-like one. Below PC, where the minor region consists of a dispersion of finite-size clusters with quasistable internal structure, a so-called "fractal-like" regime is identified where ω_T is stable. Above PC, the clusters have coalesced into a tree-like continuum, and ω_T has turned smoothly x-dependent up to the parent limit. This is referred to as the "normal" regime. There, the ω_T versus x-curve is modeled by applying a MREI approach to the restricted x-domain covered by the continuum. Such rescaled-MREI description comes, in fact, to view each percolation

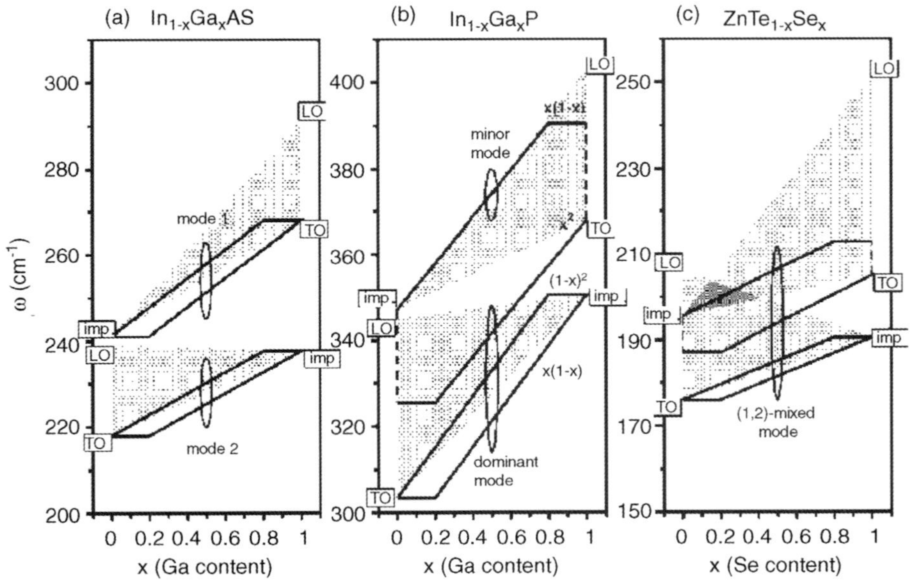

Figure 2.43 Schematic 1-bond → 2-mode (TO) percolation schemes of InGaAs (a), InGaP (b), and ZnTeSe (c), with rescaled-MREI oblique segments simplified to straight lines for more clarity. The intensity of each TO mode in the Raman spectra scales as indicated within brackets in inset (b). The optical (TO–LO) bands used for the Elliott-CPA criterion are shown as shaded areas. Reprinted figure with permission from Ref. [159]. Copyright 2010 by the American Physical Society.

continuum at PC – a pure fractal – as a pseudoparent/binary. Two input parameters only are required per bond to implement the percolation model, that is, ω_{imp}, as for the MREI model, plus Δ, the splitting between the like phonon modes from the AB-like and AC-like regions in the dilute limit.

The 1-bond → 2-mode (TO) percolation schemes for InGaAs, InGaP, and ZnTeSe are schematically reproduced in Figure 2.43. An obvious analogy exists between the three schemes, indicating that the traditional MREI-VCA/Elliott-CPA classification has in fact no *raison d'être*, and can be unified within the percolation scheme (see also Ref. [159]).

On the practical side, the VCA → percolation shift of paradigm reveals that zone-center phonons provide a natural insight into the alloy disorder at the unusual mesoscopic scale: a terminology is introduced that such phonons do act as a "mesoscope." As such they may be used to study how a pressure-induced structural phase transition is initiated in an alloy. It is a matter to decide whether the AB- and AC-like regions, as identified by distinct TO modes *per bond*, are affected in the same way by pressure, or whether one region is given a specific role. This remains unexplored.

2.3.2.2 Elucidation of the Mechanism of the Pressure-Induced Phase Transformations

One of the most important phenomena to understand is the determination of the phase transformation path. Indeed, for example, for SiO_2 α-quartz, every possible mechanism has been proposed for the pressure-induced amorphization: soft acoustical phonon, soft optical phonon, B_1 or B_2 Born stability criteria, and so on.

Performing at the same time experimental studies and *ab initio* calculation is now mandatory to improve our knowledge. One good example of what should be done is given by the wurtzite → rock salt transformation in ZnO and III-N compounds. In a recent paper [160], a general path of transformation (see Figure 2.44) is deduced from *ab initio* calculation performed in the density-functional theory (DFT) framework and the local density approximation (LDA). *Ab initio* phonon dispersion curves are determined through the density-functional perturbation theory.

This work shows that the fourfold coordinated wurtzite to the sixfold coordinated rocksalt phase transformation occurs through a quasitetragonal intermediate structure, which bridges in a crystallographic simple way, the two structures. This hypothesis is confirmed in the compounds with d-electrons ZnO, GaN, and InN and strongly supported by the experimental and *ab initio* determination of the shear constants and phonons under pressure. The other path through the hexagonal intermediate structure seems possible only in compounds containing d-electron-free cations, such as AlN and *w*-SiC. Important result of this study concern the predictions on the dynamical and elastic pressure properties of studied wurtzite compounds, such as the softening of TA modes and of the C_{44} and C_{66} shear elastic moduli in ZnO, GaN, and InN, or the independence of the wurtzite internal parameter u on pressure. The prediction is validated by the experimental results in ZnO, but are clearly needed for GaN and InN. In conclusion, many different types of experiments combined with *ab initio* calculations are needed for better understanding of the properties of compounds under extreme conditions.

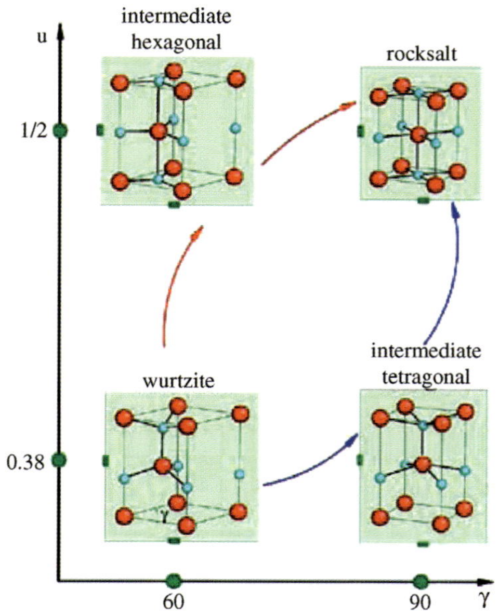

Figure 2.44 *Ab initio* calculation of the wurtzite → rock salt transformation in ZnO and III-N compounds. Reprinted figure with permission from Ref. [160]. Copyright 2010 by the American Physical Society.

2.3.2.3 **Glasses**

Glasses (and their associated liquids) were cited above as presumably the domain with the highest number of reports in the high-temperature range. Clearly, this enthusiasm will continue, as data in this field are essential for many different scientific communities: earth sciences (with underlining once more that many major experimental improvements were done by them), condensed matter physicists, with opportunity to attain pertinent data about glassy transition and structural relaxation mechanisms on annealing (the simplest glasses on the structural point of view have very high T_g and melting temperatures), chemists with access to high-temperature molten states (structure and composition), and material science people with many applications (vitroceramics, refractories, and so on.).

The solid–liquid (melting from crystal) transition remains also a domain with open questions, as the exact mechanisms of melting. As for glasses, some beyond the structurally simplest materials have the highest melting temperature, as MgO. The report of some electronic-like relaxation in this temperature range cited above has no explanation till now and deserves further investigations.

A particular field of applications will also certainly devote numerous studies in the next future: the need of knowledge for materials constituting the future Generation IV nuclear reactors. The temperature range of interest of these materials will be 1200–2000 °C and obviously this will necessitate a large effort of research to develop methods to be adapted for these extreme conditions: temperature, pressure (nevertheless, largely below what is reported here), but also under irradiation. In particular, one can think of important developments with respect to *in situ* characterizations, which can be compatible with optical spectroscopies.

Acknowledgements

Thanks are due to Dr. De Sousa Meneses, for fruitful discussions and careful reading of the infrared-data part of this chapter.

References

1 Martin, R.M. (1970) *Phys. Rev. B*, **1**, 4005.
2 Weber, W. (1977) *Phys. Rev. B*, **15**, 4789.
3 Rustagi, K.C. and Weber, W. (1976) *Solid State Commun.*, **18**, 673.
4 Born, M. and Huang, K. (1954) *Dynamical Theory of Crystal Lattices*, Oxford University Press, Oxford, England.
5 Poulet, H. and Mathieu, J.-P. (1970) *Spectres de Vibration et Symétrie des Cristaux*, Gordon & Breach Science Publishers, New York.
6 Fröhlich, H. (1958) *Theory of Dielectrics*, Oxford University Press, Oxford, p. 21.
7 Kittel, C. (1996) *Introduction to Solid State Physics*, 7th edn, John Wiley & Sons, Inc., New York, p. 347.
8 Henry, C.H. and Hopfield, J.J. (1965) *Phys. Rev. Lett.*, **15**, 964.
9 Lockwood, D.J., Yu, G., and Rowell, N.L. (2005) *Solid State Commun.*, **136**, 404.
10 Faust, W.L., Henry, C.H., and Eick, R.H. (1968) *Phys. Rev.*, **173**, 781.
11 Gervais, F. (1983) *Infrared and Millimeter Waves* (ed. K.J. Button) Academic Press, New York, p. 279.
12 Born, M. and Wolf, E. (1964) *Principles of Optics: Electromagnetic Theory of Propagation: Interference and Diffraction of Light*, Pergamon Press, Oxford.
13 Barker, A.S. and Loudon, R. (1972) *Rev. Mod. Phys.*, **44**, 18.
14 Hon, D.T. and Faust, W.L. (1973) *Appl. Phys.*, **1**, 241.
15 Cardona, M., Etchegoin, P., Fuchs, H.D., and Molinàs-Mata, P. (1993) *J. Phys. Condens. Matt.*, **5**, A61.
16 Cardona, M. (1982) Chapter 2, *Light Scattering in Solids II, Topics in Applied Physics*, vol. 50 (eds M. Cardona and G. Güntherodt), Spinger-Verlag, Berlin.
17 Yu, P.Y. and Cardona, M. (2001) *Fundamentals of Semiconductors*, Springer-Verlag, Heidelberg.
18 Sandercock, J.R. (1982) *Light Scattering in Solids III: Recent Results, Topics in Applied Physics*, vol. 51 (M. Cardona and G. Guntherodt), Springer-Verlag, Berlin, p. 176.
19 Pagès, O., Postnikov, A.V., Kassem, M., Chafi, A., Nassour, A., and Doyen, S. (2008) *Phys. Rev. B*, **77**, 125208.
20 Mintairov, A.M., Mazurenko, D.M., Sinitsin, M.A., and Yavich, B.S. (1994) *Semiconductors*, **28**, 866.
21 Pedrotti, F.L. and Pedrotti, L.S. (1993) *Introduction to Optics*, Prentice-Hall, Englewood Cliffs.
22 Berthier, S. (1993) *Optique des Milieux Composites*, Polytechnica, Paris.
23 Garnett, J.C.M. (1906) *Philos. Trans. R. Soc. Lon.*, **205**, 237.
24 Bruggeman, D.A. (1935) *G. Ann. Phys. (Leipzig)*, **24**, 636.
25 Cummins, H.Z. and Schoen, P.E. (1972) *Laser Handbook* (eds F.T. Arecchi and E.O. Schulz-Dubois), North-Holland, Amsterdam, p. 1029.
26 McMahan, A.K. and LeSar, R. (1985) *Phys. Rev. Lett.*, **54**, 1929.
27 Mailhiot, C., Yang, L.H., and McMahan, A.K. (1992) *Phys. Rev. B*, **46**, 14419.
28 Eremets, M.I., Gavriliuk, A.G., Trojan, I.A., Dzivenko, D.A., and Boehler, R. (2004) *Nat. Mater.*, **3**, 558.
29 Eremets, M.I., Gavriliuk, A.G., and Trojan, I.A. (2007) *Appl. Phys. Lett.*, **90**, 171904.
30 Lipp, M.J., Park Klepeis, J., Baer, B.J., Cynn, H., Evans, W.J., Iota, V., and Yoo, C.-S. (2007) *Phys. Rev. B*, **76**, 014113.

31 Kruger, M.B. and Jeanloz, R. (1990) *Science*, **249**, 647.

32 Polian, A., Grimsditch, M., and Philippot, E. (1993) *Phys. Rev. Lett.*, **71**, 3143.

33 Gillet, P., Badro, J., Varrel, B., and McMillan, P. (1995) *Phys. Rev. B*, **51**, 11262.

34 Sharma, S., Garg, N., and Sikka, S.K. (2000) *Phys. Rev. B*, **62**, 8824.

35 Pellicer-Porres, J., Saitta, A.M., Polian, A., Itié, J.P., and Hanfland, M. (2007) *Nat. Mater.*, **6**, 698.

36 Eremets, M. (1996) *High Pressure Experimental Methods*, Oxford Science Publicaions, Oxford.

37 Jayaraman, A.J. (1983) *Rev. Mod. Phys.*, **55**, 65.

38 Boehler, R. and De Hantsetters, K. (2004) *High Pressure Res.*, **24**, 391.

39 Mao, H.K. and Hemley, R.J. (1998) *Ultrahigh-Pressure Mineralogy: Physics and Chemistry of the Earth's Deep Interior*, vol. 37 (ed. R.J. Hemley), Reviews in Mineralogy, p. 1.

40 Piermarini, G.J. and Block, S. (1975) *Rev. Sci. Instrum.*, **46**, 973.

41 Merrill, L. and Bassett, W.A. (1974) *Rev. Sci. Instrum.*, **45**, 290.

42 LeToullec, R., Loubeyre, P., and Pinceaux, J.P. (1988) *High Pressure Res.*, **1**, 77.

43 Fourme, R. (1968) *J. Appl. Cryst.*, **1**, 23.

44 Herchen, H. and Cappelli, M.A. (1993) *Phys. Rev. B*, **47**, 14193.

45 Hennet, L., Thiaudière, D., Gailhanou, M., Landron, C., Coutures, J.P., and Price, D.L. (2002) *Rev. Sci. Instrum.*, **73**, 124.

46 Massiot, D., Trumeau, D., Touzo, B., Farnan, I., Rifflet, J.C., Douy, A., and Coutures, J.P. (1995) *J. Phys. Chem.*, **99**, 16455.

47 Rousseau, B., Brun, J.F., De Sousa Meneses, D., and Echegut, P. (2005) *Int. J. Thermophys.*, **26**, 1277.

48 Mermoux, M., Fayette, L., Marcus, B., Rosman, N., Abello, L., and Lucazeau, G. (1996) *Phys. Status Solidi A Appl. Res.*, **154**, 55.

49 Fujimori, H., Kakihana, M., Ioku, K., Goto, S., and Yoshimura, M. (2001) *Appl. Phys. Lett.*, **79**, 937.

50 Gillet, P. (1996) *Phys. Chem. Miner.*, **23**, 263.

51 McMillan, P.F., Poe, B.T., Gillet, P., and Reynard, B. (1994) *Geochim. Cosmochim. Acta*, **58**, 3653.

52 Yashima, M., Kakihana, M., Shimidzu, R., Fujimori, H., and Yoshimura, M. (1997) *Appl. Spectrosc.*, **51**, 1224.

53 Farrow, R.L. and Nagelberg, A.S. (1980) *Appl. Phys. Lett.*, **36**, 945.

54 Fujimori, H., Yashima, M., Kakihana, M., and Yoshimura, M. (2001) *J. Am. Ceram. Soc.*, **84**, 663.

55 Zouboulis, E., Renusch, D., and Grimsditch, M. (1998) *Appl. Phys. Lett.*, **72**, 1.

56 Simon, P., Moulin, B., Buixaderas, E., Raimboux, N., Herault, E., Chazallon, B., Cattey, H., Magneron, N., Oswalt, J., and Hocrelle, D. (2003) *J. Raman Spectrosc.*, **34**, 497.

57 Exarhos, G.J., Frydrych, W.S., Walrafen, G.E., Fischer, M., Pugh, E., and Garofalini, S.H. (1988) *Proceedings of the 11th International Conference of Raman Spectroscopy* (eds R.J.H. Clark and D.A. Long), John Wiley & Sons, New York, p. 503.

58 Exarhos, G.J. and Schaaf, J.W. (1991) *J. Appl. Phys.*, **69**, 2543.

59 Bernardez, L., McCarty, K.F., and Yang, N. (1992) *J. Appl. Phys.*, **72**, 2001.

60 Fayette, L., Marcus, B., Mermoux, M., Rosman, N., Abello, L., and Lucazeau, G. (1994) *J. Appl. Phys.*, **76**, 1604.

61 Fayette, L., Marcus, B., Mermoux, M., Rosman, N., Abello, L., and Lucazeau, G. (1997) *J. Mater. Res.*, **12**, 2686.

62 Mermoux, M., Fayette, L., Marcus, B., Rosman, N., Abello, L., and Lucazeau, G. (1995) *Analusis*, **23**, 325.

63 Rosman, N., Abello, L., Chabert, J.P., Verven, G., and Lucazeau, G. (1995) *J. Appl. Phys.*, **78**, 519.

64 Herchen, H. and Cappelli, M.A. (1991) *Phys. Rev. B*, **43**, 11740.

65 Herchen, H. and Cappelli, M.A. (1994) *Phys. Rev. B*, **49**, 3213.

66 Herchen, H., Cappelli, M.A., Landstrass, M.I., Plano, M.A., and Moyer, M.D. (1992) *Thin Solid Films*, **212**, 206.

67 Voronko, Y.K., Sobol, A.A., and Shukshin, V.E. (2007) *Phys. Solid State*, **49**, 1963.

68 You, J.L., Jiang, G.C., Hou, H.Y., Wu, Y.Q., Chen, H., and Xu, K.D. (2002) *Chinese Phys. Lett.*, **19**, 205.

69 You, J.L., Jiang, G.C., and Xu, K.D. (2001) *J. Non-Cryst. Solids*, **282**, 125.

70 You, J.L., Jiang, G.C., Yang, S.H., Ma, J.C., and Xu, K.D. (2001) *Chinese Phys. Lett.*, **18**, 991.

71 Goncharov, A.F. and Crowhurst, J.C. (2005) *Rev. Sci. Instrum.*, **76**, 063905.

72 Maehara, T., Yano, T., and Shibata, S. (2005) *J. Non-Cryst. Solids*, **351**, 3685.

73 Maehara, T., Yano, T., Shibata, S., and Yamane, M. (2004) *Philos. Mag.*, **84**, 3085.

74 Slotznick, S.P. and Shim, S.H. (2008) *Am. Miner.*, **93**, 470.

75 Wan, S.M., Zhang, X., Zhao, S.J., Zhang, Q.L., You, J.L., Chen, H., Zhang, G.C., and Yin, S.T. (2007) *J. Appl. Crystallogr.*, **40**, 725.

76 Wan, S.M., Zhang, X., Zhao, S.J., Zhang, Q.L., You, J.L., Lu, L., Fu, P.Z., Wu, Y.C., and Yin, S.T. (2008) *Cryst. Growth Des.*, **8**, 412.

77 Yano, T., Shibata, S., and Maehara, T. (2006) *J. Am. Ceram. Soc.*, **89**, 89.

78 You, J.L., Jiang, G.C., Hou, H.Y., Chen, H., Wu, Y.Q., and Xu, K.D. (2006) *J. Phys.: Conf. Ser.*, **28**, 25.

79 Sova, R.M., Linevsky, M.J., Thomas, M.E., and Mark, F.F. (1998) *Infrared Phys. Technol.*, **39**, 251.

80 Rozenbaum, O., De Sousa Meneses, D., Auger, Y., Chermanne, S., and Echegut, P. (1999) *Rev. Sci. Instrum.*, **70**, 4020.

81 De Sousa Meneses, D., Brun, J.F., Rousseau, B., and Echegut, P. (2006) *J. Phys. Condens. Matt.*, **18**, 5669.

82 De Sousa Meneses, D., Brun, J.F., Echegut, P., and Simon, P. (2004) *Appl. Spectrosc.*, **58**, 969.

83 Spetzler, H., Chen, G., Whitehead, S., and Getting, I.C. (1993) *Pure Appl. Geophys.*, **141**, 341.

84 Jacobsen, S.D., Spetzler, H., Reichmann, H.J., and Smyth, J.R. (2004) *Proc. Nat. Acad Sci.*, **101**, 5867.

85 Thomsen, C., Strait, J., Vardeny, Z., Maris, H.J., Tauc, J., and Hauser, J.J. (1984) *Phys. Rev. Lett.*, **53**, 989.

86 Decremps, F., Belliard, L., Perrin, B., and Gauthier, M. (2008) *Phys. Rev. Lett.*, **100**, 035502.

87 Decremps, F., Gauthier, M., and Ricquebourg, F. (2006) *Phys. Rev. Lett.*, **96**, 105501.

88 Polian, A. and Grimsditch, M. (1983) *Phys. Rev.*, **B27**, 6409.

89 Munro, R.G., Block, S., Mauer, F.A., and Piermarini, G. (1982) *J. Appl. Phys.*, **53**, 6174.

90 Polian, A. and Grimsditch, M. (1984) *Phys. Rev. Lett.*, **52**, 1312.

91 Polian, A. and Grimsditch, M. (1984) *Phys. Rev. B*, **29**, 6362.

92 Shimizu, H., Ohnishi, M., Sadaki, S., and Ishibashi, Y. (1995) *Phys. Rev. Lett.*, **74**, 2820.

93 Shimizu, H., Nabetani, T., Nishiba, T., and Sasaki, S. (1996) *Phys. Rev. B*, **53**, 6107.

94 Tulk, C.A., Gagnon, R.E., Kiefte, H., and Clouter, M.J. (1996) *J. Chem. Phys.*, **104**, 7854.

95 Tulk, C.A., Gagnon, R.E., Kiefte, H., and Clouter, M.J. (1997) *J. Chem. Phys.*, **107**, 10684.

96 Shimizu, H. (1998) *Rev. High Pressure Sci. Technol.*, **7**, 1124.

97 Hemley, R.J. (1987) *High-Pressure Research in Mineral Physics* (eds M.H. Manghnani and Y. Syono), Terra Scientific, Tokyo, p. 347.

98 Hemley, R.J., Jephcoat, A.P., Mao, H.K., Ming, L.C., and Manghnani, M.H. (1988) *Nature*, **334**, 52.

99 McNeil, L.E. and Grimsditch, M. (1992) *Phys. Rev. Lett.*, **68**, 83.

100 Kingma, K.J., Meade, C., Hemley, R.J., Mao, H.K., and Veblen, D.R. (1993) *Science*, **259**, 666.

101 Kingma, K.J., Hemley, R.J., Mao, H.K., and Veblen, D.R. (1993) *Phys. Rev. Lett.*, **70**, 3927.

102 Gregoryanz, E., Hemley, R.J., Mao, H.K., and Gillet, P. (2000) *Phys. Rev. Lett.*, **84**, 3117.

103 Carpenter, M. and Salje, E. (1998) *Eur. J. Mineral.*, **10**, 693.

104 Calderon, E., Gauthier, M., Decremps, F., Hamel, G., Syfosse, G., and Polian, A. (2007) *J. Phys. Condens. Matt.*, **19**, 436228.

105 Polian, A., Vo-Thanh, D., and Richet, P. (2002) *Europhys. Lett.*, **57**, 375.

106 Iota, V., Yoo, C.S., Klepeis, J.H., Jenei, Z., Evans, W., and Cynn, H. (2007) *Nat. Mater.*, **6**, 34, and references therein.

107 Giordano, V.M. and Datchi, F. (2007) *Europhys. Lett.*, **77**, 46002.

108 Grimsditch, M., Polian, A., and Vogelgesang, R. (2003) *J. Phys. Condens. Matt.*, **15**, S2335.

109 Shintani, H. and Tanaka, H. (2008) *Nat. Mater.*, **7**, 870.
110 Andrikopoulos, K.S., Christofilos, D., Kourouklis, G.A., and Yannopoulos, S.N. (2006) *J. Non-Cryst. Solids*, **352**, 4594.
111 Niss, K., Begen, B., Frick, B., Ollivier, J., Beraud, A., Sokolov, A., Novikov, V.N., and Alba-Simonesco, C. (2007) *Phys. Rev. Lett.*, **99**, 055502.
112 Mishima, O., Calvert, L.D., and Whalley, E. (1985) *Nature*, **314**, 76.
113 Klotz, S., Strässle, T., Nelmes, R.J., Loveday, J.S., Hamel, G., Rousse, G., Canny, B., Chervin, J.C., and Saitta, A.M. (2005) *Phys. Rev. Lett.*, **94**, 025506.
114 Mishima, O., Calvert, L.D., and Whalley, E. (1984) *Nature*, **310**, 393.
115 Nguyen, J.H., Kruger, M.B., and Jeanloz, R. (1997) *Phys. Rev. Lett.*, **78**, 1936.
116 Arora, A.K., Nithya, R., Yagi, T., Miyajima, N., and Mary, T.A. (2004) *Solid State Commun.*, **129**, 9.
117 Cappelli, M.A. and Herchen, H. (1991) *Abstr. Pap. Am. Chem. Soc.*, **202**, 35.
118 Fayette, L., Marcus, B., Mermoux, M., Abello, L., and Lucazeau, G. (1994) *Diam. Relat. Mater.*, **3**, 438.
119 Mermoux, M., Fayette, L., Marcus, B., Rosman, N., Abello, L., and Lucazeau, G. (1995) *Diam. Relat. Mater.*, **4**, 745.
120 Mermoux, M., Marcus, B., Abello, L., Rosman, N., and Lucazeau, G. (2003) *J. Raman Spectrosc.*, **34**, 505.
121 Voevodin, A.A. and Laube, S.J.P. (1995) *Surf. Coat. Technol.*, **77**, 670.
122 Li, X.J., Zhou, C.D., Jiang, G.C., and You, J.L. (2006) *Mater. Charact.*, **57**, 105.
123 Cynn, H., Sharma, S.K., Cooney, T.F., and Nicol, M. (1992) *Phys. Rev. B*, **45**, 500.
124 Gajovic, A., Furic, K., Stefanic, G., and Music, S. (2005) *J. Mol. Struct.*, **744**, 127.
125 Lughi, V. and Clarke, D.R. (2007) *J. Appl. Phys.*, **101**, 053524.
126 Boucher, S., Piwowarczyk, J., Marzke, R.F., Takulapalli, B., Wolf, G.H., McMillan, P.F., and Petuskey, W.T. (2005) *J. Eur. Ceram. Soc.*, **25**, 1333.
127 Cormier, L., Meneses, D.D., Neuville, D.R., and Echegut, P. (2006) *Eur. J. Glass Sci. Technol. B*, **47**, 430.
128 Daniel, I., McMillan, P.F., Gillet, P., and Poe, B.T. (1996) *Chem. Geol.*, **128**, 5.
129 Farber, D.L. and Williams, Q. (1996) *Am. Miner.*, **81**, 273.
130 Kalampounias, A.G., Yannopoulos, S.N., and Papatheodorou, G.N. (2006) *J. Chem. Phys.*, **125**, 164502.
131 Malfait, W.J. and Halter, W.E. (2008) *Phys. Rev. B*, **77**, 014201.
132 Mysen, B. (1996) *Geochim. Cosmochim. Acta*, **60**, 3665.
133 Mysen, B. (1997) *Contrib. Mineral. Petrol.*, **127**, 104.
134 Mysen, B.O. and Toplis, M.J. (2007) *Am. Miner.*, **92**, 933.
135 Pedeche, S., Simon, P., Matzen, G., Moulin, B., Blanchard, K., and Querel, G. (2003) *J. Raman Spectrosc.*, **34**, 248.
136 Reynard, B. and Webb, S.L. (1998) *Eur. J. Mineral.*, **10**, 49.
137 Richet, P. and Mysen, B.O. (1999) *Geophys. Res. Lett.*, **26**, 2283.
138 Sharma, S.K., Cooney, T.F., Wang, Z.F., and vanderLaan, S. (1997) *J. Raman Spectrosc.*, **28**, 697.
139 Shimodaira, N. (2006) *Phys. Rev. B*, **73**, 214206.
140 Voronko, Y.K., Sobol, A.A., and Shukshin, V.E. (2006) *Inorg. Mater.*, **42**, 981.
141 Wang, Y.Y., You, J.L., and Jiang, G.C. (2008) *Chinese J. Inorg. Chem.*, **24**, 765.
142 You, J.L., Jiang, G.C., Chen, H., and Xu, K.D. (2006) *Rare Metals*, **25**, 431.
143 You, J.L., Jiang, G.C., and Xu, K.D. (2000) *Spectrosc. Spect. Anal.*, **20**, 797.
144 Huang, S.P., Yoshida, F., You, J.L., Jiang, G.C., and Xu, K.D. (1999) *J. Phys. Condens. Matt.*, **11**, 5429.
145 Kalampounias, A.G., Yannopoulos, S.N., and Papatheodorou, G.N. (2006) *J. Non-Cryst. Solids*, **352**, 4619.
146 Kalampounias, A.G., Yannopoulos, S.N., and Papatheodorou, G.N. (2006) *J. Chem. Phys.*, **124**, 014504.
147 McMillan, P.F. and Wolf, G.H. (1995) *Structure, Dynamics and Properties of Silicate Melts (Reviews in Mineralogy,* Vol. 32*)*, pp. 247.
148 Grzechnik, A. and McMillan, P.F. (1998) *Am. Miner.*, **83**, 331.
149 Auguste, F., Tkatcheva, O., Mediaas, H., Ostvold, T., and Gilbert, B. (2003) *Inorg. Chem.*, **42**, 6338.
150 Bessada, C., Lacassagne, V., Massiot, D., Florian, P., Coutures, J.P., Robert, E., and

Gilbert, B. (1999) *Z. Naturfors. A J. Phys. Sci.*, **54**, 162.

151 Robert, E. and Gilbert, B. (2000) *Appl. Spectrosc.*, **54**, 396.

152 Sinogeikin, S., Bass, J., Prakapenka, V., Lakshtanov, D., Shen, G., Sanchez-Valle, C., and Rivers, M. (2006) *Rev. Sci. Instrum.*, **77**, 103905.

153 Mikkelsen, J.C. and Boyce, J.B. (1983) *Phys. Rev. B*, **28**, 7130.

154 Chang, I.F. and Mitra, S.S. (1971) *Adv. Phys.*, **20**, 359.

155 Genzel, L., Martins, T.P., and Perry, C.H. (1974) *Phys. Stat. Sol. (b)*, **62**, 83.

156 Elliott, R.J., Krumhansl, J.A., and Leath, P.L. (1974) *Rev. Mod. Phys.*, **46**, 465.

157 Taylor, D.W. (1988) *Optical Properties of Mixed Crystals* (eds R.J. Elliott and I.P. Ipatova), Elsevier Science, Amsterdam, p. 35.

158 Postnikov, A.V., Pagès, O., and Hugel, J. (2005) *Phys. Rev. B*, **71**, 115206.

159 Pagès, O., Souhabi, J., Postnikov, A.V., and Chafi, A. (2009) *Phys. Rev. B*, **80**, 035204.

160 Saitta, A.M. and Decremps, F. (2004) *Phys. Rev. B*, **70**, 035214.

3
Inelastic Neutron Scattering, Lattice Dynamics, Computer Simulation and Thermodynamic Properties

Ranjan Mittal, Samrath L. Chaplot, and Narayani Choudhury

3.1
Introduction

In the study of solids at the microscopic level, it is important to know the crystal structure and the dynamics of the atoms. Many important physical properties such as phase transitions, thermal expansion, specific heat, and thermal conductivity originate from the dynamics of atoms [1–6]. The collective motions of atoms in crystalline solids form traveling waves (called lattice vibrations). At low temperatures, typically below the Debye temperature, when the vibrations are essentially harmonic in nature, these traveling waves are noninteracting and act as simple harmonic oscillators, which are also called normal modes of vibrations. They are quantized in terms of "phonons." The phonons are characterized by three parameters, namely, their wave vector, energy, and polarization vector. The polarization vector defines the relative atomic vibrations along three orthogonal directions. In order to understand the physical properties of any solid, it is of interest to study the energy–wavelength relation (dispersion relation) of the thermal motions of the atoms, which is determined by the interatomic interactions. These studies fall in the domain of lattice dynamics.

At high temperatures, usually well above the Debye temperature, or close to phase transition, and so on, the atomic vibrations may not be harmonic. The phonons are no more noninteracting and their frequencies may change with temperature. When the anharmonicity is small, some kind of perturbation theory may be used to study such changes. However, different methods may be needed when anharmonicity is large. Molecular dynamics simulation is one such technique wherein the dynamical equation of motion is solved by brute force and atomic trajectories are calculated.

Thermodynamic Properties of Solids: Experiment and Modeling
Edited by S. L. Chaplot, R. Mittal, and N. Choudhury

ISBN: 928-3-527-40812-2

These are used to calculate the space–time correlation function and various thermodynamic properties.

Experimental studies of lattice vibration include techniques such as Raman spectroscopy, infrared absorption, inelastic neutron scattering, inelastic X-ray scattering, and so on. Unlike Raman and infrared studies, which probe only the long wavelength excitations in one-phonon scattering, inelastic neutron and X-ray scattering can directly probe the phonons in the entire Brillouin zone. While inelastic neutron scattering is widely used for such measurements, inelastic X-ray scattering has also been used [7, 8] at intense synchrotrons sources for the study of phonons in a few materials. The neutron technique will be discussed in this chapter and the X-ray experiments are explained in another chapter.

Experimental studies at high pressures and temperatures are often limited while first-principles studies of various thermodynamic properties are possible in several cases, accurate models for theoretical studies of various materials are of utmost importance. The success of the models in predicting thermodynamic properties depends crucially on their ability to explain a variety of microscopic dynamical properties [1, 5, 6]. These include an understanding of the crystal structure, equation of state, phonon dispersion relations, and density of states. The data obtained from neutron scattering and optical experiments are used to test and validate models of interatomic potentials [1, 5, 6], which in turn could be used to predict thermodynamic properties at high pressures and temperatures. The interatomic potential model can also be used for classical molecular dynamics simulations to understand the microscopic insights in a variety of novel phenomena such as pressure induced changes in atomic coordination, bonding, elastic properties, seismic discontinuities, and so on. Simulations are also useful to understand the properties of material under extreme high-pressure and -temperature conditions that are difficult to achieve in the laboratory.

The combination of neutron scattering experiments, lattice dynamics, and molecular dynamics (MD) simulations is a powerful tool to study the structure and dynamics of complex molecular systems. Using these techniques, we have studied the phonon properties and phase transitions of a variety of solids. The experiments provide valuable information about the phonon dispersion relation and density of states, while the calculations enable microscopic interpretation of the experimental data. The materials studied find a wide range of applications and involve negative thermal expansion materials, technologically important samples, and geophysically relevant minerals.

The organization of this chapter is as follows: The theoretical formalism of lattice dynamics is given in Section 3.2. The details about the computational techniques are outlined in Section 3.3. The procedure for calculation of thermodynamic properties of solids is described in Section 3.4. Theory and details about the experimental techniques for inelastic neutron scattering are described in Sections 3.5 and 3.6, respectively. In Section 3.7, we have discussed about the technique of molecular dynamics simulation. The applications of inelastic neutron scattering, lattice dynamics, and molecular dynamics are given in Section 3.8. Finally, the conclusions are given in Section 3.9.

3.2 Lattice Dynamics

3.2.1 Theoretical Formalisms

The formalism of lattice dynamics is based on the Born–Oppenheimer or adiabatic approximation. In this approximation, it is assumed that the electronic wavefunctions change adiabatically during the nuclear motion. The electrons essentially contribute an additional effective potential for the nuclear motions and the lattice vibrations are associated only with nuclear motions. In this chapter, only summary of the mathematical formalism for a perfect crystal is described. A complete account of the formal theory of lattice dynamics can be found in literature [2–4]. For small displacements of the atoms, $\mathbf{u}\begin{pmatrix} l \\ k \end{pmatrix}$, about their equilibrium positions, $\mathbf{r}\begin{pmatrix} l \\ k \end{pmatrix}$, where l denotes the lth unit cell ($l = 1, 2, \ldots N$) and k is the kth type of atom ($k = 1,2, \ldots n$) within the unit cell, the crystal potential energy can be written as a Taylor expansion. The expansion is retained only up to the second derivative in the so-called harmonic approximation, as follows:

$$\phi = \phi_0 + \phi_1 + \phi_2 \tag{3.1}$$

where

$$\phi_0 = \phi\left(\mathbf{r}\begin{pmatrix} l \\ k \end{pmatrix}\right),$$

$$\phi_1 = \sum_{lk\alpha} \frac{\partial \phi}{\partial u_\alpha\begin{pmatrix} l \\ k \end{pmatrix}}\Bigg|_o u_\alpha\begin{pmatrix} l \\ k \end{pmatrix} = \sum_{lk\alpha} \phi_\alpha\begin{pmatrix} l \\ k \end{pmatrix} u_\alpha\begin{pmatrix} l \\ k \end{pmatrix},$$

$$\phi_2 = \frac{1}{2}\sum_{lk\alpha}\sum_{l'k'\beta} \frac{\partial^2 \phi}{\partial u_\alpha\begin{pmatrix} l \\ k \end{pmatrix}\partial u_\beta\begin{pmatrix} l' \\ k' \end{pmatrix}}\Bigg|_o u_\alpha\begin{pmatrix} l \\ k \end{pmatrix} u_\beta\begin{pmatrix} l' \\ k' \end{pmatrix} = \frac{1}{2}\sum_{lk\alpha}\sum_{l'k'\beta} \phi_{\alpha\beta}\begin{pmatrix} l & l' \\ k & k' \end{pmatrix} u_\alpha\begin{pmatrix} l \\ k \end{pmatrix} u_\beta\begin{pmatrix} l' \\ k' \end{pmatrix}$$

where the suffices α and β denote Cartesian coordinates.

In the equilibrium configuration, the force on every atom must vanish. This leads to the result

$$\phi_\alpha\begin{pmatrix} l \\ k \end{pmatrix} = 0, \quad \text{for every } \alpha, k, l$$

Hence, $\phi_1 = 0$.

Thus, in the harmonic approximation,

$$\phi = \phi_0 + \frac{1}{2}\sum_{lk\alpha}\sum_{l'k'\beta}\phi_{\alpha\beta}\begin{pmatrix} l & l' \\ k & k' \end{pmatrix} u_\alpha\begin{pmatrix} l \\ k \end{pmatrix} u_\beta\begin{pmatrix} l \\ k' \end{pmatrix}. \tag{3.2}$$

Accordingly, the equation of motion of the $(lk)^{\text{th}}$ atom becomes

$$m_k \ddot{u}_\alpha\begin{pmatrix} l \\ k \end{pmatrix} = -\sum_{l'k'\beta}\phi_{\alpha\beta}\begin{pmatrix} l & l' \\ k & k' \end{pmatrix} u_\beta\begin{pmatrix} l' \\ k' \end{pmatrix}. \tag{3.3}$$

From Equation 3.3, it is clear that $\phi_{\alpha\beta}\begin{pmatrix} l & l' \\ k & k' \end{pmatrix}$ is the negative of the force exerted on atom (lk) in the α-direction due to unit displacement of the atom $(l'k')$ in the β-direction. The quantity $\phi_{\alpha\beta}$ is referred to as the force constant.

The crystal periodicity suggests that the solutions of Equation 3.3 must be such that the displacements of atoms in different unit cells must be same apart from phase factor. The equations of motion (3.3) are solved by assuming wave-like solutions of the type

$$u_\alpha\begin{pmatrix} l \\ k \end{pmatrix} = U_\alpha(k|\mathbf{q})\exp\left\{i\left(\mathbf{q}.\mathbf{r}\begin{pmatrix} l \\ k \end{pmatrix} - \omega(\mathbf{q})t\right\}\right.. \tag{3.4}$$

Here, $\mathbf{q}$ is the wave vector and $\omega(\mathbf{q})$ is the angular frequency associated with the wave.

Substituting (3.4) in (3.3),

$$m_k\omega^2(\mathbf{q})\,U_\alpha(k|\mathbf{q}) = \sum_{k'\beta} D_{\alpha\beta}\begin{pmatrix} \mathbf{q} \\ kk' \end{pmatrix} U_\beta(k'|\mathbf{q}) \tag{3.5}$$

where m_k is the mass and $\mathbf{r}\begin{pmatrix} l \\ k \end{pmatrix}$ the position coordinate of the kth atom.

Here, the dynamical matrix $D_{\alpha\beta}\begin{pmatrix} \mathbf{q} \\ kk' \end{pmatrix}$ is given as

$$D_{\alpha\beta}\begin{pmatrix} \mathbf{q} \\ kk' \end{pmatrix} = \sum_{l'}\phi_{\alpha\beta}\begin{pmatrix} l & l' \\ k & k' \end{pmatrix}\exp\left\{i\left(\mathbf{q}\cdot\left[\mathbf{r}\begin{pmatrix} l' \\ k' \end{pmatrix} - \mathbf{r}\begin{pmatrix} l \\ k \end{pmatrix}\right]\right)\right\}. \tag{3.6}$$

The $3n$-coupled equations of motion are obtained. As stated earlier, the adiabatic approximation is assumed wherein the electrons adiabatically follow the nuclear vibrations and provide an effective nuclear potential. However, this does not imply that the atoms are rigid during vibrations. The electric field setup by the displacements of the ions is modified by their electronic polarizability, which in turn modifies the force on them and affects the phonon frequencies. This may be described by a shell model [2, 4], in which each ion is regarded to be composed of a rigid or nonpolarizable core and a charged shell with effective charges $X(k)$ and $Y(k)$, respectively. The core and shell are connected by a harmonic spring constant $K(k)$. Thus, the shell can be displaced relative to the core causing a dipole, which in turn leads to a proper description of dielectric behavior of the crystals. One can calculate the dynamical matrices between the pairs, core–core, core–shell, shell–core, and shell–shell, denoted as D^{CC}, D^{CS}, D^{SC}, and D^{SS}, respectively. We have the following equations involving the displacement vectors associated with the core (U) and polarization (W):

$$m\omega^2 U = D^{CC}U + D^{CS}W \tag{3.7}$$

$$0 = D^{SC}U + D^{SS}W \tag{3.8}$$

Eliminating W, we get,

$$m\omega^2 U = [D^{CC} - \{D^{CS}(D^{SS})^{-1}D^{SC}\}]U \tag{3.9}$$

and we obtain the dynamical matrix

$$D = D^{CC} - D^{CS}(D^{SS})^{-1}D^{SC}. \tag{3.10}$$

For simplicity, the short-range forces between atoms are included only between the shells. The frequencies of the normal modes and eigenvectors are determined by diagonalizing the dynamical matrix through a solution of the secular equation

$$\det\left| m_k\omega^2(\mathbf{q})^2\delta_{kk'}\delta_{\alpha\beta} - D_{\alpha\beta}\begin{pmatrix} q \\ kk' \end{pmatrix}\right| = 0. \tag{3.11}$$

Solving Equation 3.11, $3n$ eigenvalues are obtained, which are $\omega_j^2(\mathbf{q})$, ($j = 1, 2, \ldots, 3n$). Because, the dynamical matrix is Hermitian, the eigenfrequencies are real and its eigenvectors may be chosen as orthonormal. The components of the eigenvectors $\xi_j(\mathbf{q})$ determine the pattern of displacement of the atoms in a particular mode of vibration.

The displacements of the atoms in one of these normal modes, labeled by ($\mathbf{q}j$), correspond to a wave-like displacement of atoms and are given as

$$u_\alpha\begin{pmatrix} l \\ k \end{pmatrix} = \xi_\alpha(k|\mathbf{q}j)\exp\left\{ i\left(\mathbf{q}\cdot\mathbf{r}\begin{pmatrix} l \\ k \end{pmatrix}\right)\right\} P\begin{pmatrix} \mathbf{q} \\ j \end{pmatrix} \tag{3.12}$$

where, $P\begin{pmatrix} \mathbf{q} \\ j \end{pmatrix}$ is the normal coordinate and $\xi_\alpha(k|\mathbf{q}j)$ is the normalized eigenvector of the normal mode ($\mathbf{q}j$), where j runs from 1 to $3n$ and is used to distinguish between the $3n$ normal modes at $\mathbf{q}$. Corresponding to every direction in $\mathbf{q}$-space, there are $3n$ curves $\omega = \omega_j(\mathbf{q})$, ($j = 1, 2, \ldots 3n$). Such curves are called phonon dispersion relations. The index j, which distinguishes the various frequencies corresponding to the propagation vector, characterizes various branches of the dispersion relation. Though, some of these branches are degenerate because of symmetry, in general, they are distinct. The form of dispersion relation depends on the crystal structure as well as on the nature of the interatomic forces. However, a cyclic crystal always has three zero frequency modes at $\mathbf{q} = 0$, which correspond to lateral translation of the crystal along three mutually perpendicular directions. These three branches are referred to as acoustic branches. The remaining ($3n - 3$) branches have finite frequencies at $\mathbf{q} = 0$, which are labeled as optic branches.

The crux of the lattice-dynamical problem therefore is to calculate the dynamical matrix, which is usually done by constructing a suitable model for the force constants or by using a suitable crystal potential. Corresponding to the $3n$ degrees of freedom for any wave vector $\mathbf{q}$, there are $3n$ eigenvectors each having $3n$ components. Group-theoretical analysis at various high-symmetry points and directions in the Brillouin zone are used to derive the symmetry vectors for block diagonalization of the dynamical matrix. This allows the classification of the phonon modes into different irreducible representations enabling direct comparison with single crystal Raman, infrared, and neutron data.

The phonon density of states is defined by the equation

$$g(\omega) = D' \int_{BZ} \sum_{j} \delta(\omega - \omega_j(\mathbf{q}))d\mathbf{q} = D' \sum_{jp} \delta(\omega - \omega_j(\mathbf{q}))d\mathbf{q}_p \tag{3.13}$$

where D' is a normalization constant such that $\int g(\omega)d\omega = 1$; that is, $g(\omega)d\omega$ is the ratio of the number of eigenstates in the frequency interval $(\omega, \omega + d\omega)$ to the total number of eigenstates. p is the mesh index characterizing $\mathbf{q}$ in the discretized irreducible Brillouin zone and $d\mathbf{q}_p$ provides the weighting factor corresponding to the volume of pth mesh in $\mathbf{q}$-space.

3.3
Computational Techniques

As seen in Section 3.2, the essential part involved in lattice dynamical calculations is the setting up of the "dynamical matrix" that depends on the derivatives of the crystal potential (force constants). Ideally, one would like to develop a uniform prescription for calculating the crystal potential and hence its derivatives that is applicable to various crystals.

An approach often adopted for lattice dynamical studies involves fitting of empirical values of force constants between various pairs of atoms to available experimental data. This approach is ideally suited only for simple solids, where the effective interactions between atoms are of short range, which are negligible beyond the first, second, or third neighbors. Otherwise, the number of force constant parameters necessary for reproducing the experimental data becomes very large. As we need to calculate the anharmonic properties, thermal expansion, equation of state, and so on, an approach based on force constants would not be useful.

Lattice dynamics calculations of the equation of state and vibrational properties may be carried out using either a quantum-mechanical *ab initio* approach or an atomistic approach involving semiempirical interatomic potentials. *Ab initio* density-functional calculations [9–12] have been reported for some silicate minerals including quartz SIO_2 and perovskites. However, because of the limited computational resources for complex structures an approach based on an empirical interatomic potential function is generally used in the calculations. The form of potentials used [1, 6, 7] for the calculation is

$$V(r) = \left\{\frac{e^2}{4\pi\varepsilon_o}\right\}\left\{\frac{Z(k)Z(k')}{r}\right\} + a\exp\left\{\frac{-br}{R(k)+R(k')}\right\} - \frac{C}{r^6} \quad (3.14)$$

where r is the separation between the atoms of types k and k', $Z(k)$ and $R(k)$ are, respectively, empirical charge and radius parameter of the atom type k, $1/(4\pi\,\varepsilon_o) = 9\times10^9\,\mathrm{Nm^2\,Coul^{-2}}$, $a = 1822\,\mathrm{eV}$, and $b = 12.364$. The last term in Equation 3.14 is applied only between certain atoms.

The above potential function is used for the lattice dynamical calculations of ZrW_2O_8, HfW_2O_8 [13, 14], $MSiO_4$ (M = Zr, Hf, Th, and U) [15–18], $M_3Al_2Si_3O_{12}$ (M = Fe, Si, Ca, and Mn) [19, 20], MV_2O_7 (M = Zr and Hf) [21], MPO_4 (M = Lu and Yb) [22], $NaNbO_3$ [23], and so on.

As a further improvement, a covalent potential [1, 6, 7] is also included between certain atoms.

$$V(r) = -D\exp\left(\frac{-n(r-r_o)^2}{2r}\right) \quad (3.15)$$

where n, D, and r_o are the empirical parameters of the potential.

The parameters of the potential are determined using the conditions of the structural and dynamic equilibrium of the crystal. This procedure [24] has been successful in the case of several compounds as discussed later in Section 3.7. Structural constraints used are that at $T = 0$, the free energy is minimized with respect to the lattice parameters and the atomic positions of the crystal at zero pressure, and the structure is close to that determined by the diffraction experiments. The dynamic equilibrium requires that the calculated phonon frequencies have real values for all the wave vectors in the Brillouin zone. The parameters of potentials should also reproduce various other available experimental data, namely, elastic constants, optical phonon frequencies or the range of phonon spectrum, and so on. At high pressures, the crystal structures are obtained by minimization of the free energy with respect to the lattice parameters and the atomic positions. The equilibrium structure thus obtained is used in lattice dynamics calculations. The polarizability of certain atoms has been introduced in the framework of the shell model as discussed in Section 3.2.1.

Three-body terms are not included in the potential although their contribution is, to some extent, mimicked by the two-body potential. For example, O–Si–O bond angle is partly determined by the O–O potential. This is justified from the fact that the potential reproduced the equilibrium crystal structural parameters and other dynamical properties of many solids quite satisfactorily. The interest lies in the development of an interatomic potential that gives good description of the experimental data with the limited number of adjustable parameters. The three-body terms can be included in the potential, and the parameters of the potential can be optimized when a large amount of phonon data become available.

The interatomic potential model has been used for the calculation of vibrational and thermodynamic properties of various solids such as (i) the frequency of phonons as a function of the wave vector (i.e., the phonon dispersion relation), (ii) the polarization vector of the phonons, (iii) the frequency distribution of phonons,

(iv) the thermodynamic properties of the solid such as the equation of state, specific heat, free energies, and so on, and (v) the variation of phonon frequencies due to pressure, temperature, and so on. The calculations are carried out using the current version of computer program DISPR [25] developed at Trombay, India.

3.4
Thermodynamic Properties of Solids

The theory of lattice dynamics described in Section 3.2 allows us to determine the phonon frequencies in the harmonic approximation. The phenomena that cannot be accounted [2–4] for without going beyond the harmonic approximation are the increase in specific heat beyond the value $3Nk_B$ with increasing temperature, the thermal expansion, the multiphonon process, and so on. The number of phonons excited in thermal equilibrium at any temperature is given by Bose–Einstein distribution $n(\omega)\left[=\frac{1}{\exp(\hbar\omega/k_BT)-1}\right]$. At high temperatures, $\hbar\omega_j(\mathbf{q}) \ll k_BT$, the number of phonons in a given state is directly proportional to the temperature and inversely proportional to their energy. Anharmonic effects are relatively small at low temperatures. These effects become more important at high temperatures. This change at high temperatures affects physical properties of the crystal. In the quasiharmonic approximation [2–4], (where, the vibrations of atoms at any finite temperature are assumed to be harmonic about their mean positions appropriate to the corresponding temperature), the thermodynamic properties of a crystal are based on the averages of energies associated with the $3nN$ vibrations corresponding to the number of degrees of freedom of the n atomic constituents in the N unit cells of the crystal.

The thermodynamic properties, namely, the free energy, the specific heat and the entropy are obtained from the partition function Z defined as

$$Z = Tr\left\{\exp\left(\frac{-\mathrm{H}}{k_BT}\right)\right\}. \tag{3.16}$$

If this trace is evaluated in terms of the eigenenergies of the Hamiltonian (H), then

$$Z = \exp\left(\frac{-\phi(V)}{k_BT}\right)\prod^{\mathbf{q}j}\frac{\exp\left\{\frac{-\hbar\omega_j(\mathbf{q})}{k_BT}\right\}}{1-\exp\left\{\frac{-\hbar\omega_j(\mathbf{q})}{k_BT}\right\}}. \tag{3.17}$$

All the thermodynamic properties of the crystal derived from the partition function involve summations over the phonon frequencies in the entire Brillouin zone and can be expressed as averages over the phonon density of states. The Helmholtz free energy F and entropy S are given as

$$F = -k_BT\ln Z = \phi(V) + \int\left\{\frac{1}{2}\hbar\omega + k_BT\ln\left[1-\exp\left(\frac{-\hbar\omega}{k_BT}\right)\right]\right\}g(\omega)d\omega \tag{3.18}$$

and

$$S = -\frac{\mathrm{d}F}{\mathrm{d}T} = k_B \int \left\{ -\ln\left[1 - \exp\left(\frac{-\hbar\omega}{k_B T}\right)\right] + \frac{\left(\frac{\hbar\omega}{k_B T}\right)}{\left[\exp\left(\frac{\hbar\omega}{k_B T}\right) - 1\right]} \right\} g(\omega)\mathrm{d}\omega. \quad (3.19)$$

The energy E of the crystal with volume V is

$$E = F - T\frac{\mathrm{d}F}{\mathrm{d}T} = \phi(V) + E_{\mathrm{vib}} \quad (3.20)$$

where $\phi(V)$ is the static lattice energy and E_{vib}, the vibrational energy at temperature T.

$$E_{\mathrm{vib}} = \int \left\{ n(\omega) + \frac{1}{2} \right\} \hbar\omega g(\omega)\mathrm{d}\omega \quad (3.21)$$

where $n(\omega)$ is the population factor given as

$$n(\omega) = \frac{1}{\exp(\hbar\omega/k_B T) - 1}. \quad (3.22)$$

The specific heat $C_V(T)$ is given as

$$C_V(T) = \frac{\mathrm{d}E}{\mathrm{d}T} = k_B \int \left(\frac{\hbar\omega}{k_B T}\right)^2 \frac{e^{\left(\frac{\hbar\omega}{k_B T}\right)}}{\left(e^{\left(\frac{\hbar\omega}{k_B T}\right)} - 1\right)^2} g(\omega)\mathrm{d}\omega. \quad (3.23)$$

The calculated phonon density of states can be used to compute the specific heat. While lattice dynamical calculations yield C_V, the specific heat at constant volume, experimental heat capacity data correspond to C_P, the specific heat at constant pressure. The difference $C_P - C_V$ is given as

$$C_P(T) - C_P(T) = [\alpha_V(T)]^2 BVT \quad (3.24)$$

where α_V is the volume thermal expansion and B is the bulk modulus defined as $B = -V\frac{\mathrm{d}P}{\mathrm{d}V}$.

The volume thermal expansion coefficient in the quasiharmonic approximation is given as

$$\alpha_V = \frac{1}{BV}\sum_i \Gamma_i C_{Vi}(T) \quad (3.25)$$

where Γ_i is the mode Grüneisen parameter of the phonons in state i ($= \mathbf{q}j$, which refers to the jth phonon mode at wave vector $\mathbf{q}$), which is given as

$$\Gamma_i = -\frac{\partial \ln \omega_i}{\partial \ln V}. \quad (3.26)$$

The procedure of the calculation of thermal expansion is applicable when explicit anharmonicity of phonons is not significant, and the thermal expansion arises mainly from the implicit anharmonicity, that is, the change of phonon frequencies with volume. The higher order contribution to thermal expansion arising from variation of bulk modulus with volume [26, 27] is also included. The procedure is found satisfactory [13–22] in thermal expansion calculations of ZrW_2O_8, HfW_2O_8, ZrV_2O_7, HfV_2O_7, and $MSiO_4$ (M = Zr, Hf, Th, and U) and aluminosilicate garnets. Due to very large Debye temperatures in most of these systems, the quasiharmonic approximation seems to be suitable up to fairly high temperatures.

3.5
Theory of Inelastic Neutron Scattering

The inelastic scattering of any radiation from a system involves exchange of energy and momentum between the system and the probing radiation. Thermal neutrons can exchange part of their energy or momentum with an excitation in the system. They may lose part of their energy in creating an excitation in the system or may gain energy by annihilation. Thus, the nature of the excitation can be probed by measuring the energy and momentum of the neutrons both before and after the scattering event from a system. The fundamental equations [28] describing the conservation of momentum and energy when a neutron is scattered from a crystal are

$$E_i - E_f = \hbar\omega(\mathbf{q}, j) \tag{3.27}$$

and

$$\hbar(\mathbf{k}_i - \mathbf{k}_f) = \hbar\mathbf{Q} = \hbar(\mathbf{G} \pm \mathbf{q}) \tag{3.28}$$

where k_i and k_f are incident and the scattered neutron wave vectors, respectively, and $\mathbf{Q}$ is the wave vector transfer (scattering vector) associated with the scattering process. $\mathbf{q}$ is the wave vector of the excitation with energy $\hbar\omega(= E)$, $\mathbf{G}$ is a reciprocal lattice vector, E_i and E_f are the incident and scattered neutron energies, respectively, and $\hbar\omega$ is the energy transfer to the system in the scattering process. The + (−) sign indicates that the excitation is absorbed(created) in the scattering process. Hence, the experimental technique of neutron scattering to determine the nature of excitations in the system involves study of the inelastic spectrum of scattered neutrons. The energy and wave vector of neutrons are measured using a spectrometer.

In the scattering process, the inelastic scattering cross section is directly proportional to the dynamical structure factor $S(\mathbf{Q}, \omega)$ (characteristic of the system), which is the double Fourier transform of the space–time correlation function of the constituents of the system including the phonon. Peaks in $S(\mathbf{Q},\omega)$ correspond to these elementary excitations [28, 29]. The measurements on single crystals give information about the $\mathbf{q}$ dependence of phonon (phonon dispersion relation), while polycrystalline samples provide frequency distribution of the phonons (phonon density of states $g(\omega)$). The measurement of phonon dispersion relation is not always possible because a suitable single crystal may not be available. It is difficult to determine all the phonon branches in the phonon dispersion relation separately. The complete phonon dispersion relation is often available only along high symmetry directions of the Brillouin zone. Therefore, in order to obtain a complete picture of the dynamics, it is useful to determine the phonon density of states.

3.5.1
Inelastic Neutron Scattering from Single Crystals: Phonon Dispersion Relations

The neutron scattering structure factor [28, 29] due to a one-phonon inelastic process is given as

$$S_{\mathrm{coh}}^{(1)}(\mathbf{Q},\omega) = A' \sum_{\mathbf{q}j} \frac{\hbar}{2\omega(\mathbf{q}j)} \left\{ n(\omega) + \frac{1}{2} \pm \frac{1}{2} \right\} \left| F_j^{(1)}(\mathbf{Q}) \right|^2 \delta(\mathbf{Q}-\mathbf{G} \pm \mathbf{q})\delta(\omega \mp \omega(\mathbf{q}j)) \tag{3.29}$$

where

$$F_j^{(1)}(\mathbf{Q}) = \sum_k b_k^{\mathrm{coh}} \frac{\mathbf{Q}\cdot\xi(\mathbf{q}j,k)}{\sqrt{m_k}} \exp(-W_k(\mathbf{Q}))\exp(i\mathbf{G}\cdot\mathbf{r}(k)), \tag{3.30}$$

A' is the normalization constant, and b_k, m_k, and $\mathbf{r}(k)$ are neutron scattering length, mass, and the coordinate of the kth atom, respectively. ξ is eigenvector of excitation, $F_j^{(1)}(\mathbf{Q})$ is one-phonon structure factor, $\exp(-W_k(\mathbf{Q}))$ is the Debye–Waller factor. $\hbar\mathbf{Q}$ and $\hbar\omega$ are the momentum and energy transfer on scattering of the neutron, respectively, while $n(\omega)$ is the phonon-population factor given by Equation 3.22.

The upper and lower signs $\pm$ and $\mp$ in Equation 3.29 correspond to loss and gain of the energy of the neutrons, respectively. The two delta functions in Equation 3.29 stand for the conservation of momentum and energy. These two conditions allow the determination of the phonon dispersion relation $\omega_j(\mathbf{q})$. From a large number of such measurements on a single crystal, one can identify several points of the phonon dispersion relations.

From Equation 3.30, it is clear that phonon cross sections depend strongly on $\mathbf{Q}$ and ξ, apart from the atomic structure of the solid itself. For measuring a phonon having the polarization (eigen) vector ξ, the scattering vector $\mathbf{Q}$ should be chosen such that it is aligned parallel to ξ as much as possible. Since $\mathbf{Q}=\mathbf{G}\pm\mathbf{q}$ and for longitudinal mode $\boldsymbol{q} /\!/ \xi$, one should choose $\boldsymbol{G} /\!/ \boldsymbol{q}$ for observation of a longitudinal mode. For transverse modes, $\mathbf{q} \perp \xi$ and one requires $\mathbf{G} \perp \mathbf{q}$.

For simple structures, the eigenvectors may be determined entirely from the symmetry of the space group. Thus, the structure factors $F_j(\mathbf{Q})$ may be entirely determined from the crystal structure. For more complex structures, the space–group symmetry only classifies the phonons into a number of irreducible representations. The number of phonons associated with each representation is same as that of number of symmetry vectors. The eigenvectors could be any linear combinations of the symmetry vectors associated with the irreducible representation. Calculation of individual structure factors can be done from the knowledge of the eigenvectors that may be calculated on the basis of a reasonable model of the lattice dynamics. These calculations help in identifying the regions in reciprocal space, where the neutron-scattering cross sections are large. The procedure has been used to calculate (Section 3.7.2) the neutron scattering structure factors for the measurement of the phonon dispersion relations from the single crystal of zircon, $ZrSiO_4$.

3.5.2
Inelastic Neutron Scattering from Powder Samples: Phonon Density of States

In Section 3.4.1, the scattering experiment from a single crystal is considered, in which the scattering vector has a definite orientation with respect to the reciprocal space of the single crystal. In the case of scattering from a powder sample, the reciprocal axes belonging to the different grains of the powder have different orientations, and ideally all possible orientations exist with equal probability. Thus, averaging over the various grains is equivalent to averaging over all orientations of the scattering vector **Q**. Further averaging over the magnitude of **Q**, or in the limit of large **Q** where the correlations between atomic motions become small and $S(\mathbf{Q},\omega)$ becomes independent of **Q** (except for some smoothly varying **Q** dependent factors), one gets $S(\omega)$, the density of excitations at the frequency ω, weighted with the neutron scattering lengths. Since the peaks in $S(\mathbf{Q},\omega)$ would normally be **Q** dependent, due to coherent scattering effects, the averaging over **Q** should be carried out over a suitable chosen range of **Q**.

The expression for coherent inelastic neutron scattering from a powder sample due to one-phonon excitation is obtained from the one-phonon expression for a single crystal given in Section 3.4.1 by the directional averaging of **Q**. However, this procedure is usually not followed, since one needs the dispersion relation as well as the eigenvectors over the entire Brillouin zone for performing the average. Hence, the coherent inelastic neutron scattering data from a powder sample are usually analyzed in the incoherent approximation. In this approximation, one neglects the correlations between the motions of atoms and treats the scattering from each atom as incoherent with the scattering amplitude b_k^{coh}. However, this is valid only for large **Q**. In the incoherent approximation, the cross section will not contain the phase term and the directional averaging over **Q** of the term $|\mathbf{Q}\cdot\xi|^2$ gives $Q^2\xi^2/3$. Therefore, the expression [28, 29] for coherent one-phonon scattering from a powder sample is given as

$$S_{\text{coh}}^{(1)}(\mathbf{Q},\omega) = A'' \sum_k \exp(-2W_k(\mathbf{Q})) \frac{(b_k^{\text{coh}})^2}{m_k} \sum_{\mathbf{G}\mathbf{q}j} \frac{1}{3} Q^2 |\xi(\mathbf{q}j,k)|^2 \frac{\hbar}{2\omega(\mathbf{q},j)} \times \left(n(\omega) + \frac{1}{2} \pm \frac{1}{2} \right) \delta(\mathbf{Q}-\mathbf{G}\pm\mathbf{q})\delta(\omega \mp \omega(\mathbf{q}j)) \tag{3.31}$$

where A'' is a normalization constant. Incoherent inelastic scattering also contributes to the scattering from a powder sample in almost the same way as coherent scattering in the incoherent approximation. Thus, the so-called neutron-weighted density of states involves weighting by the total scattering cross section of the constituent atoms. The measured scattering function [19, 20] in the incoherent approximation is therefore given as

$$S_{\text{inc}}^{(1)}(\mathbf{Q},\omega) = \sum_k \frac{b_k^2}{\langle b^2 \rangle} e^{-2W_k(Q)} \frac{Q^2}{2m_k} \frac{g_k(\omega)}{\hbar\omega} \left(n(\omega) + \frac{1}{2} \pm \frac{1}{2} \right) \tag{3.32}$$

where the partial density of states $g_k(\omega)$ is given as

$$g_k(\omega) = \int \sum_j |\xi(\mathbf{q}j,k)|^2 \delta(\omega - \omega_j(\mathbf{q})) d\mathbf{q}. \tag{3.33}$$

Thus, the scattered neutrons provide the information on the density of one-phonon states weighted by the scattering lengths and the population factor. The observed neutron-weighted phonon density of states is a sum of the partial components of the density of states due to the various atoms, weighted by their scattering length squares.

$$g^n(\omega) = B' \sum_k \left\{ \frac{4\pi b_k^2}{m_k} \right\} g_k(\omega) \tag{3.34}$$

where B' is a normalization constant. Typical weighting factors $\frac{4\pi b_k^2}{m_k}$ for the various atoms in the units of barns/amu are Al: 0.055; Mg: 0.150; Fe: 0.201; Ca: 0.075; Si: 0.077; and O: 0.265. By comparing the experimental phonon spectra with the calculated neutron-weighted density of states obtained from a lattice-dynamical model, the dynamical contribution to frequency distribution from various atomic and molecular species can be understood.

Any inelastic neutron scattering spectrum from a powder sample also contains a contribution from multiphonon scattering [30]. Since one is interested to measure only the one-phonon spectrum, the multiphonon component has to be estimated and subtracted from the measured spectrum. Usually, when the one-phonon spectra is a broad function, the multiphonon scattering contributes a continuous spectrum and effectively increases the background. The scattered intensity represents the scattering function in terms of $\mathbf{Q}$ and ω which, in the conventional harmonic phonon expansion, can be written as

$$S(\mathbf{Q},\omega) = S^{(0)} + S^{(1)} + S^{(n)} \tag{3.35}$$

where, $S^{(0)}$, $S^{(1)}$, and $S^{(n)}$ represent elastic, one-phonon, and multiphonon scattering, respectively. The multiphonon contribution to the total scattering is usually estimated in the incoherent approximation using Sjolander's formalism [30] in which the total contribution is treated as a sum of the partial components of the density of states from various species of atoms. The coherent scattering due to multiphonon excitations involves the rules of energy and wave vector conservation similar to those for one-phonon excitations, namely,

$$E_i - E_f = \pm\hbar\omega(\mathbf{q}1, j1) \pm \hbar\omega(\mathbf{q}2, j2) \pm \ldots \tag{3.36}$$

and

$$k_i - k_f = \mathbf{G} \pm \mathbf{q1} \pm \mathbf{q2} - \ldots \tag{3.37}$$

where the $+$ or $-$ corresponds to either creation or annihilation of the phonons. Note that, for any of the parameters (E_i, E_f, k_i, and k_f) it would be usually possible to obtain several combinations of $\mathbf{q}$ and ω satisfying the above conservation rules. Hence, it turns out, unlike the one-phonon contribution, usually the multiphonon contribution to the inelastic neutron spectrum does not give rise to sharp peaks but gives only a continuous spectrum.

It is important to note that the multiphonon contribution is quite important even at low temperatures since the phonon creation is always possible for sufficiently large incident neutron energy. The coherent cross section for the multiphonon process is more difficult to evaluate theoretically than the one-phonon spectrum. It is convenient to resort to the incoherent approximation, which may be more justified when more phonons are involved in the scattering process since the scattering is expected to be a smoothly varying function of $\mathbf{Q}$ and ω, and that the interatomic correlations are expected to be small. In this approximation,

$$S^{(m)}(\mathbf{Q}, \omega) = \sum_k \frac{A_k b_k^2}{m_k} S_k^m(\mathbf{Q}, \omega) \tag{3.38}$$

where A_k is a normalization constant. The total multiphonon scattering cross section is a weighted sum of the multiphonon contribution from each atomic species. Contribution from the incoherent multiphonon scattering is taken into account by using the total scattering length squares of the atoms. The computation of $S_k^m(\mathbf{Q}, \omega)$ is carried out using the Sjolander's formalism. $S_k^m(\mathbf{Q}, \omega)$ is given as

$$S_k^m(\mathbf{Q}, \omega) = \exp(-2W_k) \sum_{n=2}^{\infty} G_n(\omega) \frac{(2W_k)^n}{n!} \tag{3.39}$$

with $G_o(\omega) = \delta(\omega)$ and $G_1(\omega) = g_k(\omega)$ where, $g_k(\omega)$ is the partial density of states, and

$$G_n(\omega) = \int_{-\infty}^{\infty} g_k(\omega - \omega') G_{n-1}(\omega') d\omega' \tag{3.40}$$

gives the higher order terms. The multiphonon contribution may be estimated on the basis of a lattice-dynamical model. The experimental one-phonon spectrum is obtained by subtracting the calculated multiphonon contribution from the experimental data.

3.6 Experimental Techniques for Inelastic Neutron Scattering

A number of different techniques [28, 29, 31] have been developed to determine the change in energy and momentum of the scattered neutrons, but only the measurements using the triple-axis spectrometer and neutron time-of-flight technique are described in this chapter.

3.6.1 Measurements Using Triple-Axis Spectrometer

Scanning of the ω-**q** plane along any path is most conveniently carried out with triple-axis spectrometer [28]. The monochromatization of the primary neutron beam from the reactor and analysis of the energy of the scattered neutrons are obtained by neutron diffraction using a single crystal (monochromator and analyzer). The resolution of the spectrometer can be changed by replacing the crystals and collimators.

Figure 3.1 shows the schematic diagram of the triple-axis spectrometer at Trombay, India [32]. As name implies, this instrument comprises three axes. A crystal monochromator situated at the first axis helps to select a monochromatic neutron beam from the Maxwellian spectrum of neutrons from the reactor. These monochromatic neutrons are incident on a sample mounted on the second axis. As the neutrons undergo energy and momentum exchange during the scattering, one can measure the spectrum of neutrons scattered at any scattering angle by means of an analyzing crystal mounted on the third axis followed by a neutron detector. The simplest method of analyzing the spectrum is by a $(\theta_A - 2\theta_A)$scan of analyzer and detector combination, in which one may observe a neutron group at some $2\theta_A$. The measured intensity of the peak corresponds to an arbitrary locus in the (**Q**,ω) space. For ease in interpretations of data, preferred scans correspond to situations where wave vector transfer **Q** or energy transfer $\hbar\omega$ is a constant. The measurements of neutron intensities are carried out at a series of E_i and/or E_f in such a way that all the time either **Q** or ω is held constant. These are therefore referred to as constant **Q** or constant ω scans.

At Dhruva reactor, Trombay, India, the neutrons of fixed final energy E_f are observed using pyrolytic graphite (0 0 2) analyzer, while the incident energy is varied using a copper (1 1 1) monochromator. A pyrolytic graphite filter can be used to reduce the contribution from the second-order reflection. The spectrometer covers scattering angles ranging from 10 to 100°. The intensity of the scattered neutrons is measured by a BF_3 gas counter. All the measurements are carried out in the energy loss mode with constant momentum transfer (**Q**). The elastic energy resolution is approximately 15% of the final energy.

3.6.1.1 Phonon Density of States

Measurement of density of states for coherent samples requires data collection over a wide range of **Q** and ω. For good averaging over the Brillouin zone, the **Q** values are kept as large as possible. The experimental data obtained for different fixed energies E_f and momentum transfer (**Q**) values are averaged over **Q** to obtain the neutron-weighted phonon density of states $g^{(n)}(E)$. As these experiments are time consuming, in view of intensity limitations, medium flux reactors (such as Dhruva reactor) have been used to obtain density of states of coherent scatterers with range of

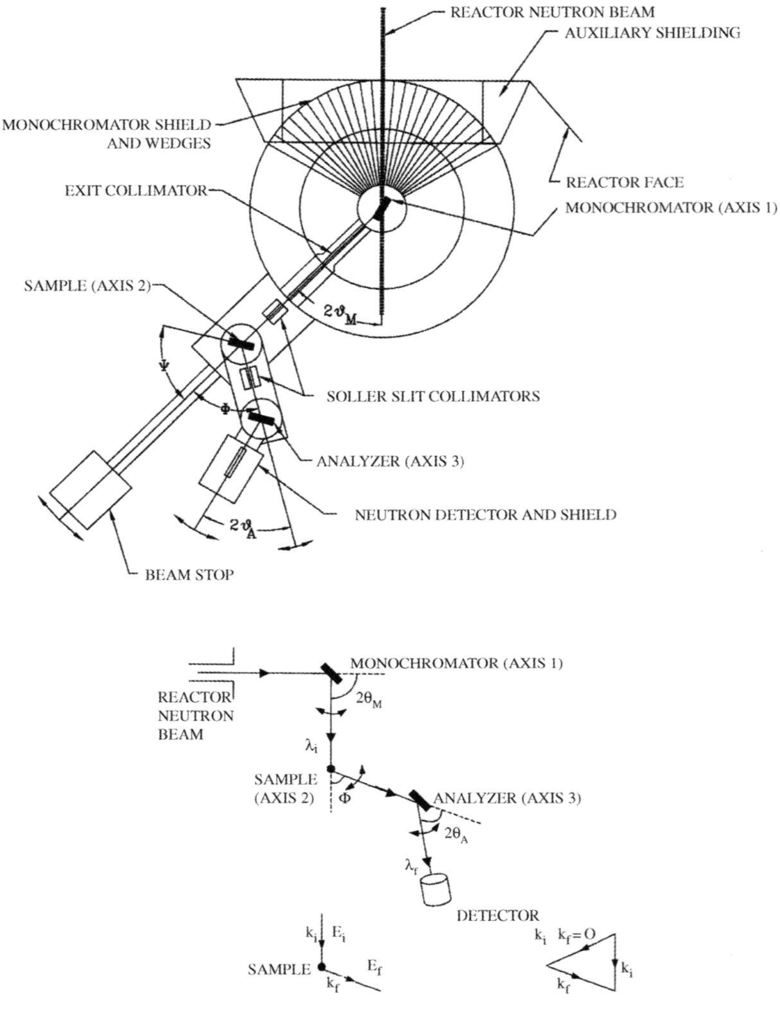

Figure 3.1 (Upper) Layout and (lower) schematic diagram of the triple-axis spectrometer at Dhruva reactor, Trombay, India (after Ref. [1]).

phonon spectrum extending upto 50 meV. At Dhruva reactor, the triple-axis spectrometer has been used for the measurements of phonon density of states of X-ray image storage material BaFCl and Cu_2O. Such measurements require approximately 10 cc of the polycrystalline sample. Several scans are recorded in the constant momentum transfer (**Q**) and the neutron energy loss mode. **Q** values from 4 to 6 $Å^{-1}$ and E_f values from 20 to 30 meV were used in different scans.

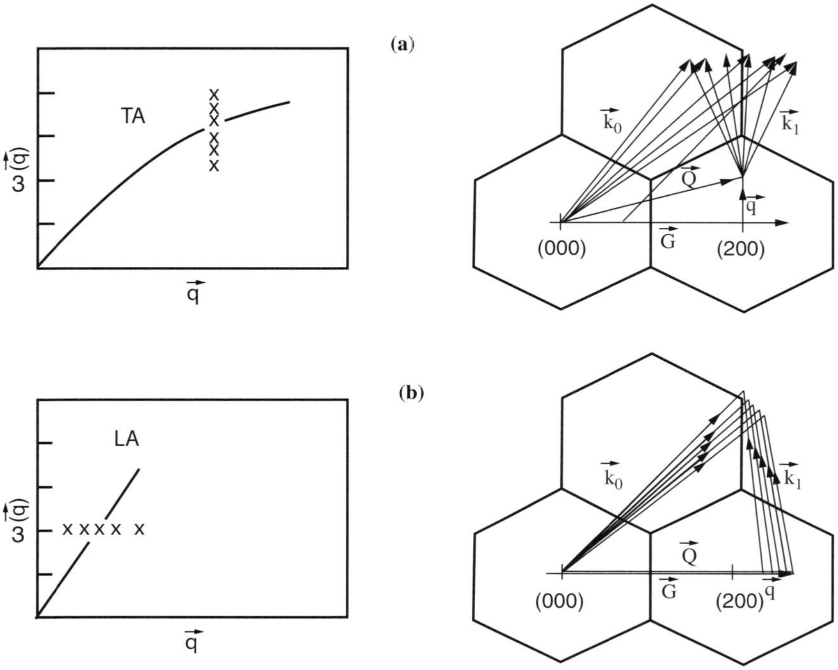

Figure 3.2 The representation of scans in (**Q**,ω) space and corresponding scattering diagrams for measurements using triple-axis spectrometer: (a) constant **Q** measurement of a transverse acoustic (TA) phonon with fixed analyzer energy and (b) constant ω measurement of a longitudinal acoustic (LA) phonon. (After Ref. [28].)

3.6.1.2 Phonon Dispersion Relations

Figure 3.2a shows the vector diagrams of scattering geometry for constant **Q** scans [19]. For a selected wave vector **q** with momentum transfer $\hbar\mathbf{Q}$, the neutron intensity is recorded as a function of energy transfer and the observed peak is assigned to the energy of the excitation. Constant **Q** mode is best suited for the measurement of excitations with weak dispersion (Figure 3.2a, whereas for a strong dispersion Figure 3.2b constant ω technique gives best results. A large number of such measurements of peaks in the $S(\mathbf{Q},\omega)$ can be used to identify several points on the phonon dispersion relation $\omega(\mathbf{q})$. The calculated one-phonon structure factors (Section 3.4.1) are generally used as guide for the selection of Bragg points for the measurements of phonon dispersion relation.

3.6.2 Measurements Using Time-of-Flight Technique

The time-of-flight (TOF) technique [31] is different from the reactor based triple-axis spectroscopy where the (**Q**,ω) space is scanned pointwise. On the other hand, time-of-flight spectrometer detects a large (**Q**,ω) space volume in a single run. This is done by the simultaneous use of several detectors equipped with the respective TOF electronics. The change in energy and the scattering vector **Q** is obtained by measuring the flight time and the scattering angle of the neutrons from a beam pulsing device (chopper) to the detectors. The energy of the neutrons is fixed before or after the scattering process.

Therefore, during an experiment, one obtains the dependence of the double differential cross section upon the energy transferred in a neutron scattering event at the sample and after introducing corrections the neutron-weighted phonon density of states $g^{(n)}(E)$ can be obtained using Equation 3.35.

3.6.2.1 Phonon Density of States

For measurement of phonon density of states the scattered neutrons from the sample are collected over a wide range of scattering angles. By choosing a suitable high incident neutron energy, measurement of the scattering function $S(\mathbf{Q},\omega)$ over a wide range of momentum and energy transfers can be undertaken and the data can be averaged over a wide range of **Q**. High-pressure inelastic neutron scattering experiments [33, 34] were carried out on polycrystalline samples of ZrW_2O_8 (Section 3.7.3.2) and $ZrMo_2O_8$ using the time-of-flight IN6 spectrometer (Figure 3.3) at the Institut Laue Langevin (ILL), Grenoble, France. The angular range of the spectrometer is from 10 to 113°. Pyrolitic graphite (0 0 2) is used as the monochromator. The second-order reflection from the graphite monochromator is removed by a beryllium filter cooled at liquid nitrogen temperature. An incident energy of 3.12 meV with an elastic resolution of 80 μeV was chosen and the measurements

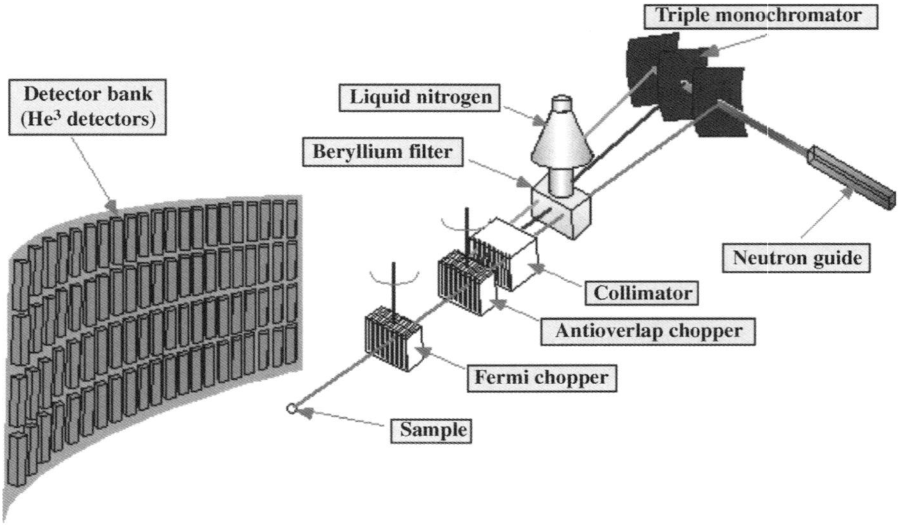

Figure 3.3 Schematic diagram of the IN6 spectrometer at ILL, Grenoble, France. (After www.ill.fr.)

were performed in the energy gain mode. The samples were compressed using argon gas in a pressure cell available at ILL.

3.6.2.2 Phonon Dispersion Relations

For measurements of phonon dispersion relation using time-of-flight technique, the incident neutron energy is measured by the time-of-flight technique and the scattered energy is analyzed simultaneously by several crystal analyzers at various scattering angles. While the triple-axis spectrometers in reactors allow for scans with constant **Q** or ω, such scans are not efficiently carried out at a pulsed source where one obtains the complete (time-of-flight) energy spectrum of the neutrons scattered from the sample along a given direction. The trajectory of the scan in (**Q**,ω) space is not along constant **Q** or constant ω, but is determined by the geometrical and other instrumental settings. The layout and schematic diagram of the PRISMA spectrometer [35] are shown in Figure 3.4. In the PRISMA spectrometer (Figure 3.4) a number of **Q**-ω scans can be simultaneously obtained using several independent analyzers and detectors. Thus, the inability to carry out constant **Q** scans has been partly compensated. While in general, these **Q**-ω scans have severe limitations as observations at arbitrary **Q** and ω (e.g., a pure longitudinal scan) are not possible. The PRISMA spectrometer at ISIS, UK has been used for the measurements of phonon dispersion relation of zircon $ZrSiO_4$ upto the energy of 70 meV.

3.7 Molecular Dynamics Simulation

Molecular dynamics simulation (MDS) is the name of a technique in which the trajectories of a system of interacting particles are calculated by numerically solving their classical equations of motion [36–42]. A wide range of applications, for example, the study of phase transitions and dynamics of microscopic defects in solids, has been developed over the past few decades. In fact, this rapid development has kept pace with the growing availability of computing power.

An essential requirement for solving the classical equations of motion is the ability to calculate the forces on the particles. There are many ways to do this. The forces may be obtained either from an interparticle potential or from the force constants in a coupled system. Alternatively, the forces in an atomic system could be evaluated from the first principles quantum mechanical calculation of the electronic energies [10, 43]. Accordingly, the MDS may be called the classical MDS or the *ab initio* MDS. The two are complementary while the former is useful for extensive studies and for relatively more complex systems.

Starting with the early simulations by Alder and Wainright [36] using hard-core potentials and Rahman [37] using Lennard–Jones potentials, there has been a great deal of progress in terms of various applications in the last four decades. However, the progress was relatively slow in the 1960s and 1970s as extensive computing facilities were not widely available. In late 1970s, we studied the defect vibrational modes in rare gas solids and identified them as local modes or the resonance

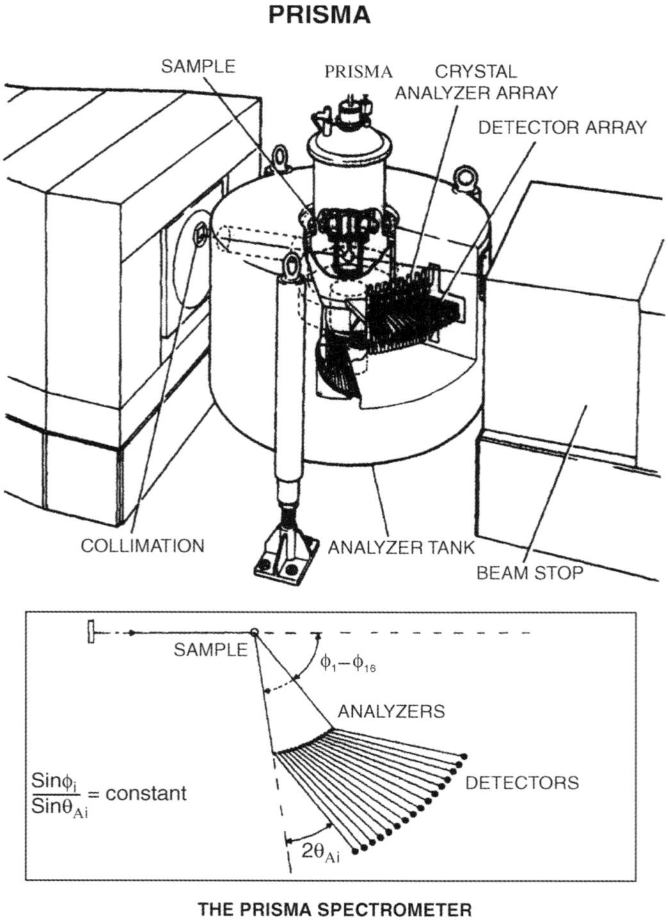

Figure 3.4 Layout (upper) and schematic diagram (lower) of the PRISMA spectrometer at ISIS, UK. (After www.isis.rl.ac.uk.)

modes [44]. A major milestone was the development of the constant-pressure technique by Parrinello and Rahman [45, 46], which facilitated studies of phase transitions involving changes in the crystallographic unit cells.

Simulations relevant to materials science involve study of a variety of physical systems, such as the bulk solids and liquids, clusters and nanoparticles, surfaces and interfaces, solid lattices with point and extended defects, and so on. The simulations may involve the characteristic parameters of temperature (T), volume (V), total energy (E), external stress (S), total number of particles (N), and so on. The quantities such as T, S, V, and E, are easily calculated at any instant of time using the atomic positions and velocities. In turn, any of these quantities (T, S, E, and V) can be controlled by suitably constraining the atomic trajectories in the simulations [45, 46]. Simulations commonly employed are those in which various combinations, for example, (NVT), (NPT), (NVE), or (NE) are kept fixed. It is now convenient to perform simulations with N about 10 000 to more than a million atoms. Bulk properties are generated by assuming periodic boundary conditions. However, the time scale of the microscopic MDS remains limited to about a few nanoseconds, that is, a few million time steps of a few femtoseconds each. This time sale is sufficient for a large variety of dynamical phenomena at the atomic level including critical propagation of extended defects and many of the phase transitions.

It is easy to make a contact between the MDS and the various scattering experiments, such as the Raman and neutron scattering experiments from various materials. Such experiments essentially measure the correlations among the positions of atoms at the same time or at different times. These correlations can be directly calculated from the simulated atomic trajectories and compared with experiments.

The simulations are validated when one obtains a good agreement between the simulated results and the experiments. Then, the simulations can be used to visualize and study the microscopic details that are not easily accessible from the experimental observations. For example, one could study the details of the collective phenomena associated with phase transitions, defect dynamics, plastic deformation, or damage. The simulations may also be extended to various sample environments, which may not have been achieved in the experiments (with due caution about the applicability of the simulations). The simulations can be only as good as the interatomic potential employed. Usually, one identifies the nature of the chemical bonds and uses suitable functional forms with adjustable parameters, which may be derived from the available experimental data or results from *ab initio* calculations. It is advisable to test the validity of the potential for the conditions, such as the interparticle distances and the coordinations that are likely to be encountered in the simulations.

We have developed empirical interatomic potentials in variety of complex ionic and molecular solids (Section 3.8), tested their lattice dynamical predictions against neutron scattering and other experimental data, and used them in molecular dynamics simulations at high temperature and pressure. At Trombay, we have developed the necessary software for molecular dynamics simulations (MOLDY), which may be applicable to systems of arbitrary size and symmetry. In particular, these softwares are also applicable to systems containing molecular ions for which the long-range Coulomb interaction is handled via the Ewald technique. Our in-house code MOLDY has been parallelized and installed on the BARC parallel computer.

3.8 Applications of Inelastic Neutron Scattering, Lattice Dynamics, and Computer Simulation

3.8.1 Phonon Density of States

Although, in principle, both reactor and spallation sources can be used for measurements of the phonon density of states, the time-of-flight method using a pulsed

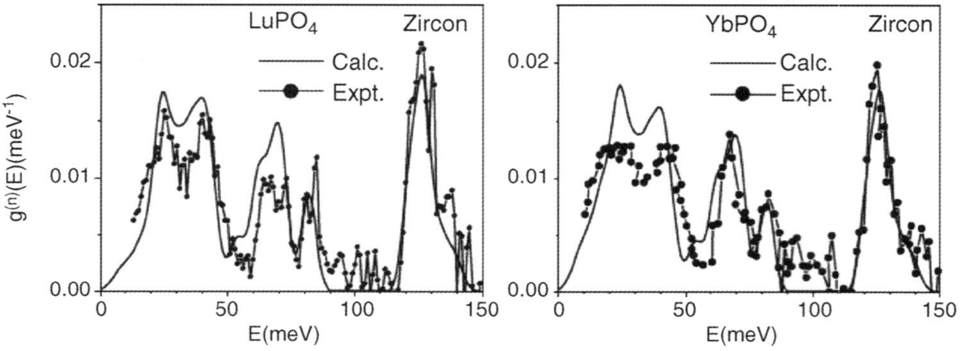

Figure 3.5 Comparison between the experimental and the calculated phonon density of states of $LuPO_4$ and $YbPO_4$ in the zircon phase. A Gaussian of FWHM 6 meV has been used for the smoothing in order to correspond to the energy resolution in the experiment (which varies between 2 and 4% of the incident energy of 200 meV). (After Ref. [22].)

spallation source is the most commonly adopted technique for the measurements of the phonon density of states, as it can cover the wide range of momentum and energy transfers. The measured phonon density of states of the rare-earth phosphates ($LuPO_4$ and $YbPO_4$) are compared with shell model calculations in Figure 3.5. The computed partial density of states [22] provides the dynamical contribution to frequency distribution from various species of atoms and has been useful in the interpretation of the observed data.

The contribution from 729 wave vectors has been included in calculation of the phonon density of states (Figure 3.5) of $LuPO_4$ and $YbPO_4$. The histogram sampling of frequencies is carried out in a frequency interval of 1.0 meV. The calculated neutron-weighted one-phonon density of states (Figure 3.5) is in good agreement with our measured data on $YbPO_4$ and the reported data [47] for $LuPO_4$. The spectra consist of phonon bands centered at about 24, 40, 68, 83, and 125 meV. There is a band gap in the energy range of 90–115 meV. These can be interpreted in terms of the partial density of states contributed from various atomic species. The calculated partial density of states (Figure 3.6) shows that rare-earth atoms contribute below 50 meV. The phonon band around 24 meV is broader in $YbPO_4$ as compared to that in $LuPO_4$ due to a dominant broad contribution from the Yb atom. The vibrations of oxygen and phosphate atoms span the entire 0–145 meV range. Above 115 meV, the contributions are mainly due to P−O stretching modes. The total phonon densities of states for $LuPO_4$ and $YbPO_4$ are similar due to the fact that the compounds are isostructural and the masses of the rare-earth elements (Yb and Lu) are nearly same.

Due to the structural complexity involved, an approach based on interatomic potentials has been used to derive phonon density of states and various microscopic and macroscopic properties for several minerals including forsterite [48, 49], fayalite [50], enstatite [51–53], $MgSiO_3$ perovskite [54, 55], almandine, pyrope, grossular, spessartine, [19, 20], zircon [16–18], sillimanite, andalusite, kyanite [56], and so on.

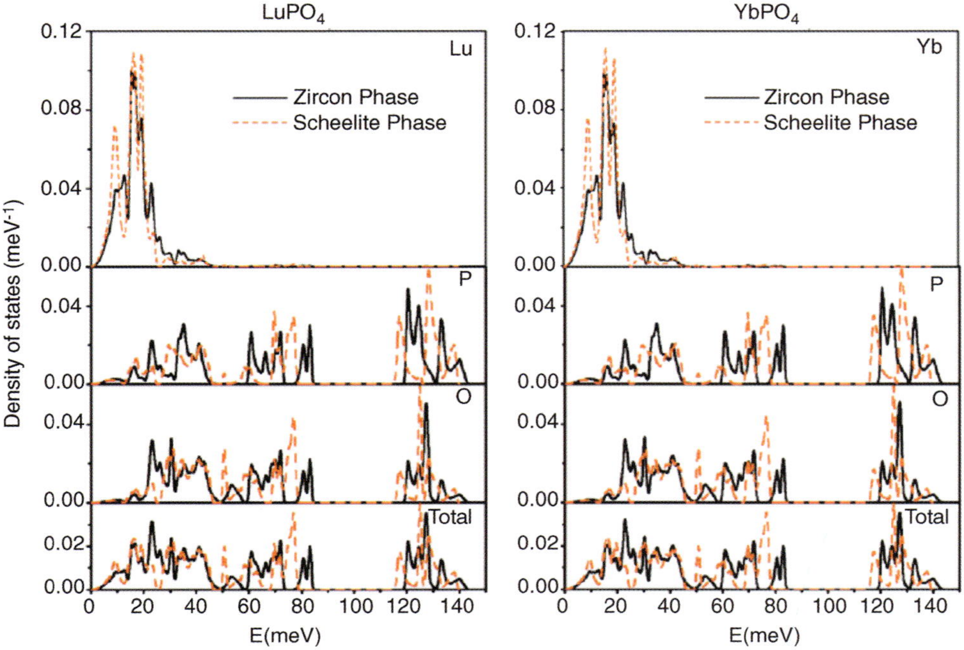

Figure 3.6 Calculated partial densities of states of various atoms in $LuPO_4$ and $YbPO_4$. (After Ref. [22].)

3.8.2 Raman and Infrared Modes, and Phonon Dispersion Relation

The study of orthosilicates, zircon ($ZrSiO_4$), hafnon ($HfSiO_4$), thorite ($ThSiO_4$), and coffinite ($USiO_4$) are of particularly importance, since these compounds are effective radiation resistant materials suitable for fission reactor applications and for storage of nuclear waste. These compounds have the zircon structure [57–59] with the space group $I4_1/amd$ (D_{4h}^{19}) and four formula units in the tetragonal unit cell. The zircon structure compounds are known to transform [60–62] to the scheelite phase ($I4_1/a$) at about 20 GPa. However, thorium silicate, $ThSiO_4$ has a zircon structure ($I4_1/amd$) at low temperature [25], whereas the high temperature form of $ThSiO_4$ has huttonite structure ($P2_1/n$). Zircon to huttonite transition is unusual [63–65] since a less dense phase usually occurs at high temperature. High-pressure studies have not been reported for $USiO_4$. Due to the various applications of these compounds, several groups [66–75] have reported experimental and theoretical studies of its structural and vibrational properties, including the specific heat and elastic constants.

The calculated phonon frequencies at the zone center for $MSiO_4$ (M = Zr, Hf, Th, and U) in the zircon phase [18] are compared in Figure 3.7. The calculations are compared with the experimental Raman data and the *ab initio* calculations.

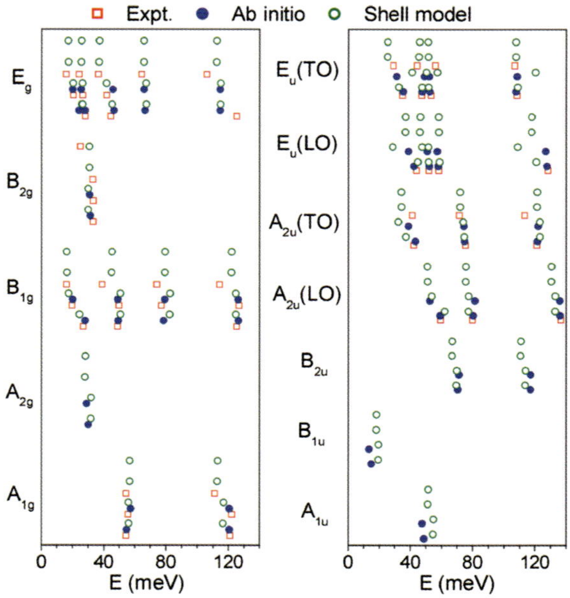

Figure 3.7 The comparison between the calculated ($T = 0$ K) and the experimental [68–73] ($T = 300$ K) zone center phonon frequencies for zircon phase of $MSiO_4$ (M = Zr, Hf, Th, and U). The *ab initio* calculations [67] for $ZrSiO_4$ and $HfSiO_4$ are also shown. The A_{2g}, A_{1u}, B_{1u}, and B_{2u} are optically inactive modes. The frequencies are plotted in the order of $ZrSiO_4$, $HfSiO_4$, $ThSiO_4$, and $USiO_4$ from below. (After Ref. [18].)

The average deviation between the calculated and the experimental frequencies is within 4–5%.

Inelastic neutron scattering is used to obtain information about the phonons in the entire Brillouin zone. In order to refine the interatomic potential for $ZrSiO_4$ we have carried out extensive inelastic neutron scattering experiments, in particular at high energy transfer up to 85 meV. The data have been obtained [15–17] using spallation and steady sources and the observed phonon dispersion relations are compared with lattice dynamical calculations. Lattice dynamical calculations are essential for the planning of neutron experiments, that is, for the calculation of the one-phonon structure factors in order to select the most appropriate Bragg points for the detection of particular phonons. They are further important for assignments of the various inelastic signals to specific phonon branches.

The phonon measurements for $ZrSiO_4$ were performed in three rounds using different techniques. Our early measurements of the phonon dispersion relation in zircon were performed on the medium resolution triple-axis spectrometer at Trombay, India [15]. Due to the relatively low neutron flux of this instrument, these measurements were restricted to an energy range up to 32 meV. Subsequently, further measurements [16] were done on the PRISMA spectrometer at ISIS using time-of-flight technique. The maximum phonon energies recorded on this instrument were 65 meV. These measurements yielded a large number of phonon frequencies because TOF techniques allow a simultaneous measurement of several phonons along specific paths in *Q–E* space. We performed additional measurements [17] using the 1T1 triple-axis spectrometer at the Laboratoire Léon Brillouin, Saclay.

The symmetry assignments and phonon dispersion relation results from the inelastic neutron scattering experiments are shown in Figure 3.8. Due to the involved nature of these experiments, such extensive measurements are often not available despite their importance. The measured data are compared with the computed first principles results and the fitted results obtained using the shell model. The experimental results are generally in good agreement with the first principles calculations. The quality of the fit is also satisfactory with approximately 4% deviation between the experimental and the calculated shell model values. The overall agreement between the experiment and the two calculations appears satisfactorily although some differences could arise due to inherent limitations in theory and experiments. The observed phonon intensities are found to be in a good qualitative agreement with the calculated one-phonon structure factors, which are quite satisfactory considering the many corrections involved in the experimental intensities and the difficulties generally encountered when making predictions from models.

The extensive phonon data have been used to refine the interatomic potential for zircon. The shell model calculations produce a good description of the available data

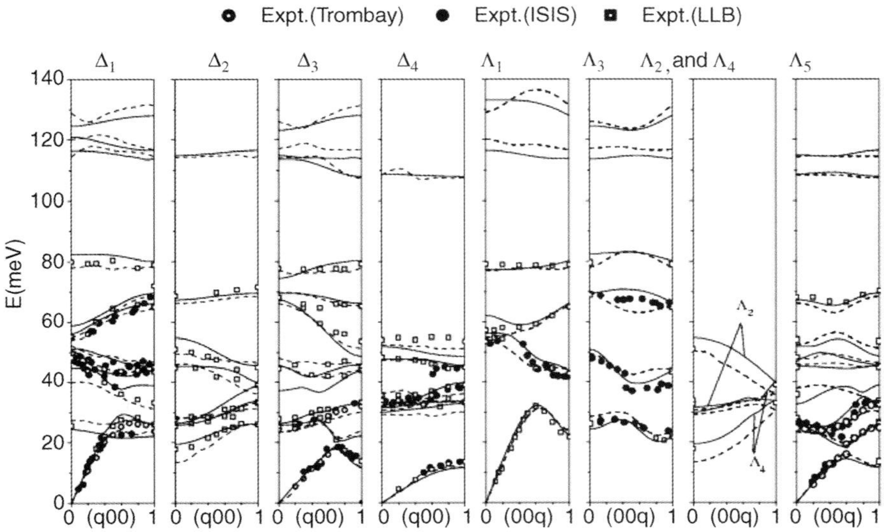

Figure 3.8 The experimental phonon dispersion curves in zircon along with the lattice dynamical calculations (solid lines: shell model; dashed lines: *ab initio*) for zircon. The open rectangles, solid circles, and open circles give the phonon peaks identified in the experiments at LLB [17],France; ISIS1 [16],UK; and Dhruva reactor [15], India, respectively. (After Ref. [16].)

on the phonon density of states measured on a polycrystalline sample. The potential derived is also fruitfully used [17] to understand the high-pressure and -temperature phase diagram, thermodynamic properties, and various microscopic and macroscopic properties of zircon.

The calculated structure factors have also been used as guides for our measurements of phonon dispersion relations from single crystals of the minerals forsterite [48], fayalite [76], and the aluminum silicate mineral andalusite [77].

3.8.3
Elastic Constants, Gibbs Free Energies, and Phase Stability

The computed elastic constants for the $MSiO_4$ (M = Zr, Hf, Th, and U) are compared in Table 3.1. The calculated bulk modulus value of $ZrSiO_4$ is 22% higher than the experimental [61] value. However, the calculated acoustic phonon branches for $ZrSiO_4$ are found to be in good agreement with the calculations [17]. Therefore, the bulk modulus should also be well reproduced. Perhaps, the measurement of the bulk modulus of $ZrSiO_4$ from natural single crystals may have been influenced [78] by the presence of known radiation damage due to radioactive impurities. This may be one of the reasons for difference between the experimental and the calculated values of bulk modulus of $ZrSiO_4$. The calculated bulk modulus values of the zircon and scheelite phases of $HfSiO_4$ are 260 and 314 GPa, respectively. These values are approximately 3.5% higher in comparison of the $ZrSiO_4$. The calculated bulk moduli for $ThSiO_4$ and $USiO_4$ in their zircon phase are nearly same. These values are approximately 80% of the bulk modulus values of $HfSiO_4$.

The phase diagram of a compound can be calculated by comparing the Gibbs free energies in various phases. In quasiharmonic approximation, Gibbs free energy of nth phase is given as

$$G = \Phi_n + PV_n - TS_n \tag{3.41}$$

Table 3.1 The elastic constants and bulk modulus [17, 18] in zircon and scheelite phase of $ZrSiO_4$ and $HfSiO_4$ and zircon phase of $ThSiO_4$ and $USiO_4$ (in GPa units).

Elastic constant	Expt. ZrSiO (Zircon)	Calc. $ZrSiO_4$ (Zircon)	Calc. $ZrSiO_4$ (Scheelite)	Calc. $HfSiO_4$ (Zircon)	Calc. $HfSiO_4$ (Scheelite)	Calc. $ThSiO_4$ (Zircon)	Calc. $USiO_4$ (Zircon)
C_{11}	424.4	432	470	441	477	334	370
C_{33}	489.6	532	288	537	282	453	483
C_{44}	113.3	110	74	107	72	78	89
C_{66}	48.2	39	133	41	136	11	20
C_{12}	69.2	73	241	77	247	38	48
C_{13}	150.2	180	255	192	274	144	159
B	205[a)]	251	303	260	314	197	217

a) Expt. Ref. [66] for elastic constants and Ref. [61] for bulk modulus for zircon.

Where, Φ_n, V_n, and S_n refer to the internal energy, lattice volume, and the vibrational entropy of the *n*th phase, respectively. The vibrational contribution is included by calculating the phonon density of states in all the phases of compound to derive the free energy as a function of temperature at each pressure. Then, the Gibbs free energy has been calculated as a function of pressure and temperature.

3.8.3.1 Zircon Structured Compound

The calculated phonon dispersion relation show [18] that for $ThSiO_4$, there is a greater density of low frequency modes in the huttonite phase in comparison of the zircon phase. This result in larger vibrational entropy in the huttonite phase, which favors [79] this phase at high temperatures.

Our Gibbs free energy calculations show that the zircon phase transforms (Figure 3.9) to the scheelite and huttonite phases at high pressure for $HfSiO_4$ and $ThSiO_4$, respectively, which is in good agreement with the experimental observations [60, 63–65]. It is likely that the phase transition pressure of $ThSiO_4$ in experiments is overestimated as these were performed with only increasing pressure and some hysteresis is expected [63–65]. Our calculated transition pressure agrees with that estimated from an analysis of the measured enthalpies [63–65]. For $ThSiO_4$, at further high pressure, the scheelite phase is found (Figure 3.9) to be stable. Experimentally, however, transformation to an amorphous phase is found (coexisting with the huttonite phase) instead of the scheelite phase, which might be due to kinetic hindrance. The free energy calculations in the zircon, scheelite and huttonite phases

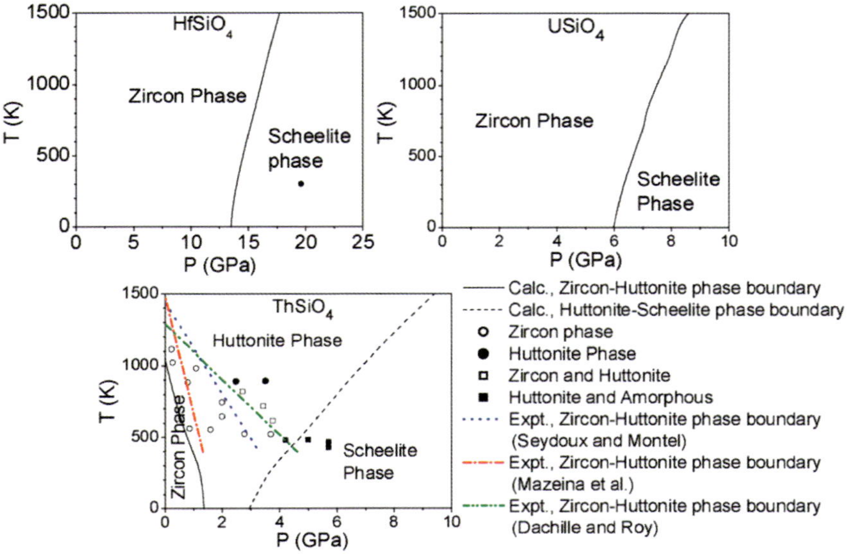

Figure 3.9 The calculated phase diagram of $HfSiO_4$, $ThSiO_4$, and $USiO_4$ as obtained from the free energy calculations. For $ThSiO_4$, experimental data are taken from Refs. [63–65]. The experimental data for $HfSiO_4$ are from Ref. 60. (After Ref. [18].)

of $USiO_4$ suggest that scheelite is the stable phase of $USiO_4$ at high pressures. The free energy changes due to volume are important in zircon to scheelite phase transition in $HfSiO_4$ and $USiO_4$, while vibrational energy and entropy play an important role in zircon to huttonite phase transition in $ThSiO_4$. It is important and satisfying to note that the free energy calculation with the present model is able to distinguish between the phases and reproduce their relative stability over a range of pressure and temperature. This is probably the most stringent test of the interatomic potentials.

3.8.3.2 **Sodium Niobate**

Sodium niobate based ceramics exhibit interesting electrical and mechanical properties that find important technological applications. Neutron diffraction studies using powder samples have been used to understand the complex sequence of low temperature phase transitions of $NaNbO_3$ in the temperature range from 12 to 350 K. Detailed Rietveld analysis of the diffraction data [23] reveals that the antiferroelectric (AFE) to ferroelectric (FE) phase transition occurs on cooling around 73 K, while the reverse ferroelectric to antiferroelectric transition occurs on heating at 245 K. However, the former transformation is not complete until it reaches 12 K and there is unambiguous evidence for the presence of the ferroelectric $R3c$ phase coexisting with an antiferroelectic phase ($Pbcm$) over a wide range of temperatures.

Using the theoretical lattice dynamical model, we have calculated [23] the energy barriers between the paraelectric $Pm\bar{3}m$, ferroelectric $R3c$, and antiferroelectric $Pbcm$ phases. The paraelectric cubic $Pm\bar{3}m$ structure has a higher energy than the ferroelectric $R3c$ and antiferroelectric $Pbcm$ phases. Further, we have calculated the double wells corresponding to the ferroelectric and antiferroelectric distortions (Figure 3.10 (a)). Our calculations reveal that both the ferroelectric and antiferroelectric distortions in $NaNbO_3$ yield similar lowering of the energy as compared to the higher energy paraelectric phase, although the ferroelectric phase has a slightly lower energy.

The ferroelectric $R3c$ phase has a slightly lower internal energy, the slightly higher vibrational entropy of the antiferroelectric $Pbcm$ phase causes the free energy crossover at $Tc \sim 50$ K. The free energy differences of the antiferroelectric $Pbcm$ and ferroelectric $R3c$ structures (inset of Figure 3.10(b)) are, however, within thermal fluctuations in the 0–400 K temperature range, which explain the coexisting ferroelectric and antiferroelectric structures over a wide range of temperature in $NaNbO_3$ observed in neutron diffraction experiments. The small energy difference between the two phases is of interest, as it would make it possible to easily switch from the antiferroelectric to ferroelectric state using realizable electric fields, which in turn would determine the potential use of this material for applications.

3.8.4 Negative Thermal Expansion from Inelastic Neutron Scattering and Lattice Dynamics

Large isotropic negative thermal expansion (NTE) from 0.3 to 1050 K was discovered in cubic ZrW_2O_8 [80]. Since then, many experimental and theoretical simulation

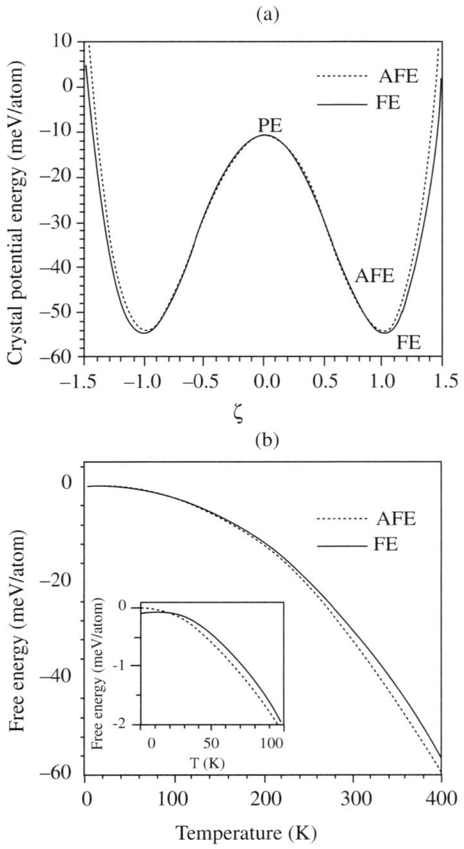

Figure 3.10 (a) Energy barriers from the paraelectric $Pm\bar{3}m$ to antiferroelectric (space group Pbcm) phase and ferroelectric (space group R3c) phase. ς corresponds to the symmetry lowering AFE and FE distortions of the ideal cubic paraelectric structure for the dashed and full lines, respectively. (b) Calculated free energy (including vibrational contributions) for the antiferroelectric and ferroelectric phases of $NaNbO_3$. (After Ref. [23].)

studies [81, 82] have been carried out to determine phonon spectrum and its relevance to NTE in framework solids. We have carried out [13, 14, 21, 33, 34, 83–85] inelastic neutron scattering and lattice dynamical calculations to understand NTE in ZrW_2O_8, HfW_2O_8, $ZrMo_2O_8$, ZrV_2O_7, HfV_2O_7, Cu_2O, and Ag_2O. Thermal expansion in insulating materials is related to the anharmonicity of lattice vibrations. The key parameters, known as Grüneisen parameters are obtained from the volume dependence of phonon frequencies. In the quasiharmonic approximation each of the phonon modes $\omega(q,j)$, (where, $\omega(q,j) =$ frequency, $q =$ wave vector, $j =$ mode index, $j = 1,3N$; N being the number of atoms in the crystallographic primitive unit cell), contribute to the thermal expansion [2, 4] equal to $\frac{1}{BV}\Gamma(\mathbf{q},j)C_V(\mathbf{q},j,T)$ (where $\Gamma(\mathbf{q},j) = -\partial \ln\omega(\mathbf{q},j)/\partial \ln V$ is the mode Grüneisen parameter, V is the cell volume, B is the bulk modulus, $C_V(\mathbf{q},j,T) = \partial E(\mathbf{q},j)/\partial T$ is the contribution of the phonon mode to the specific heat, $E(\mathbf{q},j) = \left(n + \frac{1}{2}\right)h\omega(\mathbf{q},j)$, $\left(n = \left[\exp\left(\frac{h\omega(\mathbf{q},j)}{kT}\right) - 1\right]^{-1}\right)$. Since $C_V(\mathbf{q},j,T)$ is positive for all modes at all temperatures, it is clear that the NTE would result only from large negative values of the Grüneisen parameter for certain phonons; the values should be large enough to compensate for the normal positive values of all other phonons. The frequencies of such phonons should decrease on compression of the crystal rather than increase, which is the usual behavior.

3.8.4.1 Negative Thermal Expansion Calculation

Our calculation [21] of the temperature dependence of the volume thermal expansion coefficient (Figure 3.11a) indicates that in cubic ZrV_2O_7 almost all the NTE (approximately 95%) is contributed from the phonon modes below 9 meV, among which nearly 50% of the NTE arises from just two lowest modes. The comparison between the calculated and the experimental data for cubic ZrV_2O_7 [86] is shown in Figure 3.11b. The agreement between our calculations and experimental data is excellent in the high-T phase between 400 and 900 K. In the low-T phase below 400 K, the soft phonons of the high-T phase would freeze and may no longer have the negative Grüneisen parameters. The low-T phase has positive thermal expansion coefficient. Above 900 K, the experimental data show a sharp drop in the volume at approximately 900 K, which probably signifies another phase transition.

Our estimates of NTE coefficient agree well with available experimental data. The calculations show that phonon modes of energy from 4 to 7 meV are major contributors to NTE. These important phonon modes involve translations and librations of ZrO_6 octahedral and VO_4 tetrahedral units, which significantly soften (Figure 3.12) on compression of the lattice and lead to the thermal compression.

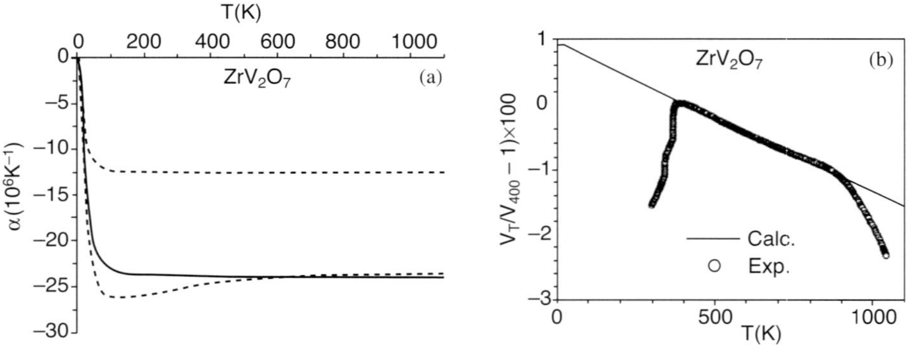

Figure 3.11 (a) The calculated volume thermal expansion (solid line) in the cubic ZrV_2O_7 along with separate contributions from the two lowest phonon branches (dotted line) and all the phonons below 9 meV (dashed line). (b) Comparison between the calculated and the experimental [86] thermal expansion behaviors of ZrV_2O_7. The low temperature phase of ZrV_2O_7 below about 400 K has positive thermal expansion coefficient. The calculations have been carried out in high-temperature phase. (After Ref. [21].)

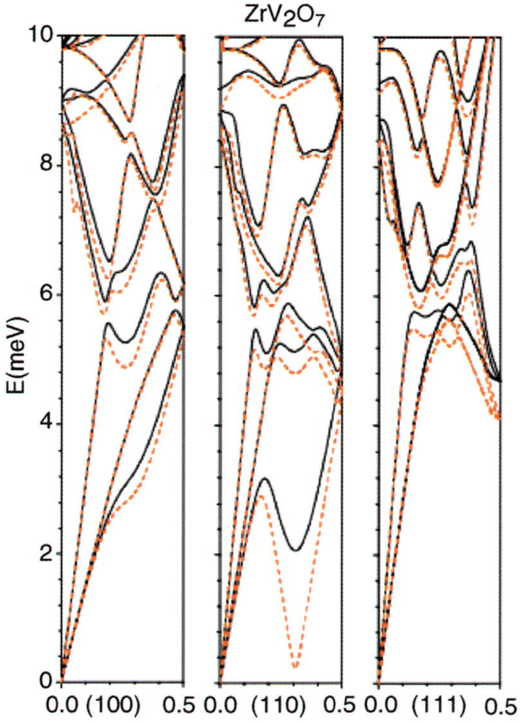

Figure 3.12 The calculated phonon dispersion relation up to 10 meV for cubic ZrV_2O_7 along the (1 0 0), (1 1 0), and (1 1 1) directions. The solid and dashed lines correspond to ambient pressure and 3 kbar, respectively (After Ref. [21].)

The maximum softening in the high-T phase occurs at $0.31\langle 1, 1, 0\rangle$, which is near $\langle 1/3, 1/3, 0\rangle$, and the freezing of these modes at low temperature could lead to the known incommensurate phase [86] below 375 K, and then below 350 K to the $3 \times 3 \times 3$ superstructure (low temperature phase) [87]. We note that the calculated soft-mode wave vector is an excellent agreement with the observed [86] incommensurate modulation. The phonon modes involved in NTE in ZrV_2O_7 are found to be quite different from those involved in cubic ZrW_2O_8.

3.8.4.2 Thermal Expansion from Experimental High-Pressure Inelastic Neutron Scattering

The measured neutron cross-section-weighted phonon density of states $g^{(n)}(E)$ for ZrW_2O_8 at 160 K and different pressures after subtracting the contributions from argon gas, absorption from the sample and empty cell background are shown in Figure 3.13. The ambient pressure results are in agreement with the previous measurements [88]. The spectra at high pressures show an unusually large softening. In conformity with the predictions, the phonon modes of energy below approximately 5 meV soften by approximately 0.15 meV at 1.7 kbar with respect to ambient

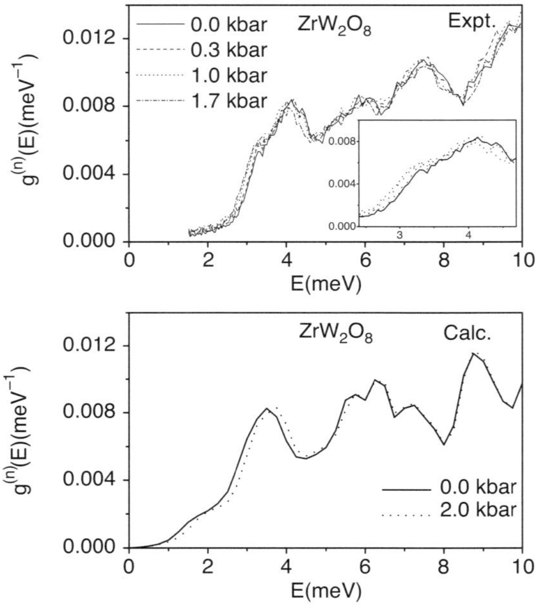

Figure 3.13 The experimental (160 K) and calculated (0 K) phonon spectra for cubic ZrW_2O_8 as a function of pressure. Inset (upper) shows the spectra approximately 4 meV on an expanded scale. (After Ref. [33].)

pressure. At energies above 5 meV, the shift of the spectrum is much less than that at lower energies. These observations (Figure 3.13) are in good agreement with the lattice dynamical calculations considering the known inherent limitations in experiment (e.g., incoherent approximation) and theory.

The Grüneisen parameter $\left[\Gamma(E) = \frac{B}{E}\frac{dE}{dP}\text{, where } B = -V\frac{dP}{dV}\right]$ for phonons of energy E has been obtained (Figure 3.14a) using the cumulative distributions for the density of states. The experimental results of Grüneisen parameters (Figure 3.14b) are in good agreement with predictions from lattice dynamics. The α_V thus derived from the neutron inelastic scattering data is in good agreement (Figure 3.14b) with that directly observed by diffraction [89]. The analysis shows that the large negative Grüneisen parameters of modes below 10 meV are able to explain the low temperature thermal expansion coefficient and its nearly constant value above 70 K [80, 89].

3.8.5
Thermodynamic Properties

The silicate perovskite, $MgSiO_3$, is a principal constituent of the Earth's lower mantle. The physical and thermodynamic properties of this mineral are of importance for interpreting the physics of the Earth's interior. In view of its geophysical importance, extensive experimental studies of its crystal structure [91], elastic constants, phonon frequencies [92, 93], equation of state [94, 95], specific heat [96], thermal expansion,

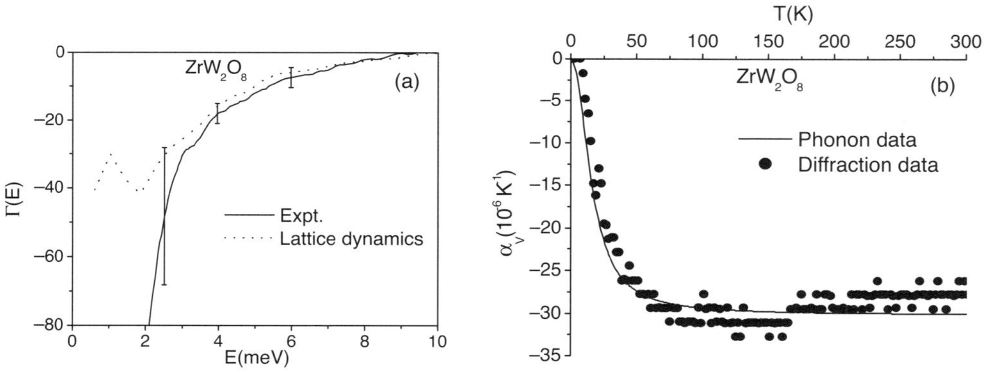

Figure 3.14 (a) The experimental Grüneisen parameter [$\Gamma(E)$] for cubic ZrW_2O_8 as a function of phonon energy (E) (averaged over the whole Brillouin zone). The Grüneisen parameter has been determined using the density of states at $P=0$ and 1.7 kbar that represents the average over the whole range in this study. The calculated $\Gamma(E)$ from the lattice dynamical calculations is shown by a dotted line. (b) The comparison between the volume thermal expansion (α_V) for cubic ZrW_2O_8 derived from the present high-pressure neutron inelastic scattering experiment (full line) and that obtained using neutron diffraction [70] (filled circles). The experimental bulk modulus value [90] of 72.5 GPa is used in the calculation of thermal expansion. (After Ref. [33].)

melting [97–100] and so on have been reported. The phase transitions of $MgSiO_3$ silicate perovskite have involved several controversies [54, 101–104] and are of interest, as they may contribute to observed seismic discontinuities in the Earth's mantle [54, 105]. Several workers have also undertaken extensive theoretical lattice dynamics and molecular dynamics studies on $MgSiO_3$ silicate perovskite using *ab initio* [10, 104, 106–110] and atomistic approaches [54, 111–114].

We had carried out detailed lattice-dynamics studies and molecular dynamics simulations of $MgSiO_3$ silicate perovskite [54, 115]. The computed equation of state, specific heat, and thermal expansion of enstatite and silicate $MgSiO_3$ perovskite [55] are shown in Figure 3.15. The calculations are in good agreement with experimental data and available *ab initio* calculations validating the model and demonstrating the role of these models in predicting high-pressure and -temperature thermodynamic properties. Similar calculations have been reported for forsterite [49], the aluminum silicate minerals sillimanite, andalusite, and kyanite [56], the garnet minerals almandine, pyrope, grossular, and spessartine [19, 20], and so on.

3.8.6
Phase Transitions in Magnesium Silicate, $MgSiO_3$

Magnesium silicate in various polymorphic forms constitutes a major component of the Earth's mantle. The upper mantle, which extends to a depth of 440 km, contains olivine, pyroxene and garnet phases, whereas the lower mantle below 660 km depth is largely made of the perovskite phase [120]. Apparently, the most important difference

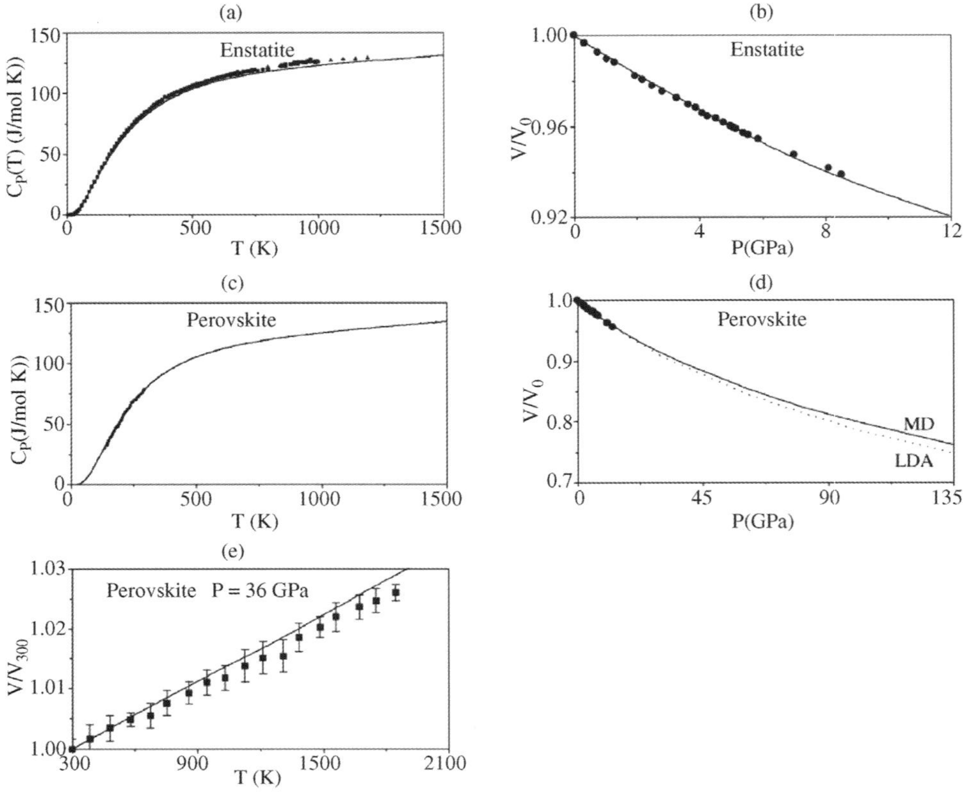

Figure 3.15 Comparisons of the calculated [55] (full line) specific heat (a) and equation of state (b) of enstatite with available experimental data [116–119] (symbols). ((c), (d), and (e)) Comparisons of the calculated [55] (full line) specific heat, equation of state and high-pressure thermal expansion of $MgSiO_3$ silicate perovskite with available experimental data [94, 96, 102] (symbols) and *ab initio* LDA calculations [10] (dotted lines). (After Ref. [55].)

between the upper and the lower mantle silicates is that the silicon coordination changes from 4 to 6, respectively, whereas the Mg coordination also increases from 6 to 8. This kind of information is indirectly inferred from seismic observations and compositional modeling of the Earth's interior based on accurate information about the structure and thermodynamic properties of the constituent phases [121–126]. Accurate modeling of mantle minerals is therefore of utmost importance, and simultaneously is also a major challenge in condensed matter physics.

We shall discuss here the results of MD simulations in Magnesium silicate as a function of pressure at high temperature starting from an important upper mantle phase, orthoenstatite. The present studies have used well-tested interatomic potentials comprising long-range Coulomb and short-range Born–Mayer type interactions. The MD simulations were performed on systems of approximately 4000 atoms using

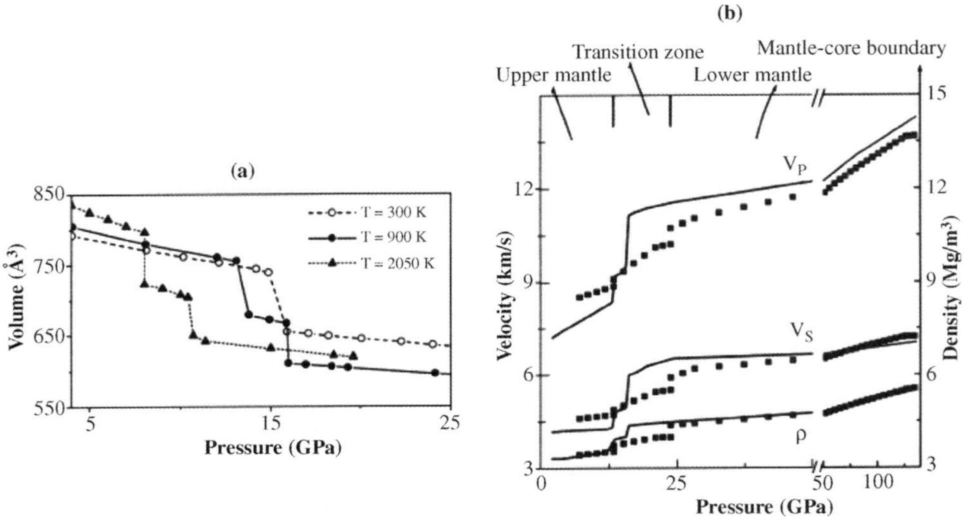

Figure 3.16 (a) The simulated volumes as a function of increasing pressure at various temperatures. (b) The solid line corresponds to the simulated longitudinal (V_P) and transverse (V_S) wave velocities and density (ϱ) as a function of pressure up to the mantle-core boundary at $T = 900$ K. The wave velocities V_P and V_S are obtained by averaging over a large number of acoustic mode frequencies along various directions. The discontinuities in the simulated *VP*, *VS*, and ϱ represent phase transitions. (After Ref. [55].)

macrocells of size (2*a*, 4*b*, and 6*c*), where *a*, *b*, and *c* are approximately 18, 9, and 5 Å, respectively. The results have been cross-checked using macrocells of sizes (2*a*, 2*b*, and 4*c*) and (4*a*, 4*b*, and 6*c*) and these simulations took several weeks on a parallel supercomputer. At each pressure and temperature the structure is equilibrated for duration of about 20 ps using a time step of 1 fs, while the system completed a phase transition is less than 50 ps.

Figure 3.16a shows the simulated change of volume as a function of pressure at three fixed temperatures of 300, 900, and 2050 K, which clearly reveal first-order transitions. The transition pressures appear to decrease with increasing temperature.

The wave velocities are calculated from the simulation of the dynamical structure factors $S(\mathbf{Q}, \omega)$ of fairly long wavelength (of approximately 40 Å) acoustic phonons in several propagation directions. The structure factor $S(\mathbf{Q},\omega)$ is related to the time correlation function $F(\mathbf{Q},t)$ of the density operator $\varrho_{\mathbf{Q}}(t)$, which is given [41] as

$$S(\mathbf{Q}, \omega) = \int_{-\infty}^{+\infty} e^{i\omega t} F(\mathbf{Q}, t)\,dt \tag{3.42}$$

where

$$F(\mathbf{Q}, t) = \frac{1}{N}\langle \varrho_{\mathbf{Q}}(t)\varrho_{-\mathbf{Q}}(0)\rangle \tag{3.43}$$

For an atomic system, the density operator is given as

$$\varrho_Q(t) = \sum_{i=1}^{N} e^{iQ \cdot r_i(t)} \tag{3.44}$$

where $\mathbf{r}_i(t)$ is the instantaneous position of the ith atom at time t.

In Figure 3.16b, we have compared the simulated seismic longitudinal and transverse velocities (V_P and V_S, respectively) and densities (ϱ) with the preliminary reference earth model (PREM) estimates [121]. The PREM results are based on seismic observations and estimates of the physical and thermodynamic properties of mantle forming phases corresponding to an actual mantle composition. The PREM results reveal several discontinuities in the observed longitudinal and transverse wave velocities (V_P and V_S, respectively) and densities (ϱ), which define the boundaries between the upper mantle, transition zone, and the lower mantle. The transition zone is defined by the observed seismic discontinuities at depths of 440 and 660 km, which on a pressure scale corresponds to approximately 13 and 24 GPa, respectively. The simulations involve only one initial phase while the PREM results correspond to an actual mantle composition.

The simulated structure factors $S(\mathbf{Q}, \omega)$ of a typical longitudinal and transverse acoustic phonon mode in the enstatite, intermediate, and perovskite phases are shown in Figure 3.17. Both the longitudinal and transverse acoustic phonon frequencies shift to higher energies upon transformation from the enstatite to the perovskite phase (via the intermediate phase). This "hardening" of frequencies in turn results in significant changes in the elastic properties, which finally manifests as discontinuities in the simulated seismic velocities (Figure 3.16b).

Figure 3.18 shows the crystal structures of the three simulated phases. Orthoenstatite [127] is orthorhombic (having the space group *Pbca* with 80 atoms in the unit cell) and its structure comprises of MgO_6 octahedral bands and single silicate tetrahedral chains. For clarity, only the coordination polyhedra around the Si atoms are shown, with the Mg atoms depicted as circles. In enstatite, there are two crystallographycally distinct MgO_6 octahedra (which are shown as circles of varying shades) and the silicate SiO_4 tetrahedra share corners to form chains parallel to the *c*-axis. The perovskite structure, on the other hand, comprises of a network of corner-shared SiO_6 octahedra, with the Mg atoms occupying the interstices. The coordination number of Mg in $MgSiO_3$ perovskite, depends on the degree of distortion from the cubic, varying from 12 (in the ideal cubic) to 8 (in the distorted orthorhombic). X-ray diffraction studies indicate that $MgSiO_3$ perovskite has the orthorhombic structure (with space group *Pnma* having 20 atoms/unit cell) over a wide range of pressures and temperature [102]. The simulated structures of enstatite and perovskite are overall consistent with these experimental observations.

The density of states is determined by Fourier transforming the velocity–velocity autocorrelation function $\langle \mathbf{v}(0).\mathbf{v}(t) \rangle$ and is given as

$$g(\omega) = \int_{-\infty}^{+\infty} \frac{v(0) \cdot v(t)}{v^2} e^{-i\omega t} dt \tag{3.45}$$

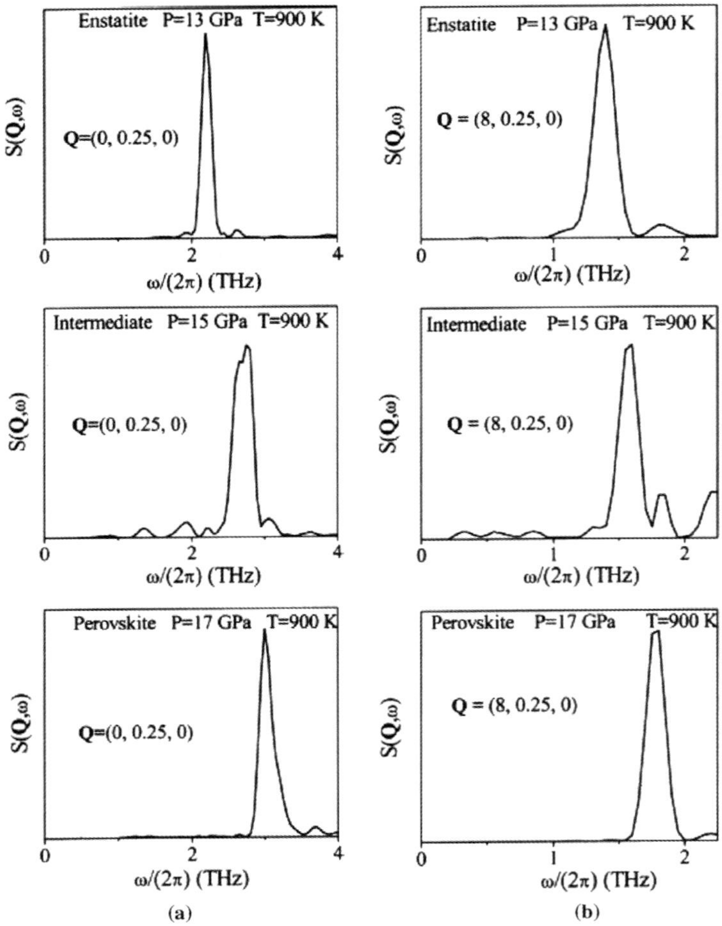

Figure 3.17 The simulated structure factor of a typical (a) longitudinal acoustic phonon mode and (b) transverse acoustic phonon mode in the enstatite, intermediate and perovskite phases. (After Ref. [55])

In Figure 3.19, the calculated vibrational density of states in enstatite phase along with the partial contributions from the Mg, Si, and O atoms is shown. The simulated density of states of enstatite is found to be in overall agreement with results obtained from lattice dynamics calculations and available inelastic neutron scattering data [53].

3.8.7
Fast Ion Diffusion in Li_2O and U_2O

Uranium, thorium, and lithium oxides are superionic conductors whose solid-state diffusion coefficients are comparable to that of liquids. They allow macroscopic

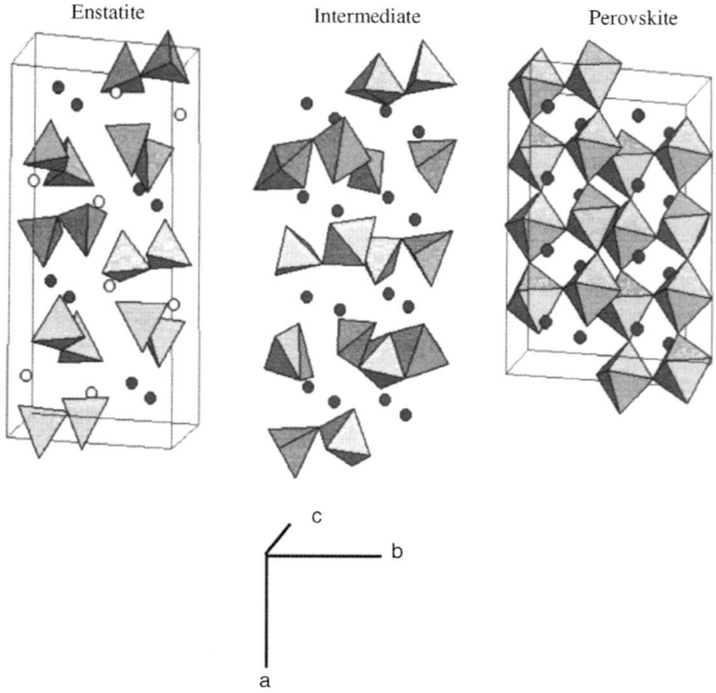

Figure 3.18 Simulated high-pressure structures in the enstatite, intermediate, and perovskite phases. The Mg atoms (circles) and the coordination polyhedra around the Si atoms are displayed. The intermediate phase (characterized by 5-coordinated silicon) is an orientationally disordered crystalline phase with space group Pmna where the Mg and Si atoms occupy the 2b and 2d sites, respectively. (After Ref. [55].)

movement of ions through their structure. A rapid diffusion of a significant fraction of one of the constituent species within an essentially rigid framework of the other species occurs· Uranium and thorium oxides are of technological importance to the nuclear industry. Li_2O finds several technological applications ranging from miniature light -weight high-power density lithium-ion batteries for heart pacemakers, laptop computers to name a few. A detailed study [128–130] of their physical and thermodynamic properties is of great interest for enhanced understanding of these systems. We have used a combination of lattice dynamics and molecular dynamics studies to understand UO_2, ThO_2 [131], and Li_2O in their normal as well as superionic phase. With the help of molecular dynamics simulations, the transition from normal to superionic phase has been studied. These oxides exhibit a sublattice melting-like behavior at temperatures around $0.8T_m$, where T_m is the melting temperature. In actinides UO_2 and ThO_2, it is the oxygen sublattice that exhibits premelting phenomena, while in antifluorite Li_2O, it is the lithium atom that exhibits superionic behavior. The melting temperatures of Li_2O, UO_2, and ThO_2 are 1705, 3310, and 3400 K, respectively. Our studies infer that superionicity sets around 1000, 2400, and

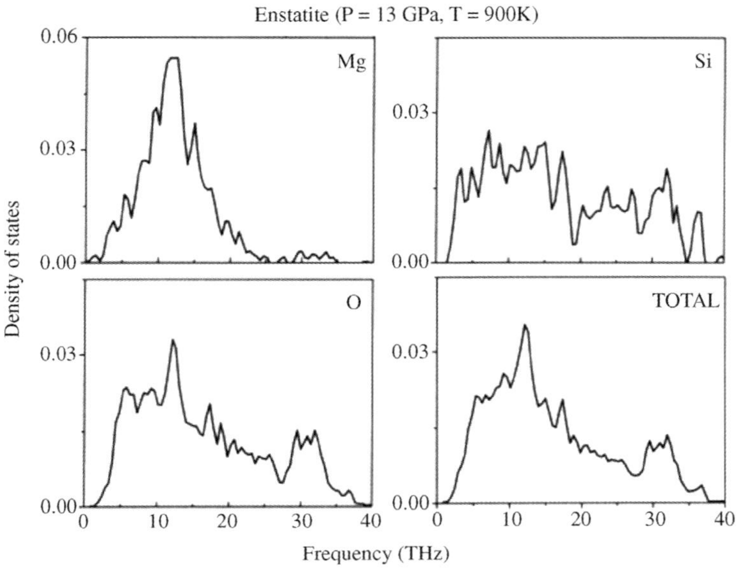

Figure 3.19 Calculated partial and total density of states (arbitrary units) of enstatite at $P = 13$ GPa and $T = 900$ K. (After Ref. [55].)

2700 K in the three oxides. Figure 3.20 gives the comparison of our calculations with the reported experimental data.

The microscopic snapshot of the processes occurring in the sublattice around transition temperature is depicted in Figure 3.21. The position–time plots of certain Li atoms, for time duration of 10 ps at 1250 K in *XY* and *YZ* planes are given in Figure 3.21. The Li atom A at an initial position (the positions are actually multiples of the lattice parameter) of (1.25,1.25,1.25) moves to a second position (1.25,1.25,1.75) with a jump time of approximately 0.065 ps, residence time being more than 3.3 ps at the initial position. It then moves to the third position of (1.25,0.75,1.25). The residence time at the second position is 3.9 ps. Before moving to the third position, it undergoes some transit zigzag motion in the octahedral region surrounding the coordinate (1.25,1.65,1.5). On the other hand, atom B starts at an initial coordinate of (3.75,0.75,1.75) and then moves to the second position (3.75,1.25,1.75) in the time duration for which these atoms were tracked. These results indicate that the lithium atoms jump from one tetrahedral position to another, passing the octahedral interstitial regions during transit; at any given instant the probability of an atom sitting in the octahedral position is rather small. Similar results have been reported for CuI [132, 133], where ionic density distribution shows no or very little occupation of the octahedral sites with increase in temperature. Average potential energy curves obtained for CuI depicts that energy is a minimum at the tetrahedral sites and rises rapidly as the octahedral site is approached. The lithium atoms essentially move from one tetrahedral position to another randomly, yet maintaining the local order and structure to a great degree.

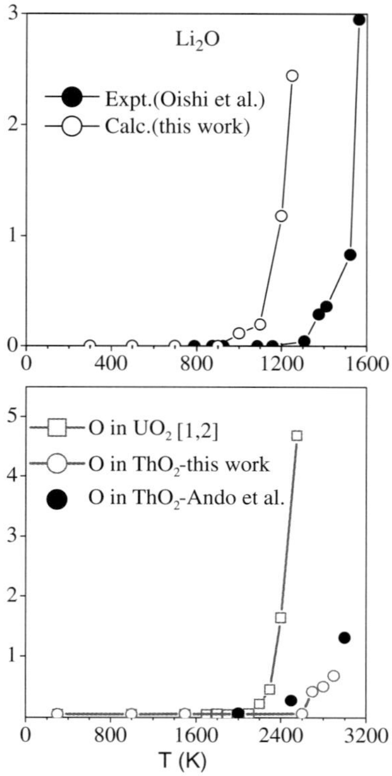

Figure 3.20 Diffusion-coefficient in different nuclear oxides (After Refs [128–131].)

It can be concluded that though considerable disturbances are occurring in one of the sublattice, the crystal structure of the system is maintained. This behavior is called fast ion conduction, and this is technologically important and used for several applications.

3.9
Conclusions

The combination of lattice dynamics calculations and inelastic neutron scattering measurements can be successfully used to study the phonon properties and their manifestations in thermodynamic quantities such as specific heat, thermal expansion, and equation of state. The measurements of phonon density of states from the powder samples and the wave vector dependence of phonon frequencies from single crystals can be made using spectrometers at both the steady state sources such as reactors and pulsed spallation sources. The experiments validate the models and the models in turn have been fruitfully used to calculate the phonon spectra and various thermodynamic properties at high pressures and high temperatures. The phonon

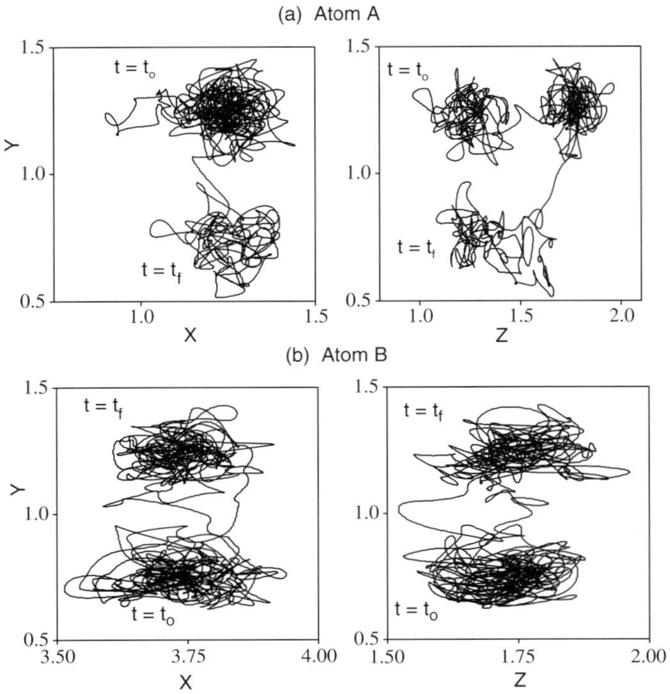

Figure 3.21 Snapshot [128] of lithium atoms at 1250 K. (After Ref. [128].)

calculations have been useful in the planning and execution of the experiments. The calculations also enabled microscopic interpretations of the observed experimental data. Molecular dynamics studies have been useful in understanding the high pressure–temperature phase transitions of solids. The simulations have enabled a microscopic visualization of the key mechanisms of phase transitions and provide useful insights about the variations in the vibrational properties.

References

1 Chaplot, S.L., Choudhury, N., Ghose, S., Rao, M.N., Mittal, R., and Goel, P. (2002) Inelastic neutron scattering and lattice dynamics of minerals. *Eur. J. Mineral*, **14**, 291–329.

2 Venkataraman, G., Feldkamp, L., and Sahni, V.C. (1975) *Dynamics of Perfect Crystals*, The MIT Press, Cambridge.

3 Born, M. and Huang, K. (1954) *Dynamical Theory of Crystal Lattices*, Oxford University Press, London.

4 Bruesch, P. (1982) *Phonons: Theory and experiments I*, Springer-Verlag, Berlin.

5 Mittal, R., Chaplot, S.L., and Choudhury, N. (2006) Modeling of anomalous thermodynamic properties using lattice dynamics and inelastic neutron scattering. *Prog. Mater. Sci.*, **51**, 211–286.

6 Choudhury, N. and Chaplot, S.L. (2009) Inelastic neutron scattering and lattice dynamics: perspectives and challenges in mineral physics, in *Neutron Applications*

in Earth, Energy and Environmental Sciences (eds L. Liang, H. Schober, and R. Rinaldi), Springer, New York, pp. 145–188.

7 Lubbers, R., Grunstendel, H.F., Chumakov, A.I., and Wortmann, G. (2000) Density of phonon states in iron at high pressure. *Science*, **287**, 1250.

8 Schwoerel-Bohning, M.W., Macrander, A.T., and Arms, D.A. (1998) Phonon dispersion of diamond measured by inelastic X-ray scattering. *Phys. Rev. Lett.*, **80**, 5572.

9 Wentzcovitch, R.M., da Silva, C., Chelikowsky, J.R., and Binggeli, N. (1998) A new phase and pressure induced amorphization in silica. *Phys. Rev. Lett.*, **80**, 2149.

10 Wentzcovitch, R.M., Martins, J.L., and Price, G.D. (1993) *Ab-initio* molecular dynamics with variable cell shape: applications to $MgSiO_3$ perovskite. *Phys. Rev. Lett.*, **70**, 3947–3950.

11 Hemley, R.J., Kresse, G., and Hafner, J. (1998) High pressure polymorphism in silica. *Phys. Rev. Lett.*, **80**, 2145.

12 Karki, B.B., Stixrude, L., Clark, S.J., Warren, M.C., Ackland, G.J., and Crain, J. (1997) Elastic properties of orthorhombic $MgSiO_3$ perovskite at lower mantle pressures. *Am. Mineral.*, **82**, 635–638.

13 Mittal, R. and Chaplot, S.L. (1999) Lattice dynamical calculation of isotropic negative thermal expansion in ZrW_2O_8 over 0–1050 K. *Phys. Rev. B*, **60**, 7234.

14 Mittal, R., Chaplot, S.L., Kolesnikov, A.I., Loong, C.-K., and Mary, T.A. (2003) Inelastic neutron scattering and lattice dynamical calculation of negative thermal expansion in HfW_2O_8. *Phys. Rev. B*, **68**, 54302.

15 Mittal, R., Chaplot, S.L., Rao, M.N., Choudhury, N., and Parthasarthy, R. (1998) Measurement of phonon dispersion relation in zircon. *Physica B*, **241**, 403–405.

16 Mittal, R., Chaplot, S.L., Parthasarathy, R., Bull, M.J., and Harris, M.J. (2000) Lattice dynamics calculations and phonon dispersion measurements of zircon, $ZrSiO_4$. *Phys. Rev. B*, **62**, 12089–12094.

17 Chaplot, S.L., Pintschovius, L., Choudhury, N., and Mittal, R. (2006) Phonon dispersion relations, phase transitions, and thermodynamic properties of $ZrSiO_4$: inelastic neutron scattering experiments, shell model and first-principles calculations. *Phys. Rev. B*, **73**, 94308.

18 Bose, P.P., Mittal, R., and Chaplot, S.L. (2009) Lattice dynamics and high pressure phase stability of zircon structured natural silicates. *Phys. Rev. B*, **79**, 174301.

19 Mittal, R., Chaplot, S.L., Choudhury, N., and Loong, C.K. (2000) Inelastic neutron scattering and lattice dynamics studies of almandine $Fe_3Al_2Si_3O_{12}$. *Phys. Rev. B*, **61**, 3983–3988.

20 Mittal, R., Chaplot, S.L., and Choudhury, N. (2001) Lattice dynamics calculations of the phonon spectra and thermodynamic properties of the aluminosilicate garnets pyrope, grossular and spessartine $M_3Al_2Si_3O_{12}$ ($M = Mg$, Ca and Mn). *Phys. Rev. B*, **64**, 94302-1-9.

21 Mittal, R. and Chaplot, S.L. (2008) Lattice dynamical calculation of negative thermal expansion in ZrV_2O_7 and HfV_2O_7. *Phys. Rev. B*, **78**, 174303.

22 Mittal, R., Chaplot, S.L., Choudhury, N., and Loong, C.-K. (2007) Inelastic neutron scattering, lattice dynamics and high pressure phase stability in $LuPO_4$ and $YbPO_4$. *J. Phys. Condens. Matt.*, **19**, 446202.

23 Mishra, S.K., Choudhury, N., Chaplot, S.L., Krishna, P.S., and Mittal, R. (2007) Competing antiferroelectric and ferroelectric interactions in $NaNbO_3$: neutron diffraction and theoretical studies. *Phys. Rev. B*, **76**, 024110.

24 Chaplot, S.L. (1988) Phonon dispersion relation in $YBa_2Cu_3O_7$. *Phys. Rev. B*, **37**, 7435.

25 Chaplot, S.L. (1978) A computer program for external modes in complex molecular-ionic crystals. Report BARC-972. Mumbai: Bhabha Atomic Research Centre.

26 Jindal, V.K. and Kalus, J. (1986) Calculation of thermal expansion and phonon frequency shift in deuterated naphthalene. *Phys. Stat. Sol. B*, **133**, 89.

27 Bhandari, R. and Jindal, V.K. (1991) Calculation of thermal expansion and implicit phonon frequency shift in deuterated anthracene. *J. Phys. Condens. Matt.*, **3**, 899.

28 Bruesch, P. (1986) *Phonons: Theory and Experiments II*, Springer-Verlag, Berlin.

29 Price, D.L. and Skold, K. (1986) *Methods of Experimental Physics: Neutron Scattering Part A*, vol. 23 (eds K. Skold and D.L. Price), Academic Press, Orlando; Carpenter, J.M. and Price, D.L. (1985) *Phys. Rev. Lett.*, **54**, 441.

30 Sjolander, A. (1958) Multiphonon processes in slow neutron scattering by crystals. *Arkiv für Fysik*, **14**, 315–371.

31 Bacon, G.E. (1975) *Neutron Diffraction*, Oxford University Press, Oxford.

32 Chaplot, S.L., Mukhopadhyay, R., Vijayaraghavan, P.R., Deshpande, A.S., and Rao, K.R. (1989) Phonon density of states of tetracyanoethylene from coherent inelastic neutron scattering at Dhruva reactor. *Pramana-J. Phys.*, **33**, 595–602.

33 Mittal, R., Chaplot, S.L., Schober, H., and Mary, T.A. (2001) Origin of negative thermal expansion in cubic ZrW_2O_8 revealed by high pressure inelastic neutron scattering. *Phys. Rev. Lett.*, **86**, 4692–4695.

34 Mittal, R., Chaplot, S.L., Schober, H., Kolesnikov, A.I., Loong, C.-K., Lind, C., and Wilkinson, A.P. (2004) Negative thermal expansion in cubic $ZrMo_2O_8$: inelastic neutron scattering and lattice dynamical studies. *Phys. Rev. B*, **70**, 214303.

35 Steigenberger, U., Hagen, M., Caciuffo, R., Petrillo, C., Cilloco, F., and Sachetti, F. (1991) The development of the PRISMA spectrometer at ISIS. *Nucl. Instrum. Meth. B*, **53**, 87–96.

36 Alder, B.J. and Wainright, T.E. (1959) Studies in molecular dynamics. I. General method. *J. Chem. Phys.*, **31**, 459.

37 Rahman, A. (1964) Correlations in the motion of atoms in liquid argon. *Phys. Rev.*, **136**, A405.

38 Chaplot, S.L. (1986) Computer simulation of dynamics in solids. *Curr. Sci.*, **55**, 949.

39 Chaplot, S.L. (1999) *Bull. Mater. Sc.*, **22**, 279.

40 Allen, M.P. and Tildesley, D.J. (1987) *Computer Simulation of Liquids*, Clarendon, Oxford.

41 Hansen, J.P. and Klein, M.L. (1976) Dynamical structure factor $S(Q,\omega)$ of rare gas solids. *Phys. Rev.*, **B13**, 878–887.

42 Chaplot, S.L. (1994) *Indian J. Pure Appl. Phys.*, **32**, 560.

43 Car, R. and Parrinello, M. (1985) Unified approach for molecular dynamics and density-functional theory. *Phys. Rev. Lett.*, **55**, 2471.

44 Rao, K.R. and Chaplot, S.L. (1979) *Current Trends in Lattice Dynamics* (ed. K.R. Rao), Indian Physics Association, Bombay, p. 589.

45 Parrinello, M. and Rahman, A. (1980) Crystal structure and pair potentials: a molecular-dynamics study. *Phys. Rev. Lett.*, **45**, 1196.

46 Martonak, R., Laio, A., and Parrinello, M. (2003) Predicting crystal structures: the Parrinello–Rahman method revisited phys. *Rev. Lett.*, **90**, 075503.

47 Nipko, J.C., Loong, C.-K., Loewenhaupt, M., Braden, M., Reichardt, W., and Boatner, L.A. (1997) Lattice dynamics of xenotime: the phonon dispersion relations and density of states of $LuPO_4$. *Phys. Rev. B*, **56**, 11584.

48 Rao, K.R., Chaplot, S.L., Choudhury, N., Ghose, S., Hastings, J.M., Corliss, L.M., and Price, D.L. (1988) Lattice dynamics and inelastic neutron scattering from forsterite, Mg_2SiO_4: phonon dispersion relation, density of states and specific heat. *Phys. Chem. Miner.*, **16**, 83–97.

49 Choudhury, N., Chaplot, S.L., and Rao, K.R. (1989) Equation of state and melting point studies of forsterite Mg_2SiO_4. *Phys. Chem. Miner.*, **16**, 599–605.

50 Price, D.L., Ghose, S., Choudhury, N., Chaplot, S.L., and Rao, K.R. (1991) Phonon density of states in fayalite, Fe_2SiO_4. *Physica B*, **174**, 87–90.

51 Ghose, S., Choudhury, N., Chaplot, S.L., Chowdhury, C.P., and Sharma, S.K. (1994) Lattice dynamics and Raman spectroscopy of protoenstatite, $Mg_2Si_2O_6$. *Phys. Chem. Miner.*, **20**, 469–477.

52 Choudhury, N. and Chaplot, S.L. (2000) Free energy and relative stability of the enstatite $Mg_2Si_2O_6$ polymorphs. *Solid State Commun.*, **114**, 127–132.

53 Choudhury, N., Ghose, S., Chowdhury, C.P., Loong, C.K., and Chaplot, S.L. (1998) Lattice dynamics, Raman spectroscopy and inelastic neutron scattering of orthoenstatite, $Mg_2Si_2O_6$. *Phys. Rev. B Condens Matt.*, **58**, 756–765.

54 Chaplot, S.L., Choudhury, N., and Rao, K.R. (1998) Molecular dynamics simulations of phase transitions and melting in $MgSiO_3$ with the perovskite structure. *Am. Mineral.*, **83**, 937–941.

55 Chaplot, S.L. and Choudhury, N. (2001) Molecular dynamics simulations of seismic discontinuities and phase transitions of $MgSiO_3$ from 4 to 6-coordinated silicon via a novel 5-coordinated phase. *Am. Mineral.*, **86**, 752–761.

56 Rao, M.N., Chaplot, S.L., Choudhury, N., Rao, K.R., Azuah, R.T., Montfrooij, W.T., and Bennington, S.M. (1999) Lattice dynamics and inelastic neutron scattering from sillimanite and kyanite Al_2SiO_5. *Phys. Rev. B*, **60**, 12061–12068.

57 Taylor, M. and Ewing, R.C. (1978) The crystal structures of the $ThSiO_4$ polymorphs: huttonite and thorite. *Acta Cryst. B*, **34**, 1074–1079.

58 Speer, J.A. and Cooper, B.J. (1982) Crystal structure of synthetic hafnon, $HfSiO_4$, comparison with zircon and the actinide orthosilicates. *Am. Mineral.*, **67**, 804–808.

59 Fuchs, L.H. and Gebert, E. (1958) X-Ray studies of synthetic coffinite, thorite and uranothorites. *Am. Mineral.*, **43**, 243–248.

60 Manoun, B., Downs, R.T., and Saxena, S.K. (2006) A high-pressure Raman spectroscopic study of hafnon, $HfSiO_4$. *Am. Mineral.*, **91**, 1888–1892.

61 Ono, S., Tange, Y., Katayama, I., and Kikegawa, T. (2004) Equations of state of $ZrSiO_4$ phases in the upper mantle. *Am. Mineral.*, **89**, 185–188.

62 Scott, H.P., Williams, Q., and Knittle, E. (2002) Ultra low compressibility silicate without highly coordinated silicon. *Phys. Rev. Lett.*, **88**, 015506.

63 Mazeina, L., Ushakov, S.V., Navrotsky, A., and Boatner, L.A. (2005) Formation enthalpy of $ThSiO_4$ and enthalpy of the thorite huttonite phase transition. *Geochim. Cosmochim. Acta*, **69**, 4675–4683.

64 Dachille, F. and Roy, R. (1964) Effectiveness of shearing stresses in accelerating solid-phase reactions at low temperatures and high pressures. *J. Geol.*, **72**, 243–247.

65 Seydoux, A.M. and Montel, J.M. (1997) Experimental determination of the thorite–huttonite phase transition. EUG IX, Terra Nova 9, Abstract, Supplement 1, p. 42119.

66 Ozkan, H. and Jamieson, J.C. (1978) Pressure dependence of the elastic constants of nonmetamict zircon. *Phys. Chem. Miner.*, **2**, 215–224.

67 Rignanese, G.M., Gonze, X., Jun, G., Cho, K., and Pasquarello, A. (2004) First-principles investigation of high-*k* dielectrics: comparison between the silicates and oxides of hafnium and zirconium. *Phys. Rev. B*, **69**, 184301.

68 Dawson, P., Hargreave, M.M., and Wilkinson, G.R. (1971) The vibrational spectrum of zircon ($ZrSiO_4$). *J. Phys. C*, **4**, 240.

69 Gervais, F., Piriou, B., and Cabannes, F. (1973) Anharmonicity in silicate crystals: temperature dependence of A_u type vibrational modes in $ZrSiO_4$ and $LiAlSi_2O_6$. *J. Phys. Chem. Solids*, **34**, 1785.

70 Lyons, K.B., Sturge, M.D., and Greenblatt, M. (1984) Low-frequency Raman spectrum of $ZrSiO_4$: V^{4+}: an impurity-induced dynamical distortion. *Phy. Rev. B*, **30**, 2127.

71 Nicola, J.H. and Rutt, H.N. (1973) A comparative study of zircon ($ZrSiO_4$) and hafnon ($HfSiO_4$) Raman spectra. *J. Phys. C*, **7**, 1381–1386.

72 Syme, R.W.G., Lockwood, D.J., and Kerr, H.J. (1977) Raman spectrum of synthetic zircon ($ZrSiO_4$) and thorite ($ThSiO_4$). *J. Phys. C*, **10**, 1335–1348.

73 Lahalle, M.P., Krupa, J.C., Lepostollec, M., and Forgerit, J.P. (1986) Low-temperature Raman study on ThSi04 single crystal and related infrared spectra at room temperature. *J. Solid State Chem.*, **64**, 181–187.

74 Kelly, K.K. (1941) The specific heats at low temperatures of ferrous silicate, manganous silicate and zirconium silicate. *J. Am. Chem. Soc.*, **63**, 2750.

75 Stull, D.R. and Prophet, H. (1971) *JANAF Thermochemical Tables*, 2nd edn, National Bureau of Standards, Washington, DC.

76 Ghose, S., Hastings, J.M., Choudhury, N., Chaplot, S.L., and Rao, K.R. (1991) Phonon dispersion relation in fayalite, Fe_2SiO_4. *Physica B*, **174**, 83–86.

77 Rao, M.N., Goel, P., Choudhury, N., Chaplot, S.L., and Ghose, S. (2002) Lattice dynamics and inelastic neutron scattering experiments on andalusite Al_2SiO_5. *Solid State Commun.*, **121**, 333–338.

78 Ozkan, H. (1976) Effect of nuclear radiation on the elastic moduli of zircon. *J. Appl. Phys.*, **47**, 4772–4779.

79 Chaplot, S.L. (1987) Free energy and the relative stability of the phases of solid tetracyanoethylene with pressure and temperature. *Phy. Rev. B*, **36**, 8471.

80 Mary, T.A., Evans, J.S.O., Vogt, T., and Sleight, A.W. (1996) Negative thermal expansion from 0.3 to 1050 Kelvin in ZrW_2O_8. *Science*, **272**, 90–92.

81 Goodwin, A.L. and Kepert, C.K. (2005) Negative thermal expansion and low-frequency modes in cyanide-bridged framework materials. *Phys. Rev. B.*, **71**, R140301.

82 Zwanziger, J.W. (2007) Phonon dispersion and Grüneisen parameters of zinc dicyanide and cadmium dicyanide from first principles: origin of negative thermal expansion. *Phys. Rev. B*, **76**, 052102.

83 Mittal, R., Chaplot, S.L., Mishra, S.K., and Bose, P.P. (2007) Inelastic neutron scattering and lattice dynamical calculation of negative thermal expansion compounds Cu_2O and Ag_2O. *Phys. Rev. B*, **75**, 174303.

84 Chatterji, T., Henry, P.F., Mittal, R., and Chaplot, S.L. (2008) Negative thermal expansion of ReO_3: neutron diffraction experiments and dynamical lattice calculations. *Phys. Rev. B*, **78**, 134105.

85 Chatterji, T., Freeman, P.G., Jimenez-Ruiz, M., Mittal, R., and Chaplot, S.L. (2009) Pressure- and temperature-induced M_3 phonon softening in ReO_3. *Phys. Rev. B*, **79**, 184302.

86 Withers, R.L., Evans, J.S.O., Hanson, J., and Sleight, A.W. (1998) An *in situ* temperature-dependent electron and X-ray diffraction study of structural phase transitions in ZrV_2O_7. *J. Solid State Chem.*, **137**, 161–167.

87 Evans, J.S.O., Hanson, J.C., and Sleight, A.W. (1998) Room-temperature superstructure of ZrV_2O_7. *Acta Cryst. B*, **54**, 705–713.

88 Ernst, G., Broholm, C., Kowach, G.R., and Ramirez, A.P. (1998) Phonon density of states and negative thermal expansion in ZrW_2O_8. *Nature*, **396**, 147–149.

89 David, W.I., Evans, J.S.O., and Sleight, A.W. (1999) Direct evidence for a low-frequency phonon mode mechanism in the negative thermal expansion compound ZrW_2O_8. *Europhys. Lett.*, **46**, 661–666.

90 Evans, J.S.O., Hu, Z., Jorgensen, J.D., Argyriou, D.N., Short, S., and Sleight, A.W. (1997) Compressibility, phase transitions, and oxygen migration in zirconium tungstate, ZrW_2O_8. *Science*, **61**, 275.

91 Yagi, T., Mao, H.K., and Bell, P.M. (1978) Structure and crystal chemistry of perovskite type $MgSiO_3$. *Phys. Chem. Minerals*, **3**, 97–110.

92 Durben, D.J. and Wolf, G.H. (1992) High T behavior of metastable $MgSiO_3$ perovskite: a Raman spectroscopic study. *Am. Mineral.*, **77**, 890–893.

93 Lu, R., Hofmeister, A.M., and Wang, Y. (1994) Thermodynamic properties of ferromagnesian silicate perovskites from vibrational spectroscopy. *J. Geophys. Res.*, **99**, 11795–11804.

94 Mao, H.K., Hemley, R.J., Fei, Y., Shu, J.F., Chen, L.C., Jephcoat, A.P., and Wu, Y. (1991) Effect of pressure, temperature and composition on the lattice parameters and density of $(Fe,Mg)SiO_3$ perovskites to 30 GPa. *J. Geophys. Res.*, **B96**, 8069–8079.

95 Knittle, E. and Jeanloz, R. (1987) Synthesis and equation of state of $(Mg,Fe)SiO_3$ perovskite to over 100 gigapascals. *Science*, **235**, 668–670.

96 Akaogi, M. and Ito, E. (1993) Heat capacity of $MgSiO_3$ perovskite. *Geophys. Res. Lett.*, **20**, 105–108.

97 Heinz, D.L. and Jeanloz, R. (1987) Measurement of the melting curve of $Mg_{0.9}Fe_{0.1}SiO_3$ at lower mantle conditions and its geophysical implications. *J. Geophys. Res.*, **92**, 11437–11444.

98 Sweeney, J.S. and Heinz, D.L. (1993) Melting of iron magnesium silicate perovskite. *Geophys. Res. Lett.*, **20**, 855–859.

99 Zerr, A. and Boehler, R. (1993) Melting of $(Mg,Fe)SiO_3$ perovskite to 625 kilobars: indication of a high melting temperature in the lower mantle. *Science*, **262**, 553–555.

100 Brown, J.M. (1993) Mantle melting at high pressure. *Science*, **262**, 529–530.

101 Meade, C., Mao, H.K., and Hu, J. (1995) High temperature phase transition and dissociation of $(Mg,Fe)SiO_3$ perovskite at lower mantle pressures. *Science*, **268**, 1743–1745.

102 Funamori, N. and Yagi, T. (1993) High pressure and high temperature in-situ X-ray observations of $MgSiO_3$ perovskite under lower mantle conditions. *Geophys. Res. Lett.*, **20**, 387–390.

103 Saxena, S.K., Dubrovinsky, L.S., Lazor, P., Cerenius, Y., Haggkvist, P., Hanfland, M., and Hu, J. (1996) Stability of perovskite $MgSiO_3$ in the Earth's mantle. *Science*, **274**, 1357–1359.

104 Dubrovinsky, L., Saxena, S.K., Ahuja, R., and Johansson, B. (1998) Theoretical study of the stability of $MgSiO_3$ perovskite in the deep mantle. *Geophys. Res. Lett.*, **25**, 4253–4256.

105 Kawakatsu, H. and Niu, F. (1994) Seismic evidence for a 920 km discontinuity in the mantle. *Nature*, **371**, 301–305.

106 Karki, B.B., Stixrude, L., Clark, S.J., Warren, M.C., Ackland, G.J., and Crain, J. (1997) Elastic properties of orthorhombic $MgSiO_3$ perovskite at lower mantle pressures. *Am. Mineral.*, **82**, 635–638.

107 Warren, M.C. and Ackland, G.J. (1996) Ab initio studies of structural instabilities in $MgSiO_3$ perovskite. *Phys. Chem. Minerals*, **23**, 107–118.

108 Stixrude, L. and Cohen, R.E. (1993) Stability of orthorhombic $MgSiO_3$ perovskite in the Earth's lower mantle. *Nature*, **364**, 613–616.

109 Wentzcovitch, R.M., Ross, N.L., and Price, G.D. (1995) Ab initio study of $MgSiO_3$ and $CaSiO_3$ perovskites at lower mantle pressures. *Phys. Earth Planet. Int.*, **90**, 101–112.

110 Wolf, G. and Bukowinski, M. (1985) Ab initio studies of the structural and thermoelastic properties of orthorhombic $MgSiO_3$ perovskite. *Geophys. Res. Lett.*, **12**, 809–812.

111 Matsui, M. and Price, G.D. (1991) Simulation of the pre-melting behavior of $MgSiO_3$ perovskite at high pressures and temperatures. *Nature*, **351**, 735–737.

112 Matsui, M. and Price, G.D. (1992) Computer simulation of the $MgSiO_3$ polymorphs. *Phys. Chem. Minerals*, **18**, 365–372.

113 Kaputsa, B. and Guillope, M. (1993) Molecular dynamics study of the perovskite $MgSiO_3$ at high temperature: structure, elastic and thermodynamic properties. *Phys. Earth Planet. Int.*, **75**, 205–224.

114 Winkler, B. and Dove, M.T. (1992) Thermodynamic properties of $MgSiO_3$ perovskite derived from large scale molecular dynamics simulations. *Phys. Chem. Minerals*, **18**, 407–415.

115 Ghose, S., Choudhury, N., Chaplot, S.L., and Rao, K.R. (1992) Phonon density of states and thermodynamic properties of minerals, in *Thermodynamic Data: Systematics and Estimation*, vol. 10, Adv. Phys. Geochem. (ed. S.K. Saxena), Springer-Verlag, New York, pp. 283–314.

116 Krupka, K.M., Hemingway, B.S., Robie, R.A., and Kerrick, D.M. (1985) High temperature heat capacities and derived thermodynamic properties of anthophyllite, diopside, enstatite, bronzite and wollastonite. *Am. Mineral.*, **70**, 261–271.

117 Krupka, K.M., Robie, R.A., Hemingway, B.S., Kerrick, D.M., and Ito, J. (1985) Low temperature heat capacities and derived thermodynamic properties of anthophyllite, diopside, enstatite,

bronzite and wollastonite. *Am. Mineral.*, **70**, 249–260.

118 Thieblot, L., Tequi, C., and Richet, P. (1999) High-temperature heat capacity of grossular ($Ca_3Al_2Si_3O_{12}$), enstatite ($MgSiO_3$) and titanite ($CaTiSiO_5$). *Am. Mineral.*, **84**, 848–855.

119 Angel, R.J. and Hugh-Jones, D.A. (1994) Equations of state and thermodynamic properties of the enstatite pyroxenes. *J. Geophys. Res.*, **B99**, 19777–19783.

120 Ringwood, A.E. (1975) *Composition and Petrology of the Earth's Mantle*, McGraw Hill, New York.

121 Dzeiwonski, A.M. and Anderson, D.L. (1981) Preliminary reference Earth model. *Phys. Earth Planet. Int.*, **25**, 297–356.

122 Duffy, T.S. and Anderson, D.L. (1989) Seismic velocities in mantle minerals and the mineralogy of the upper mantle. *J. Geophysical Res.*, **94**, 1895–1912.

123 Poirier, J.P. (1991) *Introduction to the Physics of the Earth's Interior*, Cambridge University Press, UK.

124 Hemley, R.J. (ed.) (1998) *Ultrahigh Pressure Mineralogy – Physics and Chemistry of the Earth's Deep Interior*, vol. 37, Reviews in Mineralogy, Mineralogical Society of America, Washington, DC.

125 Jeanloz, R. (1986) High pressure chemistry of the Earth's mantle and core, in *Mantle Convection: Plate Tectonics and Global Dynamics* (ed. W.R. Peltier), Gordon and Breach, New York.

126 Jeanloz, R. and Thompson, A.B. (1983) Phase transitions and mantle discontinuities. *Rev. Geophys. Space Phys.*, **21**, 51–74.

127 Ghose, S., Schomaker, V., and McMullan, R.K. (1986) Enstatite, $Mg_2Si_2O_6$: a neutron diffraction refinement of the crystal structure and a rigid body analysis of the thermal vibration. *Zeitschrift fur Kristallographie*, **176**, 159–175.

128 Goel, P., Choudhury, N., and Chaplot, S.L. (2004) Superionic behavior of lithium oxide Li_2O: a lattice dynamics and molecular dynamics study. *Phys. Rev. B*, **70**, 174307.

129 Goel, P., Choudhury, N., and Chaplot, S.L. (2007) Fast ion diffusion, superionic conductivity and phase transitions of the nuclear materials UO_2 and Li_2O. *J. Phys. Condens. Matter*, **19**, 386239.

130 Goel, P., Choudhury, N., and Chaplot, S.L. (2008) Atomistic modeling of the vibrational and thermodynamic properties of uranium dioxide, UO_2. *J. Nucl. Mater.*, **377**, 438–443.

131 Goel, P., Choudhury, N., and Chaplot, S.L. (2008) International Conference on Material Chemistry, Mumbai, India.

132 Zheng-Johannson, J.X.M. and McGreevy, R.L. (1996) A molecular dynamics study of ionic conduction in CuI. II. Local ionic motion and conduction mechanisms. *Solid State Ionics*, **83**, 35–48.

133 Ihata, K. and Okazaki, H. (1997) Structural and dynamical properties of α-CuI: a molecular dynamics study. *J. Phys. Condens Matt.*, **9**, 1477–1492.

4
Phonon Spectroscopy of Polycrystalline Materials Using Inelastic X-Ray Scattering

Alexei Bosak, Irmengard Fischer, and Michael Krisch

4.1
Introduction

The study of phonon dispersion in condensed matter at momentum transfers Q and energies E, characteristic of collective atom motions, provides insight into many physical properties such as elasticity, interatomic forces, structural phase stability and transitions, anharmonicity, and electron correlation. Phonon spectroscopy has been traditionally the domain of neutron spectroscopy. Neutrons as probing particle are particularly suitable, since (i) the neutron–nucleus scattering cross-section is sufficiently weak to allow a large penetration depth, (ii) the energy of neutrons with wavelengths of the order of interparticle distances is about 100 meV, and therefore comparable to typical phonon energies, and (iii) the momentum of the neutron allows one to probe the whole dispersion scheme out to several tens of nm^{-1}.

X-rays represent another probe that can in principle as well be used to determine the phonon dispersion throughout the Brillouin zone (BZ). However, it has been pointed out in several textbooks that this would represent a formidable experimental challenge mainly due to the fact that an X-ray instrument would have to provide an extremely high-energy resolution of at least $\Delta E/E = 10^{-7}$. For example, the following statement can be found in the textbook *Solid State Physics* by Ashcroft and Mermin about the possibility to measure phonons with X-rays: *In general the resolution of such minute photon frequency shifts is so difficult that one can only measure the total scattered radiation of all frequencies* . . . [1]. Indeed, the first attempts for the extraction of lattice dynamics from X-ray experiments were made by analyzing the diffuse scattering intensity between the diffraction spots. It was shown by Laval in 1939 [2] that the thermal movements of the atoms contributed to these intensities, which could therefore provide information on the lattice dynamics. One of the first experiments along these lines was performed in 1948 by Olmer on fcc aluminum [3]; his work was completed by Walker in 1956 [4]. Furthermore, in combination with the lattice dynamics theory of Born, interatomic force constants could be derived by Curien [5] and Jacobsen [6] for α-iron and copper. There was, however, no possibility at that time

Thermodynamic Properties of Solids: Experiment and Modeling
Edited by S. L. Chaplot, R. Mittal, and N. Choudhury

ISBN: 928-3-527-40812-2

to measure such dispersion curves directly, due to the lack of scattered intensity. In the 1980s a spectrometer setup, using a rotating anode X-ray generator, yielded an instrumental energy resolution ΔE of 42 meV, but the photon flux was not sufficient to perform an inelastic X-ray scattering (IXS) experiment [7]. Using synchrotron radiation from a bending magnet at HASYLAB provided the necessary high photon flux, and first pioneering experiments were performed on graphite and beryllium with an energy resolution of $\Delta E = 55$ meV [8, 9]. This modest energy resolution and photon flux were significantly improved by the construction of a dedicated instrument located at a wiggler source at HASYLAB, where an instrumental resolution of 9 meV could be achieved [10]. With the advent of the third-generation synchrotron radiation sources, in particular the European Synchrotron Radiation Facility (ESRF) in Grenoble (France), the Advanced Photon Source (APS) at the Argonne National Laboratory (United States) and the Super Photon Ring (SPring-8) in Japan, the IXS technique gained its full maturity within a few years. Thanks to the high brilliance of the undulator X-ray sources and important developments in X-ray optics, experiments are now routinely performed with an energy resolution of 1.5 meV. At present, there are worldwide five instruments dedicated to IXS from phonons: ID16 and ID28 at the ESRF [11, 12], 3ID and 30ID at the APS [13, 14], and BL35XU at SPring-8 [15]. Two further projects are in an advanced design stage.

Today, the use of photons complements the capabilities of inelastic neutron spectroscopy, in particular in cases where neutron techniques are difficult or impossible to apply. This concerns the study of disordered and polycrystalline systems where well-defined sound-like excitations are only clearly resolvable in the small momentum transfer (Q) limit, typically below 10 nm^{-1}. In particular, in systems with a high speed of sound ($v_{\text{sound}} \sim E/Q$), the required energy transfer and energy resolution cannot be met simultaneously. These kinematic restrictions are absent for X-rays (see Figure 4.1). Another important advantage of inelastic X-ray scattering lies in the possibility to study samples in very small quantities ($V = 10^{-4}$ -10^{-6} mm^3). This has opened new possibilities in material research and high-pressure science. IXS is applied to a large variety of very different materials ranging from quantum liquids over high T_c superconductors to biological aggregates. The interested reader is referred to several reviews that allow gaining an excellent overview of the various research fields [16–20].

The present review is focused on IXS from polycrystalline materials. In polycrystalline systems, the directional information is lost due to the random orientation of the individual crystalline grains. The information content is therefore limited, or, in the best case, involves sophisticated modeling. An approximate average longitudinal acoustic (LA) phonon dispersion can be determined, if Q is chosen to be within the first Brillouin zone. Earlier examples comprise studies of the high-pressure phases of ice [21, 22] and ice clathrates [23, 24], iron [25, 26], and other geophysically relevant materials [27–29]. The vibrational density of states (VDOS) can be determined – in analogy to coherent INS [30] – in the so-called incoherent approximation. Section 4.4 provides a rigorous theoretical treatment of IXS at low Q and discusses the relation of IXS-derived elastic properties with the macroscopic aggregate elasticity. Density of state measurements are presented in Section 4.5. The VDOS provides the link

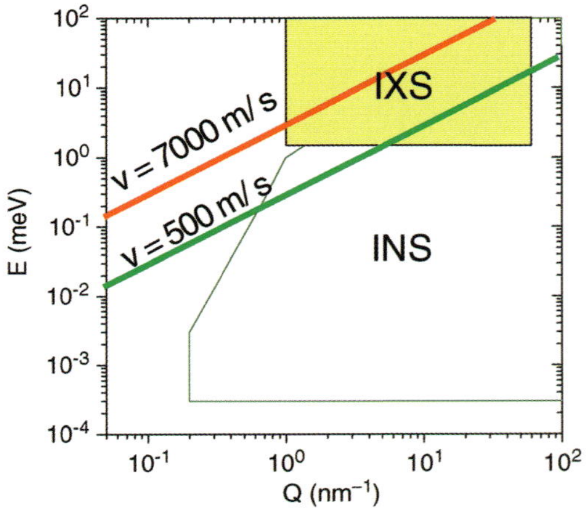

Figure 4.1 Approximate energy transfer *E*–momentum transfer *Q* range, accessible by inelastic neutron (INS) and X-ray scattering (IXS). The green and red lines represent the linear dispersion of sound waves with a velocity as indicated in the figure.

between the microscopic interatomic interactions and the macroscopic thermodynamic and elastic properties. In the case of monoatomic systems, well-established integral equations allow the derivation of the vibrational contribution to the heat capacity, free energy and entropy, and the atomic mean square displacements and mean force constants [31]. Furthermore, the Debye temperature and velocity can be derived. The extraction of these quantities for multicomponent systems is not straightforward and necessitates modeling of the lattice dynamics for each constituent atom species. Section 4.6 presents a novel methodology for the extraction of the single-crystal lattice dynamics from polycrystalline materials, exploiting the information in the intermediate *Q*-range. The review is complemented by a short overview of the theoretical background (Section 4.2) and the instrumental setup (Section 4.3). Section 4.7 contains the concluding remarks and attempts to project onto the future.

4.2 Theoretical Background

4.2.1 Scattering Kinematics and Dynamical Structure Factor

In the following discussion, we assume the validity of both (quasi)harmonic and adiabatic approximations. In this particular case, the dynamical matrix, being the Fourier transform of the force matrix, provides the complete description of the phonon system, including elasticity and lattice thermodynamics of the material.

The inelastic scattering process is depicted schematically in Figure 4.2. The momentum and energy conservation impose that

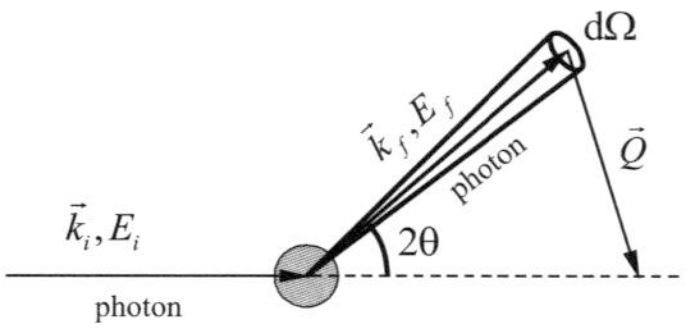

Figure 4.2 Schematics of the inelastic scattering process. E_i (E_f), and $\vec{k}_i$ ($\vec{k}_f$) denote the energy and the wave vector of the incident (scattered) photon, respectively, and 2θ is the scattering angle. $\vec{Q}$ is the momentum transfer and $d\Omega$ is the portion of solid angle within which the scattered photons are detected.

$$\vec{Q} = \vec{k}_i - \vec{k}_f, \tag{4.1a}$$

$$E = E_i - E_f. \tag{4.1b}$$

In order to reach sufficiently large momentum transfers to cover several Brillouin zones, X-rays with an energy of at least 10 keV need to be employed. Considering that typical energies of phonons are in the meV range, $E \ll E_i$, the momentum transfer depends only on the incident wave vector (energy) and the scattering angle 2θ:

$$Q = 2k_i \sin(\theta). \tag{4.2}$$

The general form of the dynamical structure factor $S(\vec{Q}, E)$ for single-phonon scattering takes the following form [17]:

$$S(\vec{Q}, E) = \sum_j G(\vec{Q}, j) F(E, T, \vec{Q}, j), \tag{4.3}$$

$$G(\vec{Q}, j) = \left| \sum_d f_d(\vec{Q}) e^{-W_d(\vec{Q}) + i\vec{Q}\cdot\vec{r}_d} (\vec{Q} \cdot \hat{\sigma}_d^j(\vec{q})) M_d^{-1/2} \right|^2, \tag{4.4}$$

and the thermal factor

$$F(E, T, \vec{Q}, j) = \frac{1}{E_j(\vec{q})} [(\langle n(E_j(\vec{q}), T)\rangle + 1)\delta(E - E_j(\vec{q})) + \langle n(E_j(\vec{q}), T)\rangle \delta(E + E_j(\vec{q}))] \tag{4.5a}$$

with thermal occupation factor $\langle n(E, T)\rangle = (1/(\exp(E/kT) - 1))$, or as

$$F(E, T, \vec{Q}, j) = \frac{1}{1 - \exp\left(-\frac{E}{kT}\right)} \frac{1}{E_j(\vec{q})} [\delta(E - E_j(\vec{q})) - \delta(E + E_j(\vec{q}))], \tag{4.5b}$$

where $E_j(\vec{q})$ is the energy of mode j at momentum transfer $\vec{q}$, f_d is the atomic scattering factor of atom d with mass M_d, $W_d(\vec{Q})$ is the Debye–Waller factor at the position $\vec{r}_d$, and $\hat{\sigma}_d^j$ is the d-site projected component of the $3N$-dimensional normalized eigenvector of the phonon mode j, defined in the periodic notation $\hat{\sigma}^j(\vec{q} + \vec{\tau}) = \hat{\sigma}^j(\vec{q})$. $\vec{\tau}$ is an arbitrary reciprocal lattice vector and $\vec{q} = \vec{Q} - \vec{\tau}$ is the reduced momentum transfer.

In polycrystalline materials, it is not possible to determine the phonon dispersion along specific directions, as only the magnitude of the momentum transfer $\vec{Q}$ is defined. The scattering process takes place on a spherical shell of radius Q, and the scattered intensity is integrated over the surface of this shell. If the Debye–Waller factors are assumed to be the same for all types of atoms $(\tilde{W})$, we obtain for the averaging over the sphere of radius $Q = |\vec{Q}|$:

$$S(Q, E, T) = g(Q, E)F(E) \cdot \exp(-2\tilde{W}), \tag{4.6}$$

$$g(Q, E) = \left\langle \left| \sum_d f_d(Q)(\vec{Q} \cdot \hat{e}_d(\vec{Q}, j)) M_d^{-1/2} \right|^2 \delta(E - E_{\vec{Q},j}) \right\rangle, \tag{4.7}$$

where $\langle \cdots \rangle$ means averaging over the sphere of radius Q and the phonon modes j.

Inside the first Brillouin zone, $\vec{Q} = \vec{q}$ and scattering from quasilongitudinal phonon modes dominates, since only phonon modes with an eigenvector component parallel to their propagation direction contribute to the $S(Q,E)$ (see Equation 4.7). In the long wavelength limit, the displacements of different atom species become collinear and parallel to the eigenvectors of the corresponding elastic waves $\hat{e}(\vec{n}, j)$ $(\vec{n} = \vec{Q}/|\vec{Q}|)$. In this case, the scattering from a polycrystal is completely defined by the macroscopic elasticity of the crystal:

$$g(Q, E) \rightarrow A \left\langle |\vec{Q} \cdot \hat{e}(\vec{n}, j)|^2 \delta(E - V_{\vec{n},j}|\vec{Q}|) \right\rangle, \tag{4.8}$$

where A is a scaling factor and $V_{\vec{n},j}$ is the sound velocity, obtained from the Christoffel's equation [32].

In the limit of large momentum transfers, the normalized $g(Q, E)$ should approach the generalized vibrational density of states:

$$g(Q, E) \rightarrow AQ^2 \sum_n \frac{G_n(E)}{M_n} f_n^2(Q), \tag{4.9}$$

where $G_n(E) = \sum_{\vec{Q},j} |\hat{e}_n(\vec{Q}, j)|^2 \delta(E - E_{\vec{Q},j})$ are the partial densities of states, and A is a scaling factor. As inelastic X-ray scattering from phonons is essentially a coherent scattering process, the same incoherent approximation as for coherent INS can be applied [30].

4.2.2 IXS Cross Section

The IXS cross section is directly proportional to the dynamical structure factor $S(\vec{Q}, E)$ associated with the collective motions of the atoms within the validity of the following assumptions [10, 33]: (i) the scattering cross section is dominated by the Thomson scattering term, and the other resonant and spin-dependent contributions to the electron–photon (e–ph) scattering cross section can be neglected, (ii) the adiabatic approximation allows separating the electronic and nuclear parts of the total wave function, and therefore the center of mass of the electronic cloud follows the nuclear motion, and (iii) there are no electronic excitations in the considered energy transfer range. Under the above conditions, the IXS double differential cross section can be written as

$$\frac{\partial^2 \sigma(\vec{Q}, E)}{\partial \Omega \partial E} = r_0^2 (\vec{\varepsilon}_i \cdot \vec{\varepsilon}_f)^2 \frac{k_f}{k_i} S(\vec{Q}, E), \tag{4.10}$$

where Ω denotes the solid angle in which the inelastic scattered X-rays are observed, r_0 is the classical electron radius, and $\vec{\varepsilon}_i(\vec{\varepsilon}_f)$ are the photon polarization vectors of the incident (scattered) photons. The prefactor to $S(\vec{Q}, E)$ is the Thomson scattering cross section that describes the coupling to the electromagnetic field.

An important experimental aspect is related to the optimum scattering intensity. The flux of I scattered photons into the solid angle $\Delta\Omega$ and energy interval ΔE is given by

$$I = I_0 \frac{\partial^2 \sigma}{\partial \Omega \partial E} \Delta\Omega \Delta E n t e^{-\mu t}, \tag{4.11}$$

where I_0 is the incident photon flux, n is the number of scattering units per unit volume, t is the sample thickness, and μ is the total absorption coefficient. The maximum IXS signal is obtained for $t = 1/\mu$. The dominating attenuation process for elements with an atomic number $Z > 3$ is photoelectric absorption for X-ray energies above 10 keV. The photoelectric absorption process is roughly proportional to Z^4, far away from electron absorption edges. As the Thomson cross section is proportional to $|f(\vec{Q})|^2 \approx Z^2$ (for small Q), I is roughly proportional to $1/Z^2$. Furthermore, an estimate for the inelastic X-ray scattering intensity I can be obtained, similar to the approach by Sinn [34]:

$$I \propto \frac{t e^{-\mu t} \varrho Z^2}{\Theta_D^2 M^2}, \tag{4.12}$$

where Θ_D and ϱ are the Debye temperature and the density, respectively. Figure 4.3 (top panel) shows the IXS signal as a function of Z for a monoatomic system with an optimum sample thickness $t = 1/\mu$ and $f(Q) = Z$, and for a X-ray energy of 17.794 keV. The alkali metals (Na, K, Rb, and Cs) form an exception, since they are very soft (low Debye velocity) and have a low density. Furthermore, Se, In, and Hg are soft materials, while Yb shows an exceptionally high scattering intensity due to an anomaly in its density. The study of materials composed of heavy atoms is more difficult than the ones of low Z materials. On the other hand, in cases where the sample thickness is limited ($t \ll 1/\mu$) either due to the availability of the material or constraints imposed by the sample environment (high pressure, high/low temperature, high magnetic field, and so on), it is an advantage to study high Z materials. This is illustrated in the bottom panel of Figure 4.3 for a sample thickness of 20 μm. It is apparent that the IXS signal for low Z materials is weak and increases with Z up to $Z = 59$, with a gain of a factor 100 in this range.

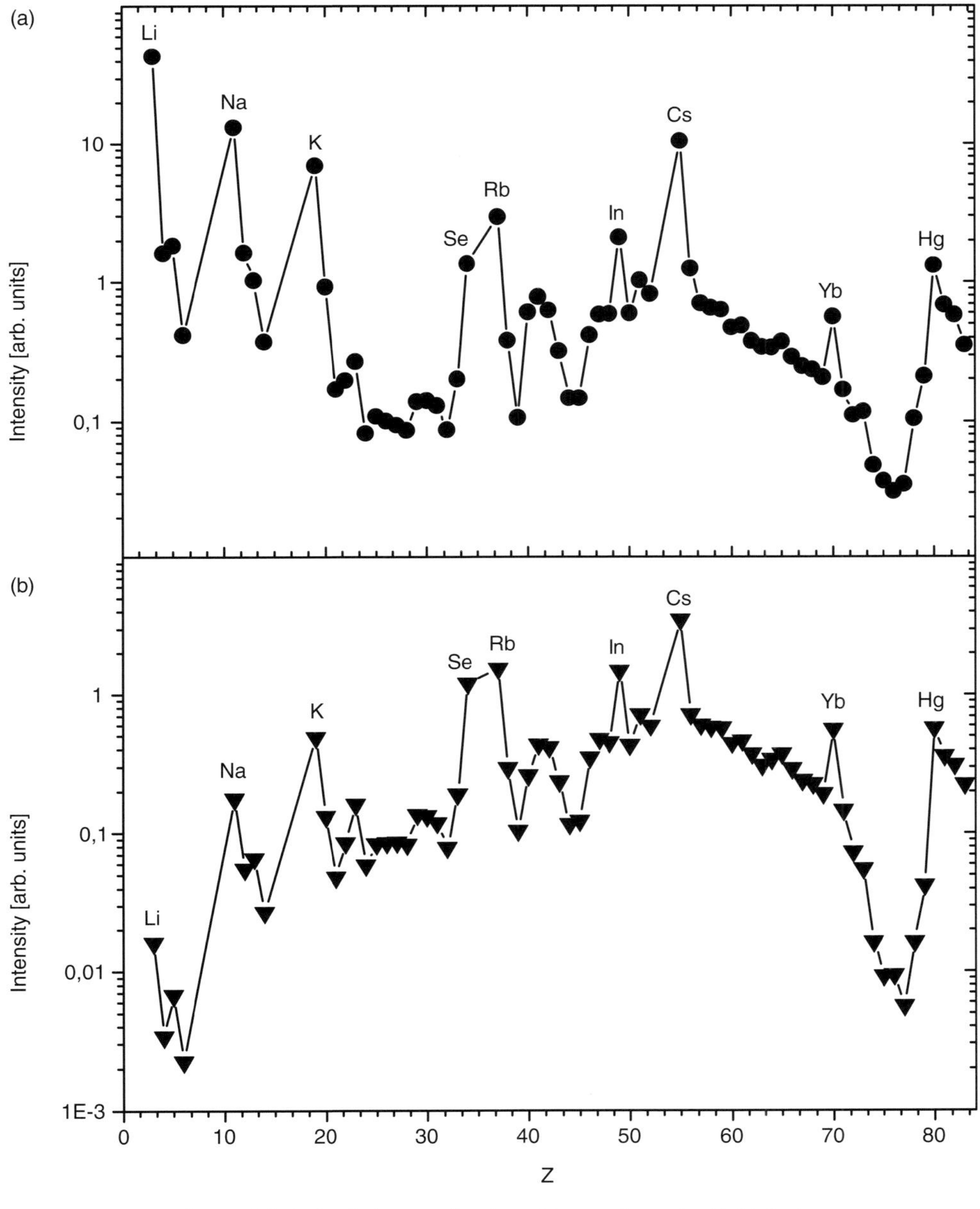

Figure 4.3 Figure of merit of the IXS technique as a function of the atomic number Z. (a) Optimum sample thickness $t = 1/\mu$. (b) $t = 20\,\mu\text{m}$.

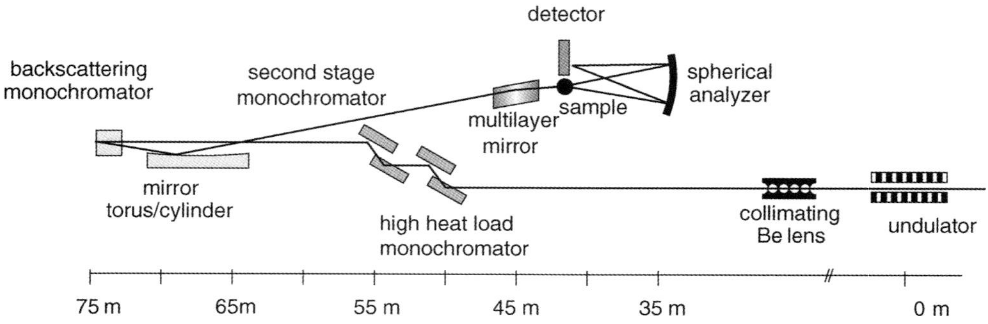

Figure 4.4 Schematic optical layout of the inelastic scattering beamline II (ID28) at the European Synchrotron Radiation Facility.

4.3 Instrumental Principles

All IXS instruments are based on the triple-axis spectrometer as developed by Brockhouse for INS [35]. The three axes comprise the very high-energy resolution monochromator (first axis), the sample goniometry and the scattering arm (second axis), and the crystal analyzer (third axis). A detailed account of the instrumental developments and the two different types of very high-energy resolution monochromators is given in Refs [17, 20]. Here, the optical layout and the working principle of the inelastic scattering beamline ID28 at the ESRF shall be described in some detail.

Figure 4.4 gives an overview of the optical layout. The X-ray source consists of three 1.6 m long undulators of 32 mm magnetic period that are operated on the third or fifth harmonics. These standard undulators are complemented by two short-period (17.6 mm period) undulators that provide a factor 2 higher flux at 17 794 eV.

A collimating refractive lens made out of beryllium can be inserted in order to provide a better match between the vertical divergence of the X-ray beam at the source point and the angular spectral acceptance of the high heat load silicon (1 1 1) monochromator. The design parameters are given by the following equation [36]:

$$R/N = 2\delta p, \tag{4.13}$$

where $1-\delta$ is the real part of the refractive index, p is the distance between the source and the optical element, R is the radius of the cylindrical lens holes, and N is the number of holes. In order to accept the full vertical profile of the X-ray beam, R was chosen to be 1 mm, leading to the number of holes varying between 13 and 34, depending on the selected photon energy. The experimentally determined residual divergence amounts to 1–2 μrad, and the increase in spectral flux is 6, 20, and 30% at 17.794, 21.747, and 25.704 keV, respectively.

The X-ray beam from the undulator is premonochromatized by a silicon (1 1 1) double-crystal monochromator to a relative bandwidth of $\Delta E/E = 1.5 \times 10^{-4}$. Due to the high heat load produced by the intense undulator beam (total power up to 200 W and a power density up to 15.2 W/mm^2), the crystal has to be cooled down to cryogenic temperatures of about 125 K where silicon displays a maximum in the thermal conductivity and a minimum in the linear thermal expansion coefficient [37]. This allows keeping the thermal deformation of the crystal below the limits where photon flux losses occur.

In order to further reduce the heat load on the very high-resolution backscattering monochromator, a silicon (3 3 1) channelcut monochromator can be inserted behind the silicon (1 1 1) premonochromator. In this case, the bandwidth amounts to $\Delta E/E = 1.5 \times 10^{-5}$.

The required energy resolution in the meV range at photon energies between 15 and 25 keV is achieved by the utilization of high-order reflections from perfect crystals. It can be shown that the resolving power ($E/\Delta E$) is given by [38]

$$\left(\frac{\Delta E}{E}\right) = \frac{d_{hkl}}{\pi \Lambda_{\text{ext}}}, \tag{4.14}$$

where d_{hkl} denotes the lattice spacing, associated with the (hkl) reflection order, and Λ_{ext} the primary extinction length, a quantity deduced within the framework of the *dynamical theory of X-ray diffraction* [38]. Λ_{ext} increases with increasing reflection order. In order to reach a high resolving power, it is therefore necessary to use high-order Bragg reflections and to have highly perfect crystals. Relative lattice spacing variations $\Delta d/d$ within the diffracting volume, induced by defects, distortions, or impurities, must be significantly smaller than the desired relative energy resolution: $\Delta d/d \ll (\Delta E/E) \approx 10^{-8}$. This stringent requirement practically limits the choice of the material to silicon. A further important consideration is related to the spectral angular acceptance (Darwin width), which is given by

$$\omega_{\text{D}} = \left(\frac{\Delta E}{E}\right) \tan \theta_B. \tag{4.15}$$

As a consequence of the above criteria, the main monochromator consists of a flat perfect single crystal, operating at a Bragg angle of 89.98° and utilizing the silicon ($n\,n\,n$) reflection orders. This close-to-exact backscattering configuration ensures that the Darwin width is larger than the X-ray beam divergence, and therefore all the photons within the desired energy bandwidth are transmitted. The monochromator has an asymmetry cut, α (angle between the lattice plane normal and the crystal surface) of 75°, with the vertical diffraction plane. In this way the power density of the incident X-rays is further reduced and undesirable thermal gradients within the diffracting volume can be avoided. The crystal is temperature controlled by a high-precision platinum 100 Ω (Pt100) thermometer bridge in closed-loop operation with a PID-controlled heater unit [39]. Further details on the monochromator can be found in Ref. [40].

The X-rays are focused by a platinum-coated toroidal mirror with a variable meridional radius and a sagittal radius of 0.101 m, operated at a glancing angle of 2.7 mrad. The meridional radius can be changed by a bending device, thus allowing the optimization of the optical scheme if additional optical elements such as the collimating lens, already discussed, are utilized. The mirror has a reflectivity of about 80% throughout the exploited energy range and provides a focus of $270 \times 60\ \mu m^2$ full-width half-maximum in the horizontal and vertical plane, respectively. Furthermore, the left and right parts of the mirror have an infinite sagittal radius, and therefore the X-ray beam is focused only in the vertical direction. This part of the mirror is utilized in conjunction with the focusing-graded multilayer described below.

If a smaller X-ray spot at the sample position is required, for example, for studies using diamond anvil cells, the beam size in the horizontal direction needs to be further reduced. This is accomplished by a focusing multilayer, located 2 m in front of the sample position. Due to the large optical demagnification (1 : 50), the Bragg angle change along the illuminated multilayer length, l, exceeds its spectral angular acceptance; the multilayer must therefore have a lateral gradient in its d-spacing. It can be shown that to the lowest order this gradient is given by [41]

$$\frac{\Delta d}{d} = \frac{l \cos \theta}{2} \left(\frac{1}{q} - \frac{1}{p} \right). \tag{4.16}$$

Here, p and q denote the source-multilayer and multilayer-focal plane distance. Such a multilayer was realized using the ESRF in-house facilities [42] with the following characteristics: 120 layers of Ru/B_4C with a period of 3 nm in the center, a nonlinear gradient of 6.1% over the 240 mm long footprint, and a reflectivity of about 70% throughout the utilized energy range.

The sample stage consists of a four-circle goniometry with the three translational degrees of freedom. Additional circles or a kappa goniostat can be mounted if it is required by a specific experimental setup. A microscopy and an optical spectrometer are mounted on-line for sample alignment and pressure determination by the ruby fluorescence method. Furthermore, various pressure cells, high-temperature oven, and cryostats allow the study of materials in a wide range of temperature and pressure.

Although the required energy resolution is the same for the monochromator and the analyzer, there is a significant difference concerning their angular acceptance. The spatial angular acceptance of the monochromator should be compatible with the divergence of the X-ray beam, whereas the analyzer must have a much larger angular acceptance, which is only restricted by the required Q-resolution. An angular acceptance up to $4 \times 10\ \text{mrad}^2$ is an adequate compromise between Q-resolution and signal maximization. These constraints necessitate focusing optics. Since it is not possible to elastically bend a crystal without introducing important elastic deformations, which in turn deteriorate the intrinsic energy resolution, one has to position small, unperturbed crystals onto a spherical substrate with a radius, fulfilling the Rowland condition. This polygonal approximation to the spherical shape yields the intrinsic energy resolution, provided the individual crystals are unperturbed and the geometrical contribution of the cube size to the energy resolution is negligible. The solution realized at the ESRF consists of gluing 12 000 small crystals of $0.6 \times 0.6 \times 3\ \text{mm}^3$ size onto a spherical silicon substrate [43]. This procedure yields very good results, and provided the record energy resolution of 0.9 meV, utilizing the silicon [13 13 13] reflection order at 25 704 eV.

Table 4.1 Instrumental energy resolution, maximum momentum transfer Q_{max}, and Q-spacing between analyzers, $Q_n - Q_m$, at the utilized photon energies.

Reflection (Si)	(8 8 8)	(9 9 9)	(11 11 11)	(13 13 13)
Wavelength (Å)	0.7839	0.6968	0.5701	0.4824
Energy (eV)	15 816	17 793	21 747	25 701
Q_{max} (nm^{-1})	67.7	76.2	93.2	110
Energy resolution (meV)	6.0	3.0	1.5	0.9–1.1
$Q_n - Q_m$ (nm^{-1})	1.1	1.2	1.5	1.75
ΔQ (nm^{-1})	0.25	0.28	0.34	0.40

ΔQ denotes the typical Q-resolution in the horizontal scattering plane (for 20 mm horizontal analyzer slit opening).

Apart from the heavy constraints imposed on the analyzer construction, another important ingredient for a good analyzer performance is the temperature stability of the analyzer. To this purpose, the analyzers are actively stabilized utilizing the same kind of thermometer–heater unit as for the high-energy resolution monochromator. As all other optical components, the analyzers are operated in high vacuum.

The energy-analyzed photons are detected by a Peltier-cooled silicon diode detector that has an intrinsic energy resolution of 800 eV [44]. The dark counts due to electronic and environmental noises amount to about 0.2 counts/min. Further components of the spectrometer are an entrance pinhole, slits in front of the analyzers in order to define the momentum transfer resolution, and a detector pinhole for aberrant ray suppression. The analyzer crystals are temperature stabilized at about 22.5 °C with a typical stability of 1 mK/24 h. The momentum transfer is selected by rotating the spectrometer around a vertical axis passing through the scattering sample in the horizontal plane. Since there are nine independent analyzer systems, spectra at nine different momentum transfers can be recorded simultaneously. Table 4.1 summarizes the main characteristics of the spectrometer.

Energy scans are done by changing the relative lattice constant between monochromator and analyzer crystals via the temperature T: $\Delta E/E = \Delta d/d = \alpha(T)\Delta T$, where $\alpha(T) = \alpha_0 + \beta\Delta T$ (with $\alpha_0 = 2.58 \times 10^{-6}\,K^{-1}$, $\beta = 1.6 \times 10^{-8}\,K^{-2}$, and $\Delta T = T - 22.5\,°C$) is the thermal expansion coefficient of silicon [45]. The validity of this conversion has been checked by comparing the experimentally determined diamond dispersion curve for longitudinal acoustic and optical phonons with well-established inelastic neutron scattering results, *ab initio* calculations, and Raman spectroscopy [46].

4.4 IXS in the Low-*Q* Limit

In the low-Q (long wavelength) limit, the IXS spectrum is governed by the macroscopic elasticity of the crystal. The $\left|\vec{Q} \cdot \hat{e}(\vec{n}, j)\right|^2$ term in Equation 4.7 implies that only components of the phonon eigenvector parallel to the propagation direction contribute to the dynamical structure factor within the first Brillouin zone. The IXS signal is therefore generally dominated by (quasi)longitudinal acoustic phonons and its dispersion can be recorded, if the probed Q-range is chosen to be within the first Brillouin zone. If sound velocities shall be derived from the orientation-averaged LA dispersion, care has to be taken concerning the Q-range over which data are considered, as shall be demonstrated using the example of bcc iron. Furthermore, if the single-crystal elastic properties significantly deviate from isotropy, nonnegligible differences can appear for the properties of the polycrystalline aggregate. This effect obviously becomes more pronounced for textured samples. Moreover, the appearance of quasitransverse excitations needs to be considered. These aspects will be discussed in the following sections [47].

4.4.1
Scattering from (Quasi)Longitudinal Phonons

It is a common practise to fit the dispersion by a sine function:

$$E = \frac{2\hbar}{\pi} V_L Q_{\max} \sin\left(\frac{\pi}{2}\frac{Q}{Q_{\max}}\right). \tag{4.17}$$

The value of $Q_{\max}$ is either left free or fixed to a value, determined approximating the volume of the Wigner–Seitz Brillouin zone by a sphere of radius $Q_{\max}$ [26]. The validity and the limitations of this approach are examined below, using a realistic lattice dynamics model.

As no analytic solution exists for arbitrary Q-values in polycrystalline materials, we proceeded with a numerical modeling, taking the example of bcc iron. Born–von Kármán coupling constants were taken up to the fifth shell [48]. The theoretical constant Q IXS spectra were convoluted with the resolution function (3 meV FWHM), and the thermal factors are calculated for 298 K. The resulting spectra were fitted using coupled pseudo-Voigt functions in order to reproduce typical data analysis procedures. The result of these simulations is shown in Figure 4.5. The left panel displays $S(Q,E)/Q$ as a function of the reduced momentum transfer. Here, the

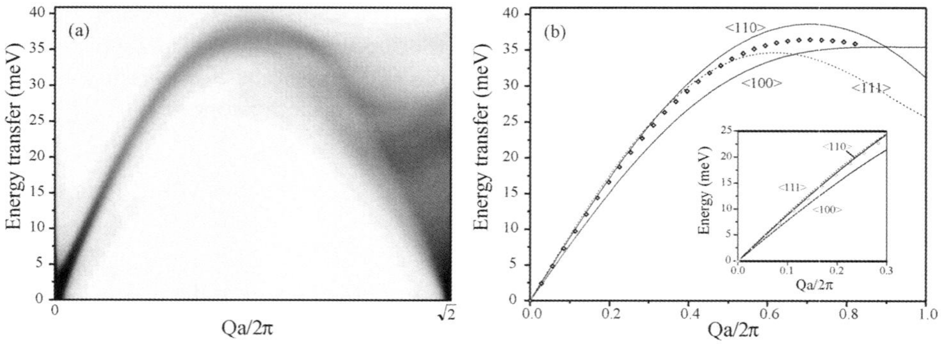

Figure 4.5 Body-centered cubic iron. (a) Stokes side of $S(Q,E)/Q$, represented as grayscale image. (b) Simulated LA polycrystalline dispersion (open diamonds) compared to the single-crystal dispersion along the main symmetry directions. Figure taken from Bosak *et al.* [47].

intensity is represented on a gray scale. The LA dispersion is well defined – as testified by the small energy spread of the LA phonon at constant Q – up to $Qa/2\pi \approx 1/\sqrt{2}$. Above this value, the dispersion is increasingly smeared out, and a phonon position can no longer be defined. The right panel shows the simulated polycrystalline LA dispersion, together with the single-crystal dispersion along the three main symmetry directions. We note that for the low-Q region, the polycrystalline LA dispersion is very close to the LA dispersion along the $\langle 110 \rangle$ direction. If the sine fit is performed in the range up to the maximum of the dispersion, the resulting sound velocity is rather sensitive to the choice of Q_{max}. If Q_{max} remains free, its resulting value is close to $0.66 \times 2\pi/a$, significantly lower than the theoretical value for a spherical approximation of the Brillouin zone for which the resulting $Q_{max} = 0.78 \times 2\pi/a$. As a result, the sound velocities derived in these two ways differ significantly.

In order to validate these findings, an IXS experiment was performed, utilizing the Si(8 8 8) configuration at an incident photon energy of 15 816 eV, with a total energy resolution of 5.5 meV. The sample consisted of a polycrystalline foil of high-purity bcc iron of 30 μm thickness, roughly corresponding to one absorption length. IXS spectra were recorded for two angular settings of the spectrometer, spanning a Q-range from 3 to 13.4 nm^{-1}. Figure 4.6 shows a set of representative spectra. The resulting dispersion of the LA phonon branch is reported in the left panel of Figure 4.7,

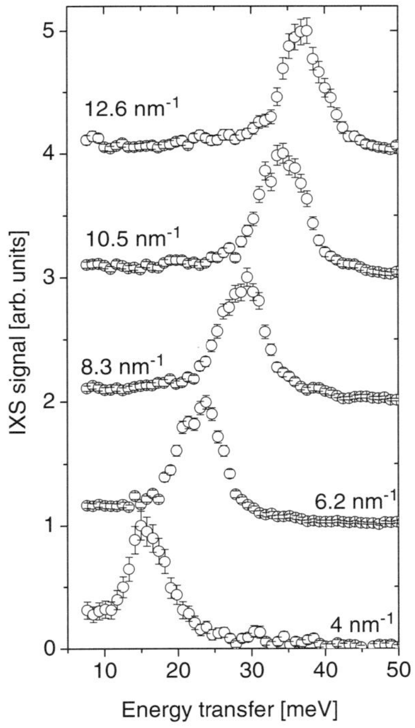

Figure 4.6 Representative IXS spectra of polycrystalline bcc iron at the indicated momentum transfers within the Brillouin zone 1. The peak intensities are scaled to the same height for clarity.

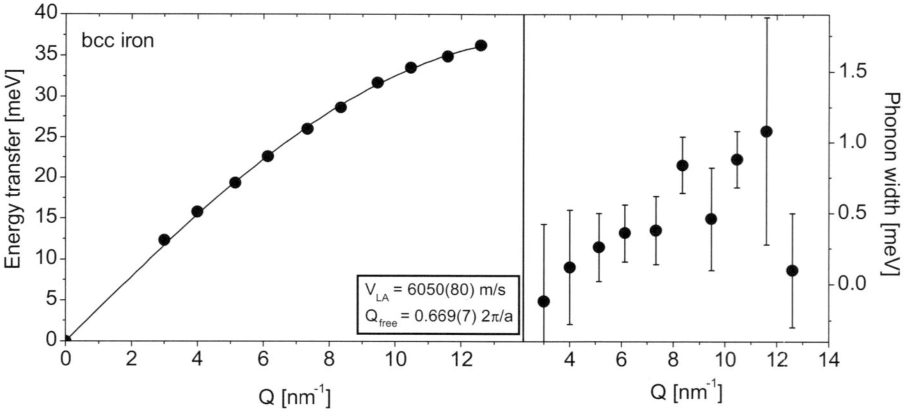

Figure 4.7 Left panel: Low-Q LA phonon dispersion of polycrystalline bcc iron (full circles). The solid line represents the best fit, using Equation 4.17, and the resulting fit parameters are indicated at the bottom of the panel. Right panel: Evolution of the phonon line width as a function of Q.

together with a sine fit to the experimental data. We note the good quality of the fit and, more important, the excellent agreement of the sound velocity V_{LA} and the parameter Q_{max} with the calculated values: $V_L = 6035\ \mathrm{m\,s^{-1}}$ and $Q_{max} = 0.66 \times 2\pi/a$. This good agreement can be expected, since the largest Q-value at which an IXS spectrum was recorded corresponds to $Q = 0.585 \times 2\pi/a$, a value that is significantly lower than $0.707 \times 2\pi/a$, identified above as the critical Q-value. The right panel of Figure 4.7 shows the Q-evolution of the excitation width. These were obtained, using a model Lorentzian function, convoluted with the experimental resolution function. Apart from the highest Q-point, we note a monotonic increase of the width, in qualitative agreement with our simulation, which yields an excess width of 0.25 meV (FWHM) at $13.4\ \mathrm{nm^{-1}}$.

4.4.2 Scattering from Quasitransverse Phonons

Contributions from quasitransverse acoustic (qTA) phonons, whose eigenvectors possess a finite component along Q, can become nonnegligible in the first BZ due to a strong elastic anisotropy and/or the low crystal symmetry. This is illustrated in a simulation for the case of sodium and α-quartz in Figure 4.8. Here, we utilized Equation 4.8 and single-crystal elastic moduli, reported in literature. In order to emphasize the spread in sound speed, we have chosen this representation rather than to report $g(Q,E)$. Another advantage of the chosen representation is that we can obtain semiquantitative estimates without involving lattice dynamics calculations. It must be kept in mind that the relation $E = VQ$ is only approximate at $Q = 5\ \mathrm{nm^{-1}}$ due to the bending of the acoustic dispersion curve. For the case of sodium, two clearly distinct and rather broad distributions can be identified, corresponding to the longitudinal and transverse acoustic (TA) phonon branches. Figure 4.8b shows the

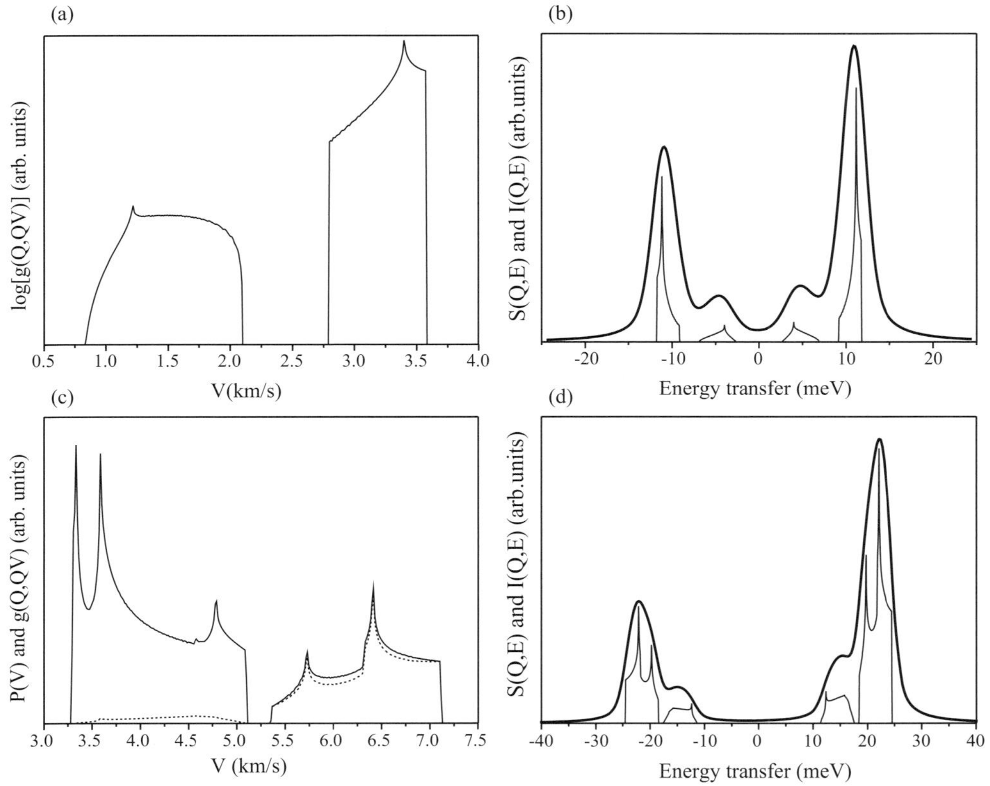

Figure 4.8 $g(Q,QV)$ for sodium (a) and α-quartz (c) at $Q = 5\,\text{nm}^{-1}$. Corresponding dynamical structure factor $S(Q,E)$ (thin line) and $I(Q,E)$ (thick line) – its convolution with a typical energy resolution function of 3 meV – as a function of energy transfer for sodium (b) and α-quartz (d). Figure adapted from Bosak *et al.* [47].

IXS signal $I(Q,E)$ (including the thermal factor $F(E)$, which enhances the lower energy qTA excitation), convoluted with a typical energy resolution of 3 meV. The estimated ratio of transverse/longitudinal peaks is as much as ~18% for sodium due to its high elastic anisotropy. For crystal symmetries lower than cubic, the qTA contribution can become even more important. This is illustrated in Figure 4.8c and d. It can be appreciated that the velocity distribution takes a more complex form; the TA and LA velocities are no longer clearly separated, and each individual distribution contains at least two distinct peaks. This is a direct consequence of the strong elastic anisotropy. Though the transverse contributions are strongly suppressed due to the $|\vec{Q}\cdot\hat{e}(\vec{n},j)|^2$term in the $S(Q,E)$, the shape of the IXS spectrum $I(Q,E)$ is composed of a main peak and a clearly visible low-energy shoulder, associated with the LA and TA phonons, respectively (see Figure 4.8d).

Figure 4.9 shows representative experimental IXS spectra for sodium, together with model calculations. These spectra clearly reveal, besides the usual Stokes and

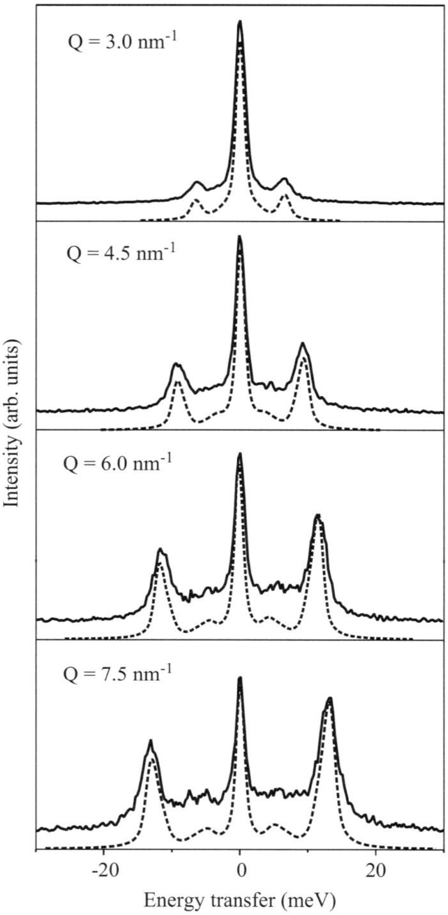

Figure 4.9 Low-*Q* IXS spectra of polycrystalline sodium, recorded with 1.5 meV energy resolution (solid lines) compared with an elasticity-based model (dashed line). Experimental and theoretical spectra are shifted along the vertical axis for clarity. Figure taken from Bosak *et al.* [47].

anti-Stokes scattering from LA phonons, some extra intensity at lower energy, due to the contribution of TA phonons. The LA peak positions of the model were slightly corrected in order to account for the bending of the acoustic dispersion, and an elastic line was added for better comparison. The excellent agreement of modeled and experimental spectra indicates undoubtedly that the low-energy transfer signal is the result of scattering from quasitransverse phonons.

4.4.3
The Aggregate Elasticity of Polycrystalline Materials

There are several approaches to calculate the isotropic elastic properties of polycrystalline materials from single-crystal elastic constants. The most common models presume either (i) constant stress throughout the body (Reuss average) [49], thus neglecting the boundary conditions for the strain, or (ii) constant strain throughout the body (Voigt average) [50], thus neglecting the boundary conditions for the stress. These two models provide the lower and upper bounds of the isotropic moduli. The Hill approximation [51] corresponds to the average of the Voigt and Reuss approximation. The most accurate results for isotropic and textured samples are obtained from self-consistent iterative approaches [52] or first-principle calculations [53], which aim at solving the problem of an elastic inclusion in a homogeneous matrix of the same elastic properties as the one of the polycrystalline material. These treatments are relatively complicated, time consuming, and the results depend on parameters like the grain shape.

The procedure, proposed by Matthies and Humbert [54], is much simpler. The resulting average has the properties of a geometrical mean and obeys the physical condition for the compliance tensor C on averaging: $\langle C\rangle = \langle C^{-1}\rangle^{-1}$. In other words, the average value must be equal to the inverse of the average of the inverse value. We note that this condition is not fulfilled for the Voigt, Reuss, and Hill approximations. For an isotropic distribution (no texture effects), analytical expressions can be obtained for crystal symmetries higher than monoclinic. For cubic symmetry, one obtains [54]

$$C_{44} = \mu = \frac{\left(C_{11}^0 - C_{12}^0\right)^{2/5} C_{44}^{3/5}}{2^{2/5}}, \tag{4.18a}$$

$$C_{11} = 2\mu + \lambda = \frac{\left(4\mu + C_{11}^0 + 2C_{12}^0\right)}{3}, \tag{4.18b}$$

$$C_{12} = \lambda = C_{11} - 2C_{44}, \tag{4.18c}$$

where μ and λ are the Lamé constants. These are linked to the bulk modulus K and shear modulus G by the following relationship: $\lambda = K - (2/3)G$ and $\mu = (3/2)(K - \lambda) = G$. It follows that longitudinal and transverse sound speeds are not necessarily exactly the same as the ones derived from the phonon dispersion measurements as they are *different* functions of the elastic moduli.

In Table 4.2 the values of aggregate sound velocities are compared to those determined from the simulated IXS spectra. Average (macroscopic) aggregate values are calculated using the principle of a geometric mean for the elastic tensor [54] as described above and available elastic moduli [55–57]. The simulation was performed for $Q = 5\ \mathrm{nm}^{-1}$ and $T = 298$ K, employing Equation 4.8, and did not take into account any energy or momentum transfer resolution effects. It must be noted that the IXS average for V_S is not accessible experimentally in most cases.

This analysis reveals that V_L and V_S, derived from IXS measurements, can differ from the aggregate macroscopic average up to several percent with exception of diamond. Na displays the largest deviation for V_L (6.2%), while the value of V_S is in reasonable agreement with the macroscopic average (−0.9%). A complementary situation occurs for Co where the difference for V_L is very small (0.3%), but very large for V_S (7.7%). Finally, the overall discrepancy is most pronounced for SiO_2 as a consequence of the low crystal symmetry.

The above analysis shows that the LA sound velocity obtained via low-Q IXS measurements from polycrystalline materials can be quite different from the

Table 4.2 Comparison of aggregate longitudinal V_L and shear sound velocities V_S with those determined from the simulated IXS spectra. Table taken from Bosak of *et al.* [47].

Material	Macroscopic average		IXS average	
	V_L (m s^{-1})	V_S (m s^{-1})	V_L (m s^{-1})	V_S (m s^{-1})
Diamond[a)]	18 175	12 351	18 219 (+0.2%)	12 238 (−0.9%)
Na (fcc)[a)]	3115	1434	3308 (+6.2%)	1421 (−0.9%)
Fe (bcc)[a)]	5916	3220	6035 (+2.0%)	3118 (−3.2%)
Fe (hcp)[b)]	8634	4709	8627 (−0.1%)	4785 (+1.6%)
Fe (hcp)[c)]	9341	5300	9344 (0.0%)	5380 (+1.5%)
Co (hcp)[a)]	5704	2934	5721 (+0.3%)	3161 (+7.7%)
Quartz[a)]	6093	4039	6318 (+3.7%)	4210 (+4.2%)

The difference in % is indicated in parentheses. See text for further details.
a) Tabulated elastic constants from Landolt Börnstein [55].
b) Elastic constants derived from radial X-ray diffraction and ultrasonic techniques [56].
c) Elastic constants obtained from *ab initio* calculations [57].

aggregate one, and its value is sensitive to the dispersion fitting procedure. In the absence of very low-Q data (well within the linear $E(Q)$ regime), the choice of Q_{max} can become crucial and attention must be paid to the data treatment.

4.4.4
Effects of Texture

Texturing of polycrystalline samples is a quite common phenomenon as a result, for example, of the synthesis procedure or if the material is submitted to nonhydrostatic pressure. As graphite is the most anisotropic among all the elementary solids, it is a good example for the demonstration of anisotropy effects in low-Q IXS spectra. The preparation of nontextured polycrystalline graphite samples presents significant difficulties. We therefore limited our study to a pyrolytic graphite sample, which was rotated with a frequency of about 10 Hz around an axis, perpendicular to the crystallographic c-axis and the horizontal scattering plane. This configuration is equivalent to the study of a textured sample with an orientation distribution function $f(g) = 4\delta(\psi)/\sin(\theta)$, where θ and ψ are the corresponding Euler angles. As input parameters, we have used the recently determined set of graphite elastic moduli [58]. Figure 4.10 shows a typical experimental IXS spectrum recorded at 6.85 nm^{-1}, together with calculations for both an isotropic model and a model that incorporates the texture as simulated in the experiment. First, we note that two distinct structures at very different energy transfers are observed. Coarsely speaking, these are due to LA and TA phonons propagating close to the c-axis (low-energy transfer) and LA phonons in the basal plane (high-energy transfer). Their large energy difference is due to the very different bonding strengths. The main effect of texture is reflected in line shape changes of the low-energy feature, while at higher energy transfers the influence of texture is only moderate.

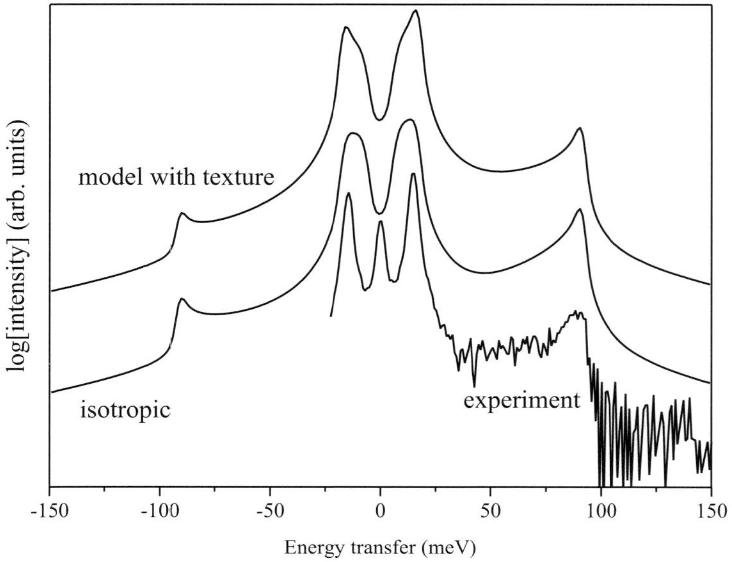

Figure 4.10 Experimental IXS spectrum of textured pyrolytic graphite at $Q = 6.85\ nm^{-1}$ compared with elasticity-based models with and without texture. Figure taken from Bosak *et al.* [47].

The position and shape of the experimental low-energy feature is not very well described by the model. Besides the spectral weight arising from LA (0 0 1) phonons in the long wave limit, the scattering from TA phonons, propagating close to the basal plane and polarized along the *c*-axis, give a significant contribution. The TA phonon energy grows faster than the LA phonon energy with increasing Q due to the parabolic-like dispersion of these modes: $E^2 = CQ^2 + DQ^4$ (C and D are constants) [59]. As a consequence, for some Q-values the low-energy feature becomes narrower than predicted within the frame of our elastic model, which assumes a linear relationship between E and Q. Exploiting the simple dynamic model of Nicklow *et al.* [60], it can be demonstrated that the apparent Q-dispersion of high- and low-energy features mimics the LA in-plane dispersion and LA dispersion along Γ–A, respectively. This is illustrated by comparing the intensity map obtained for pyrolytic graphite with the single-crystal data [61] (see Figure 4.11). This extreme case of graphite underlines that the term "average sound velocity" and even "orientationally averaged sound velocity" must be used with prudence, specifically if strongly anisotropic systems are considered.

4.5
IXS in the High-Q Limit: The Phonon Density of States

The phonon density of states (PDOS) gives the energy distribution $g(E)$ of the phonons, which can be measured using either INS or IXS. A further method is provided by nuclear inelastic scattering if the studied material contains a Möss-

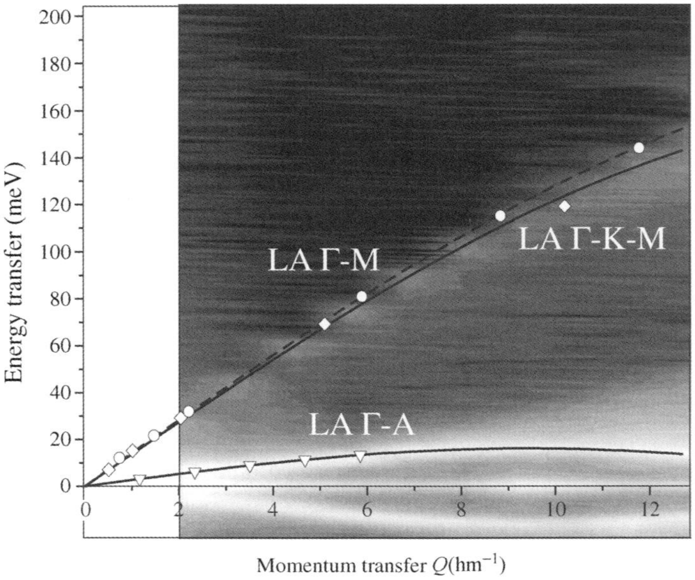

Figure 4.11 Low-Q IXS intensity map of textured pyrolytic graphite compared with single-crystal phonon dispersion data; longitudinal acoustic phonons along Γ-A (triangles), Γ-M (circles), and Γ-K-M (diamonds) [58, 61]. Solid and dashed lines are the results of a lattice dynamics calculation [60]. Positive energy transfer corresponds to the Stokes side of the spectra. Figure taken from Bosak *et al.* [47].

bauer isotope [62, 63]. In monoatomic polycrystalline samples, $g(E)$ can be determined directly, while in systems with different atomic species, only the generalized DOS can be obtained. Here the individual contributions of the different constituent atoms are weighted by their corresponding scattering length (INS) or atomic from factor (IXS). For large Q, the normalized $g(Q, E)$ should approach the generalized VDOS, (see Equation 4.9).

As inelastic X-ray scattering from phonons is essentially a coherent scattering process, the same incoherent approximation as for coherent INS needs to be applied [64–66]. Several aspects have to be considered in order to ensure a correct VDOS approximation. Intuitively, the larger the momentum transfer, the better the approximation, and in the limit of a very large Q-sphere, even one IXS spectrum will give a good VDOS approximation. However, the radius of the largest Q-sphere is given by the maximum scattering angle of the IXS spectrometer. Another aspect concerns the thickness of the integration shell. While for INS there are no constraints associated to this since the neutron scattering length b is independent of Q, for IXS the atomic form factor $f(Q)$ displays a pronounced Q-dependence with an approximately exponential decay. This decay is element dependent, and the Q-value for which $f(Q)$ has dropped by 50% from its maximum ($f(Q=0)=Z$) corresponds roughly to the inverse of the spatial extent of the atom. This leads to a distortion of the VDOS, if the integration is performed over a large Q-range.

In general, no recipe exists for the choice of the shell sampling, but it was shown that semiquantitative criteria for a uniform sampling can be established, which are independent of any specific lattice dynamics model, and only result from simple symmetry consideration [67]. In practise, the collection of typically 10 IXS spectra in the 50–70 nm^{-1} Q-range is sufficient to ensure a correct sampling. The PDOS is then determined after correcting for the crystal analyzer efficiencies, subtracting the elastic line, and following a previously established data treatment in which the multiphonon contributions are eliminated simultaneously with the deconvolution of the instrumental function [68]. A benchmark study on polycrystalline diamond has convincingly proven the validity of the approach [67]. In the following, a few representative examples of PDOS determinations with X-rays shall be presented.

4.5.1 Magnesium Oxide

MgO is regarded as the prototype oxide due to its simple structure and the large stability field (in pressure and temperature) of the NaCl structure. It is an important ceramic for industrial applications, and is of great interest for Earth sciences, since it is a major mineral in the Earth's lower mantle. Consequently, several calculations of its lattice dynamics exist in literature and allow a critical comparison with experimental results. In the present context, MgO served as a benchmark for a diatomic system.

Ten IXS spectra were recorded, covering a Q-range of about 15 nm^{-1} (approximately 60–75 nm^{-1}) with a momentum resolution of 0.7 nm^{-1}, both in the horizontal and vertical directions. The spectra were recorded with an energy resolution of 5.4 meV, employing the silicon (8 8 8) setup. For MgO we can only extract the X-VDOS for the selected Q-region. The weighting with $f(Q)^{-2}Q^{-2}$ is no more useful as the contributions for the two atoms are significantly different. On the other hand, the variation of effective scattering factors $f(Q)^2Q^2$ for Mg and O is less than 8% over the sampled Q-range, therefore no correction is needed.

Figure 4.12 shows the comparison of the experimental X-VDOS with results from different *ab initio* calculations [69–72]. As expected, the overall form of the X-VDOS is roughly the same as for the VDOS, but relative intensities of peaks are changed due to the different scattering power of Mg and O for X-rays. We note significant differences among the various VDOS. This is not surprising, since none of the calculated single-crystal dispersions match the experimental phonon dispersion perfectly [73, 74]. The best agreement between experiment and theory is obtained for calculations using density functional perturbation theory (DFPT) within the local density approximation (LDA) [69, 70]. The calculations by Drummond *et al.* [71], performed using DFT within the quasiharmonic approximation, underestimate the transverse acoustic phonon energies and overestimate the energies of the highest optical branches. The calculation by Parlinski *et al.* [72] shows the largest energy difference for the first peak in the VDOS ($\sim$4 meV) and significantly underestimates the energy of the highest optical branches.

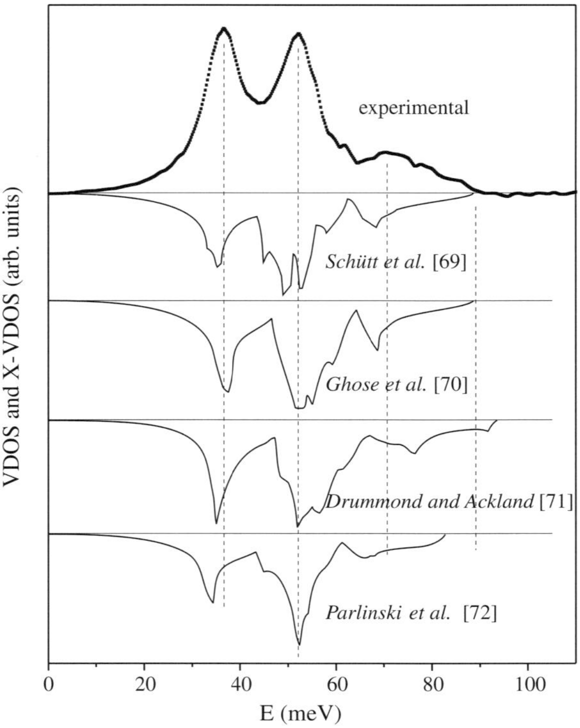

Figure 4.12 Reconstructed generalized X-VDOS of MgO versus calculated *ab initio* results [88–92]. Figure taken from Bosak *et al.* [67].

As only the X-VDOS is available, no thermodynamic data can be obtained. The present data should be completed by neutron VDOS measurements in order to obtain the partial VDOS for Mg and O, or, less directly, the present X-VDOS results should be used for the adjustment of *ab initio* models, most likely using DFPT within LDA. It can be shown that the real VDOS can be obtained at least for the low-energy part [67]. For this portion of the VDOS ($E < 20$ meV), we obtain a scaling factor α of about 0.908 and extract from the thus corrected generalized VDOS the low-temperature limit of the Debye temperature and the average sound speed by a parabolic fit to the experimental data. In Table 4.3 we compare our results with those calculated from the available elastic data. The agreement is remarkable, thus proving that even for a multicomponent system, aggregate properties can be correctly extracted.

Finally, we would like to stress that the combination of IXS–VDOS measurements (determination of the Debye velocity V_D) with results from IXS within the first Brillouin zone (determination of the aggregate compressional velocity V_P) and X-ray diffraction (determination of the bulk modulus K and the density ϱ) allows the extraction of the single-crystal elastic moduli C_{11}, C_{44}, and C_{12} for cubic systems. From low- and high-Q IXS measurements, we obtain [67, 76] $V_P = 9960\ \mathrm{m\,s^{-1}}$ and $V_D = 6630\ \mathrm{m\,s^{-1}}$, whereas K=162.5 GPa was taken from literature [77]. A set of three

Table 4.3 Selected macroscopic parameters for MgO. Table taken from Bosak *et al.* [67].

Parameter	Calculated from VDOS	Calculated from elastic data [75]
θ_D – low-temperature limit	935(20) K	940 K
Average sound speed V_D	6.63 (13) km s^{-1}	6.65 km s^{-1}

Table 4.4 MgO: comparison of elastic moduli derived from polycrystalline X-ray data with single-crystal Brillouin light scattering (BLS) data. The estimated error for X-ray data is ~3%.

	V_D (km s^{-1})	V_P (km s^{-1})	B (GPa)	C_{11} (GPa)	C_{12} (GPa)	C_{44} (GPa)
XRD [77]			162.5			
IXS (present)	6.63	9.96		299.3	159.6	94.1
BLS [75]				296.8	155.8	95.3

equations then allows the determination of the three independent elastic moduli:

$$K = \frac{1}{3}(C_{11} + 2C_{12}), \tag{4.19}$$

$$V_{P,D} = \left(\frac{1}{12\pi} \sum_{j=1}^{3} \int \frac{1}{V_j^3} \mathrm{d}\Omega \right)^{-\frac{1}{3}}. \tag{4.20}$$

V_P is computed as the average of the longitudinal contribution in Equation 4.20, while V_D is formally obtained by averaging over all directions. The results for MgO are displayed in Table 4.4. The maximum deviation with respect to single-crystal data is as low as 2.5%, which can be considered as an excellent result.

4.5.2
Boron Nitride [78]

An interesting and technologically important example concerns the superhard polymorphs of boron nitride. The physical and chemical properties of boron nitride have attracted a lot of interest due to their fascinating mechanical properties and high melting points in the zinc blende (z-BN) and wurtzite (w-BN) phases. At ambient conditions, the most usual form is the graphite-like hexagonal phase with two-layer stacking (h-BN), although the three-layer form also exists (r-BN). Surprisingly enough, in contrast to carbon, the zinc blende phase is thermodynamically the most stable one at ambient conditions, while the most common h-BN is metastable. The wurtzite phase is metastable under any condition, but can be produced via a martensitic mechanism from h-BN. Experimental data concerning the lattice dynamics of any form of boron nitride are limited to frequencies of Γ-point optical phonons obtained by Raman spectroscopy [79, 80]. The only exception is h-BN for which an IXS study on a tiny single crystal was recently performed [81].

Ten IXS were recorded, spanning the Q-range 60–73 nm^{-1} for w-BN and 52–68 nm^{-1} for z-BN, with an overall energy resolution of 3 meV. The spectra were properly weighted, summed, and treated according to the established protocol. Figure 4.13 shows the derived generalized IXS–PDOS for z-BN, compared to theoretical results. It is worth noting that among the available *ab initio* calculations, the energy position of special points varies [82–86]. The best overall agreement with our experimental data is obtained using a plane-wave pseudopotential method within the density-functional theory [82]: a linear response approach to the density functional was used to derive the Born effective charges, phonon frequencies, and eigenvectors. While the overall agreement between theory and experiment is quite remarkable, there are noticeable deviations for the high-energy cutoff. These are not due to experimental errors, since the X-VDOS cutoffs – nearly identical for both BN forms (162–164 meV) – coincide well with the zone center LO z-BN energy of ~162 meV, determined by Raman spectroscopy [79, 80]. The choice of the best model in the case of the wurtzite polymorph is less obvious (see Figure 4.14), as both

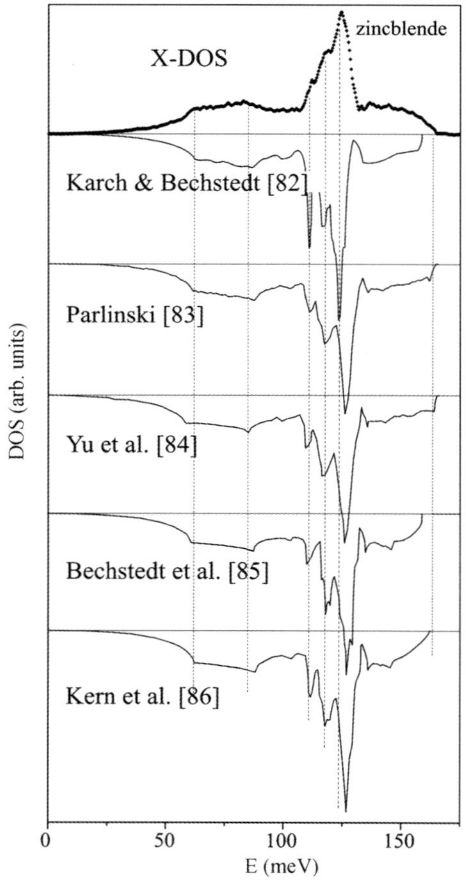

Figure 4.13 Experimental X-VDOS versus *ab initio* calculations of z-BN [82–86]. The vertical bars indicate special points.

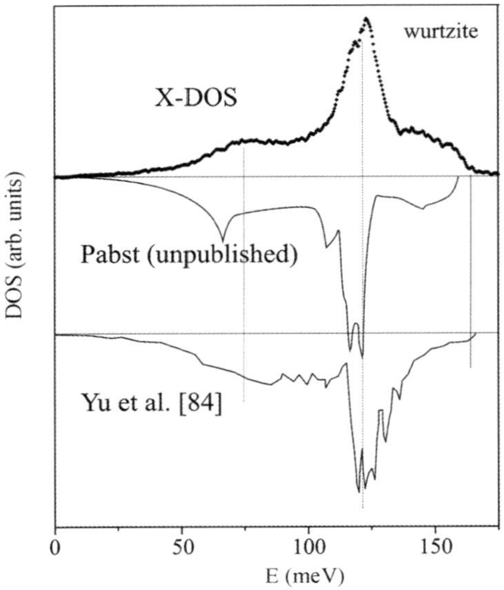

Figure 4.14 Experimental X-VDOS versus *ab initio* calculations of w-BN [84] (M. Pabst, unpublished). The vertical bars indicate special points. Figure taken from Bosak *et al.* [78].

available theoretical data sets [84] (M. Pabst, unpublished) show visible deviations from the experiment.

A parabolic fit to the low-energy part of the VDOS (up to 40–50 meV) allows the determination of the average sound speed v_D. Our value for v_D is in reasonable agreement with the one calculated from elastic constants [82, 87]. The low-temperature limit of θ_D obtained from the same fit of the VDOS is obviously very close to the calculated one, but differs significantly from the one found from specific heat measurements (1600–1800 K) [88]. Such a difference can be explained by the presence of impurities and point defects.

Tables 4.5 and 4.6 summarize the main results for the macroscopic parameters obtained combining both DOS limit and low-Q scattering results [78]. It is worth

Table 4.5 Selected macroscopic parameters for z-BN. Table taken from Bosak *et al.* [78].

Parameter	This work	Other experiments	Calculations
Longitudinal velocity ($m\,s^{-1}$)	16 000(30)	16 420(8)[a]	16 260[b], 16 350[c]
Shear velocity ($m\,s^{-1}$)	11 200(40)	10 780(3)[a]	10 675[b], 10 695[c]
Debye velocity ($m\,s^{-1}$)	12 200(20)	11 760(3)[a]	11 645[b], 11 715[c]
low-T θ_D (K)	2000(40)	~1700[d], 1930(5)[a]	1910[b], 1920[c]

a) Calculated from experimentally determined elastic moduli [87].
b) *ab initio* calculations [82] (M. Pabst, unpublished).
c) *ab initio* calculations [89].
d) Obtained by a fit to specific heat data [88].

Table 4.6 Selected macroscopic parameters for w-BN.

Parameter	This work	Other experiments	Calculations
Longitudinal velocity ($m\,s^{-1}$)	16 000(30)	—	16 205[a)], 16 660[b)]
Shear velocity ($m\,s^{-1}$)	11 450(70)	—	10 590[a)], 11 025[b)]
Debye velocity ($m\,s^{-1}$)	12 400(50)	—	11 595[a)], 12 065[b)]
low-t θ_D (K)	2030(80)	~1700[c)]	1900[a)], 1980[b)]

a) *ab initio* calculations [82], (M. Pabst, unpublished).
b) *ab initio* calculations [89].
c) Obtained by a fit to specific heat data [88].

noting that the shear velocity and, consequently, the Debye temperature are higher for w-BN. The effect is modest, but nevertheless observable.

4.5.3 Clathrate Ba_8Si_{46} [90]

Novel materials based on structures with metallic ions located in oversized crystalline cages are an intriguing class of substances. The encaged ions form a nanoscale crystalline subnetwork and influence a wide variety of physical properties. Among these materials those that become superconducting are particularly interesting since they represent model systems to study the electron–phonon interactions that mediate superconductivity.

The filled clathrates based on group IV elements (i.e., Si, Ge, and so on) also belong to this family of materials and include several superconductors [91–94]. Type-I (e.g., Ba_8Si_{46}) and type-II (e.g., Na_xSi_{136}) Si-clathrates form three-dimensional crystalline lattices based on rigid oversized 20 atom and 24 or 28 atom Si cages in which metal atoms are enclosed [95, 96].

The mechanism of superconductivity for the type-I Ba_8Si_{46} clathrate was investigated both theoretically by *ab initio* calculations and experimentally [97, 98]. The joint experimental and theoretical study [97] has shown that superconductivity is an intrinsic property of the sp^3 silicon network. A large electron–phonon coupling exists in such covalent structures. Recently, theoretical work has pointed out that the low-frequency modes in Ba_8Si_{46}, in particular those arising from the Ba vibrations in the large Si_{24} cages, contribute significantly to the electron–phonon coupling parameter λ [99].

In order to study the Ba vibrations, inelastic scattering experiments have been performed utilizing both INS and IXS. The spectrometers IN4 and IN5 at the Institut Laue Langevin (Grenoble, France) have been used for INS experiments. The incident neutron wavelength was 2.25 Å with an energy resolution of 0.8 meV. Information on the generalized VDOS was obtained from the INS spectrum using the incoherent approximation. IXS data were collected on ID28 at the ESRF with 3 meV energy resolution.

The complementarity of the two techniques results from the different atomic scattering factors for the two probes (neutrons and X-rays): the scattering probability of X-rays by the electronic shell of the sample constituents is roughly proportional to

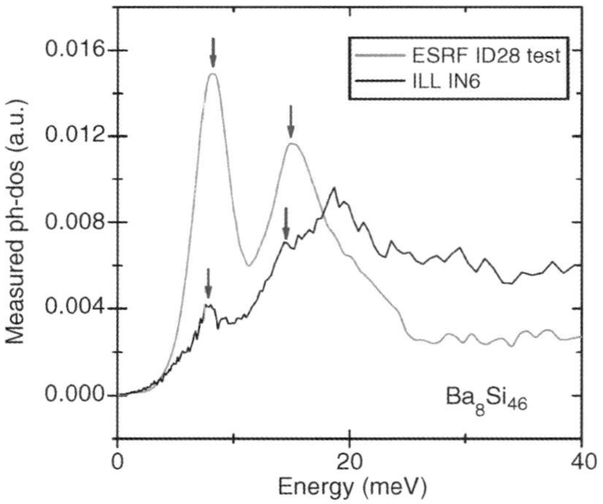

Figure 4.15 Generalized density of states of Ba_8Si_{46} as obtained by INS and IXS. Figure adapted from Lortz *et al.* [90].

Z^2, and is therefore strongly barium biased. This can be seen in Figure 4.15 where X-ray- and neutron-weighted VDOS are compared. The two Einstein-like modes associated with the Ba vibrations are much better visible in the IXS spectrum.

It is worth noting that the volume of the sample used for the IXS experiment at ambient conditions was $\sim 3 \times 10^{-3}\,\mathrm{mm}^3$. This opens up the possibility to perform VDOS measurements under high pressures using diamond anvil cell techniques.

4.6 IXS in the Intermediate *Q*-Range

As developed in the previous chapters, IXS spectra recorded at low *Q* (within the first Brillouin zone) give access to the orientation-averaged longitudinal velocity, while at high *Q* an appropriate averaging of IXS spectra permits the reconstruction of the (generalized) VDOS. Compared to single-crystal work, the lattice dynamics information content is therefore limited. This is an important constraint, since novel materials or materials studied under extreme conditions are often available only in polycrystalline form because of synthesis procedures or the presence of structural phase transitions. Attempts to overcome this limitation have been undertaken by analyzing neutron powder diffraction data. The method proposed by Dimitrov *et al.* [100] analyses the pair distribution function (PDF) obtained by a Fourier transform of the elastic scattering intensity $S(Q)$. The parameters of a phonon dispersion model are then refined with respect to the experimental PDF. Goodwin *et al.* [101] analyzed reverse Monte Carlo configurations, refined to the powder data. Here, we explore the information contained in IXS spectra recorded in the intermediate *Q*-range, where the spectral shape varies with momentum transfer due to – though relaxed – selection rules [102]. Therefore, additional constraints (beyond VDOS and averaged longitudinal sound velocity) are provided for the lattice dynamics.

The approach consists of recording IXS spectra over a large momentum transfer region (typically 2–80 nm^{-1}) and compare them with a model calculation that properly takes into account the polycrystalline state of the material and the IXS cross section. A least-squares refinement of the model IXS spectra then provides the single-crystal dispersion scheme. As a benchmark for the proposed methodology, we have chosen beryllium.

The IXS experiment was performed utilizing the Si (9 9 9) backscattering configuration with an energy resolution of 3.0 meV. A total of 90 IXS spectra were recorded, spanning a momentum transfer region from 1.9 to 79.5 nm^{-1}. We chose the exact angular positions to uniformly cover the achievable momentum transfer range and to avoid Bragg peaks. The sample was a sintered pellet of beryllium grains of 15 mm diameter, roughly corresponding to one absorption length. Due to the large grain size, it was necessary to rotate the sample with a frequency of about 10 Hz. The absence of texture was verified determining the relative intensities of the Debye–Scherrer rings and comparing them to the ideal powder X-ray diffraction spectrum.

The complete experimental momentum transfer – energy – intensity map is shown in Figure 4.16. The experimental data are normalized to the incoming X-ray intensity, the polarization of the beam, the analyzer efficiencies, and the atomic form factor. Furthermore, the elastic line was carefully subtracted using a pseudo-Voigt fit. In the low-Q limit, the dispersion of the averaged longitudinal dynamics is observable, while in the high-Q limit the VDOS is approached. In the intermediate momentum transfer range and lower energy region, acoustic phonons form the arc structure. Optical vibrations compose the band in the higher energy range. The Q-values of the IXS spectra, selected for the refinement procedure, are marked by a white line ($Q = 3.2$, 8.0, 14.4, 19.3, 22.9, 42.8, 52.2, 62.6, 69.4, and 78.5 nm^{-1}). A Born–von Kármán lattice dynamics model was refined to the 10 spectra, using the beryllium hexagonal unit cell with $a = 2.2858$ Å and $c = 3.5843$ Å and two atoms at (0, 0, 0) and (1/3, 1/3, 0.5). We have included up to the seventh next-neighbor (NN) atomic force

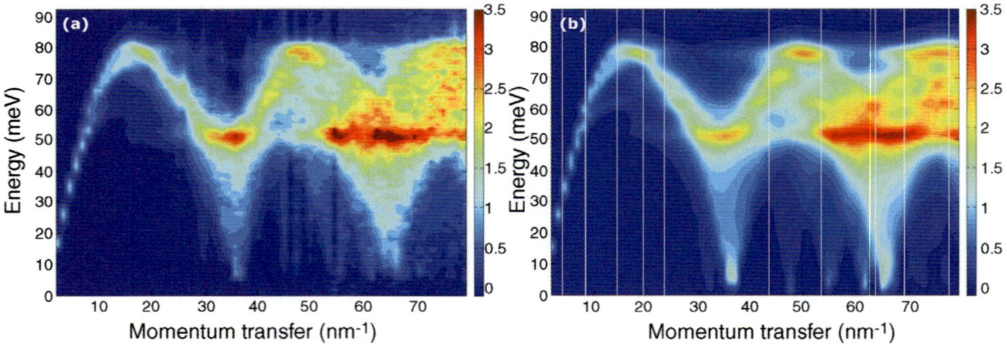

Figure 4.16 Normalized IXS intensity map for polycrystalline beryllium: (a) experimental data; (b) refined model. The vertical white lines mark the momentum transfer values corresponding to the selected spectra used for the fitting. Figure adapted from Fischer *et al.* [108].

constants (29 parameters), using a set of published force constants as the starting parameter [103]. The scattering intensity was calculated for a mesh of 43 200 points on 1/24 of the spherical shell surface, and then averaged and properly weighted as described above. The merit function, measuring the agreement between the experimental data and model calculations during the fitting routine, decreased by a factor 2.4 from $\chi^2 = 3.73$ to 1.55. Figure 4.17 shows four representative experimental IXS spectra together with the resulting computed spectra using the starting and refined models. It can be noted that the starting model already shows a remarkable agreement with the experimental data, improvements in the refined model are nevertheless evident. At low momentum transfers (3.2 and 8.0 nm^{-1}), the positions of the peaks are improved. For the spectrum at 14.4 nm^{-1}, the intensity distribution of the main feature at about 77 meV is better reproduced. Finally, a better agreement can be appreciated for the low-energy portion in the spectrum at 62.6 nm^{-1}.

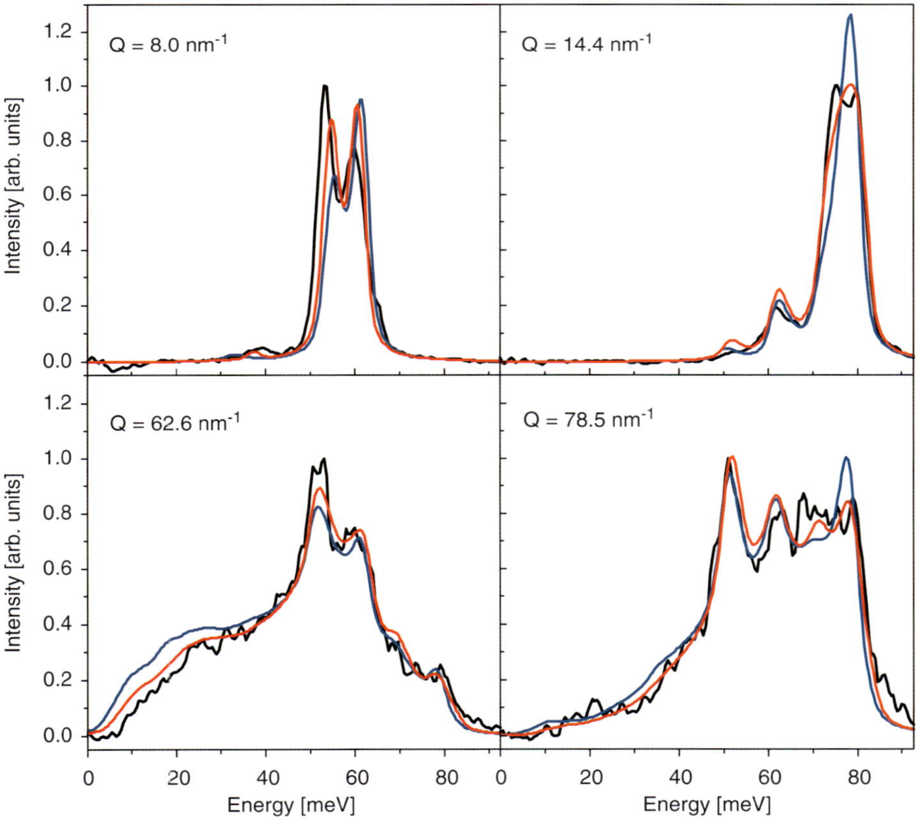

Figure 4.17 Representative IXS spectra of polycrystalline beryllium: experiment (black line), starting model spectra (blue line), and refined model spectra (red line). Figure adapted from Fischer *et al.* [102].

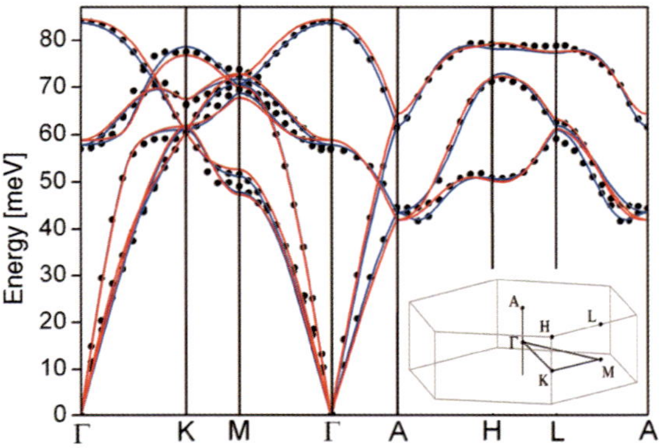

Figure 4.18 Dispersion relation for beryllium: experimental INS data [111] (black dots), starting model (blue line), and refined model calculation (red line). Figure adapted from Fischer *et al.* [102].

Figure 4.18 compares the dispersion calculated from the refined force constants (red line) with the experimental one (black dots) and the starting model (blue line). The excellent overall agreement confirms the results obtained for the polycrystalline system. Differences can be observed in the optical branches. This is due to the fact that the shape of the optic part in the IXS spectra is more sensitive to the choice of the number of neighbors and the set of force constants, illustrating the limitation of the Born–von Kármán model in this very anisotropic system. A reasonably good starting model is essential for the fit to converge. In general, there are several possibilities to produce suitable starting parameters. One is to use force constants of the material, if they are available, or, alternatively, of a system with similar physical properties. Another option is to use an empirical model such as rigid ion or shell models. Finally, a set of Hellmann–Feynman forces can be computed, using *ab initio* methods. It is important to stress that the proposed methodology is limited to relatively simple materials. As the number of atoms per unit cell increases, the density of phonon branches increases, and distinct features can no longer be observed in the polycrystalline IXS spectra, making a refinement fit impossible. A further constraint arises if the material is composed of light and heavy atoms. As the IXS cross section is roughly proportional to Z^2, where Z is the number of electrons, the weak signal from the light-atom species is masked by the strong signal arising from the heavy-atom species.

Even without a refinement as described above, a stringent discriminating test can be made if several lattice dynamics models exist. This is actually the case for beryllium, for which, besides the force constants of Hannon *et al.* [103], another set has been derived by Kwasniok [104]. Both studies have refined a Born–von Kármán model to the experimental dispersion relation, obtained by inelastic neutron scattering [105]. Kwasniok utilized 33 parameters, including up to the 8 next neighbors, while Hannon *et al.* refine 29 parameters considering 7 next neighbors. The two calculations show an equally good agreement with the experimental data for the main symmetry direction of the phonon dispersion (see panel a of Figure 4.19). In contrast to this, the computed individual IXS spectra show distinct differences as can

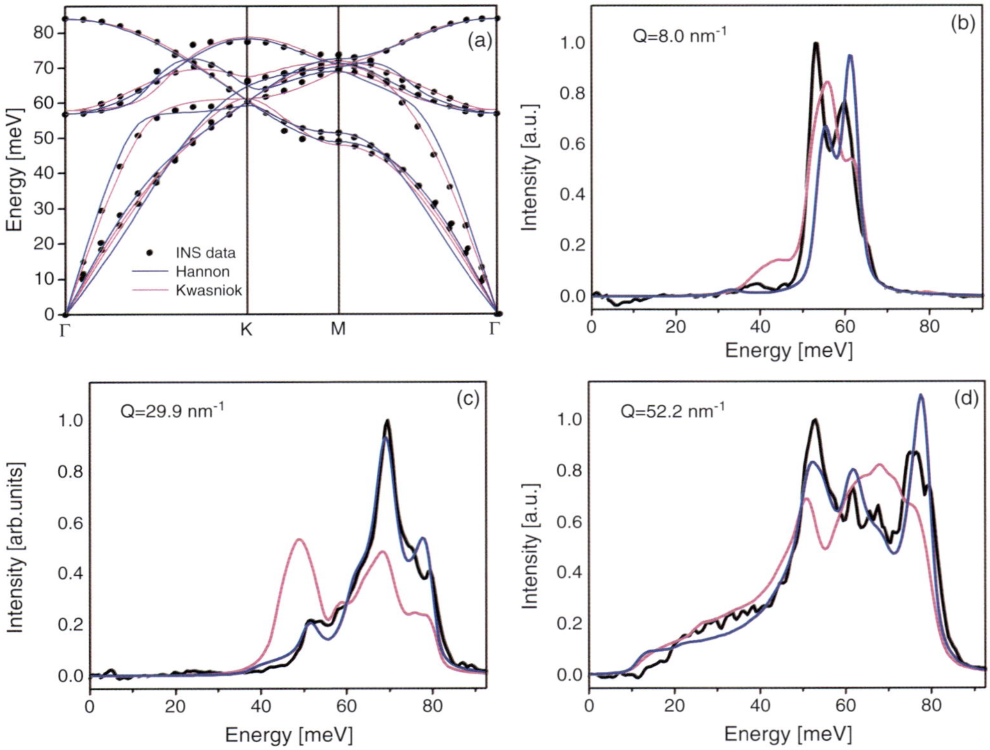

Figure 4.19 (a) Dispersion relations of Be along high symmetry directions. Polycrystalline IXS spectra at (b) $Q = 29.9\,\text{nm}^{-1}$, (c) $Q = 42.8\,\text{nm}^{-1}$, and (d) $Q = 52.2\,\text{nm}^{-1}$. INS data (black points) and IXS spectra (black lines) are compared to the results derived from model I [103] (pink lines) and model II [104] (blue lines). Figure adapted from Fischer *et al.* [102].

be appreciated in panels b–d of the same figure. Obvious discrepancies in peak positions and intensities occur for the model of Kwasniok in all spectra. On the contrary, the calculation of Hannon *et al.* reproduces the spectra quite well throughout the whole *Q*-region. This observation can be understood considering that in the INS dispersion only main symmetry directions are taken into account, whereas the calculation of the polycrystalline IXS spectra demands the averaging over a spherical surface with radius *Q*, therefore involving also nonhigh symmetry directions. The additional information from nonhigh symmetry directions allows making the appropriate choice between different models.

4.7 Concluding Remarks

The present study has focused on inelastic X-ray scattering in polycrystalline materials and has provided both the formal theoretical background and examples to illustrate the present status. It has been emphasized that attention has to be paid when orientation-averaged properties are linked to the macroscopic aggregate properties. If the single-crystal elastic properties significantly deviate from isotropy, nonnegligible differences can appear for the properties of the polycrystalline aggregate. This effect obviously becomes more pronounced for textured samples. Furthermore, if sound velocities shall be derived from the orientation-averaged LA dispersion, care has to be taken concerning the Q-range over which data are considered, as we demonstrate using the example of bcc iron.

IXS within the "incoherent approximation" approach promises to become a valuable tool in the determination of the frequency distribution function, thus complementing well-established inelastic neutron and nuclear scattering techniques. Even if only the generalized VDOS is accessible for nonmonoatomic systems, experimental results provide a discriminating test for the validity of the approximations made in the respective calculations. The technique can be applied to a very wide class of materials and furthermore opens the possibility to study systems in extreme conditions such as high pressure and/or high temperature. It is worth noting that with respect to inelastic neutron scattering, the amount of material needed is 3–5 orders less, and anomalous absorption (like for B, Cd, Gd, and so on) or anomalously high cross sections (H) are not present. For multicomponent systems, the X-VDOS is only defined for a particular spherical shell in Q-space due to the Q-dependence of the atomic scattering factor $f(Q)$. Since the scattering strengths for neutrons and X-rays are essentially different, it opens the possibility to extract directly the partial densities of states in at least binary systems from coupled N-VDOS and X-VDOS measurements.

Combining IXS data at low- and high-momentum transfers with X-ray diffraction results can provide accurate values for the average longitudinal sound speed, the Debye velocity, and the bulk modulus. These three quantities fully determine the elasticity tensor of cubic materials, thus allowing the determination of single-crystal elastic moduli in polycrystalline materials. A further step in this direction is the introduction of a new methodology for the determination of the single-crystal lattice dynamics from polycrystalline materials. It could be shown that for the chosen simple test case of beryllium, a least-squares refinement of a model calculation can be performed. It has been furthermore demonstrated that even without a fit procedure the validity of a model calculation can be assessed by inspection of individual polycrystalline inelastic spectra. As a matter of fact, a comparison (experiment versus model/theory) not only of the phonon energies but also the intensity distribution, and furthermore including all crystallographic directions (as we deal with a polycrystalline material), provides an important constraint. While the least-squares refinement procedure will only be successful for relatively simple systems, the comparative approach can be applied to more complex materials.

The study of polycrystalline materials by IXS will considerably benefit from a new generation of IXS spectrometers. In particular, a multianalyzer crystal spectrometer, covering the entire momentum transfer range, would allow the collection of all the IXS spectra simultaneously, consequently reducing the data acquisition time significantly. This will enable the investigation of materials in increasingly complex environments such as high pressure and high temperature (by laser heating), static and pulsed magnetic fields, as well as time-resolved studies.

References

1 Ashcroft, N.W. and Mermin, N.D. (1976) *Solid State Physics*, Saunders College Publishing.
2 Laval, J. (1939) *Bull. Soc. Franç. Minér.*, **62**, 137.
3 Olmer, P. (1948) *Acta Crystallogr.*, **1**, 57.
4 Walker, C.B. (1956) *Phys. Rev.*, **103**, 547.
5 (a) Curien, H. (1952) *Acta. Crystallogr.*, **5**, 393; (b) Curien, H. (1952) *Acta Crystallogr.*, **5**, 554.
6 Jacobsen, E.H. (1955) *Phys. Rev.*, **97**, 654.
7 (a) Egger, H., Hofmann, W., and Kalus, J. (1984) *Appl. Phys. A*, **35**, 41; (b) Hofmann, W.G. (1989) Thesis. University of Bayreuth.
8 Burkel, E., Peisl, J., and Dorner, B. (1987) *Europhys. Lett.*, **3**, 957.
9 Dorner, B., Burkel, E., Illini, Th., and Peisl, J. (1987) *Z. Phys. B Condens. Matter*, **69**, 179.
10 Burkel, E. (1991) Inelastic scattering of X-rays with very high energy resolution, in *Springer Tracts in Modern Physics*, vol. 125, Springer Verlag, Berlin.
11 Sette, F., Ruocco, G., Krisch, M., Masciovecchio, C., and Verbeni, R. (1996) *Phys. Scripta.*, **T66**, 48.
12 http://www.esrf.eu/UsersandScience/Experiments/DynExtrCond/ID28.
13 Schwoerer-Böhning, M. and Macrander, A.T. (1998) *Rev. Sci. Instrum.*, **69**, 3109.
14 http://www.aps.anl.gov/News/APS_News/2006/20061025.htm.
15 Baron, A.Q.R., Tanaka, Y., Goto, S., Takeshita, K., Matsushita, T., and Ishikawa, T. (2000) *J. Phys. Chem. Solids*, **61**, 461.
16 Ruocco, G. and Sette, F. (1999) *J. Phys.: Condens. Matter*, **11**, R259.
17 Burkel, E. (2000) *Rep. Prog. Phys.*, **63**, 171.
18 Ruocco, G. and Sette, F. (2001) *J. Phys.: Condens. Matter*, **13**, 9141.
19 Scopigno, T., Ruocco, G., and Sette, F. (2005) *Rev. Mod. Phys.*, **77**, 881.
20 Krisch, M. and Sette, F. (2007) Inelastic X-Ray scattering from phonons, in *Light Scattering in Solids, Novel Materials and Techniques*, Topics in Applied Physics 108, Springer-Verlag, Berlin.
21 Schober, H., Koza, M.M., Tölle, A., Masciovecchio, C., Sette, F., and Fujara, F. (2000) *Phys. Rev. Lett.*, **85**, 4100.
22 Koza, M.M., Schober, H., Geil, B., Lorenzen, M., and Requardt, H. (2004) *Phys Rev. B*, **69**, 024204.
23 Baumert, J., Gutt, C., Shpakov, V.P., Tse, J.S., Krisch, M., Müller, M., Requardt, H., Klug, D.D., Janssen, S., and Press, W. (2003) *Phys. Rev. B.*, **68**, 174301.
24 Baumert, J., Gutt, C., Krisch, M., Requardt, H., Müller, M., Tse, J.S., Klug, D.D., and Press, W. (2005) *Phys. Rev. B*, **72**, 054302.
25 Fiquet, G., Badro, J., Guyot, F., Requardt, H., and Krisch, M. (2001) *Science*, **291**, 468.
26 Antonangeli, D., Occelli, F., Requardt, H., Badro, J., Fiquet, G., and Krisch, M. (2004) *Earth Planet Sci. Lett.*, **225**, 243.
27 Fiquet, G., Badro, J., Guyot, F., Bellin, Ch., Krisch, M., Antonangeli, D., Mermet, A., Requardt, H., Farber, D., and Zhang, J. (2004) *Phys. Earth Planet. Interiors*, **143–144**, 5.
28 Antonangeli, D., Krisch, M., Fiquet, G., Badro, J., Farber, D.L., Bossak, A., and Merkel, S. (2005) *Phys. Rev. B*, **72**, 134303.
29 Badro, J., Fiquet, G., Guyot, F., Gregoryanz, E., Occelli, F., Antonangeli, D., and D'Astuto, M. (2007) *Earth Planet Sci. Lett.*, **98**, 085501.
30 Squires, G.L. (1978) *Introduction to the Theory of Thermal Neutron Scattering*, Cambridge University Press, Cambridge.
31 Jones, W. and March, N.H. (1973) *Theoretical Solid State Physics*, John Wiley & Sons, Ltd.
32 Auld, B.A. (1973) *Acoustic Fields and Waves in Solids*, vol. 1, John Wiley & Sons, Inc., New York.
33 Schülke, W. (1991) *Handbook on Synchrotron Radiation*, vol. 3 (eds G.

Brown and D.E. Moncton), Elsevier, Amsterdam.

34 Sinn, H. (2001) *J. Phys.: Condens. Matter*, **13**, 7525.

35 Brockhouse, B.N. and Hurst, D.G. (1952) *Phys. Rev.*, **88**, 542.

36 Snigirev, A.A., Snigireva, I.I., Kohn, V.G., and Lengeler, B. (1996) *Nature*, **384**, 49.

37 Bilderback, D.H. (1986) *Nucl. Instrum. Meth.*, **A246**, 434.

38 Zachariasen, H. (1944) *Theory of X-Ray Diffraction in Crystals*, Dover, New York.

39 Automatic Systems Laboratory, Milton Keynes, England.

40 Verbeni, R., Sette, F., Krisch, M.H., Bergmann, U., Gorges, B., Halcoussis, C., Martel, K., Masciovecchio, C., Ribois, J.F., Ruocco, G., and Sinn, H. (1996) *J. Synchr. Radiat.*, **3**, 62.

41 Morawe, Ch. (2006) Proc. SRI E3-014.

42 Morawe, Ch., Pecci, P., Peffen, J.-Ch., and Ziegler, E. (1999) *Rev. Sci. Instrum.*, **70**, 3227.

43 (a) Masciovecchio, C., Bergmann, U., Krisch, M., Ruocco, G., Sette, F., and Verbeni, R. (1996) *Nucl. Instrum. Meth. B*, **111**, 181; (b) Masciovecchio, C., Bergmann, U., Krisch, M., Ruocco, G., Sette, F., and Verbeni, R. (1996) *Nucl. Instrum. Meth. B*, **117** (1996) 339.

44 Canberra Eurisys, Lingolsheim, France.

45 Bergamin, A., Cavagnero, G., Mana, G., and Zosi, G. (1999) *Eur. Phys. J. B*, **9**, 225.

46 Verbeni, R., D'Astuto, M., Krisch, M., Lorenzen, M., Mermet, A., Monaco, G., Requardt, H., and Sette, F. (2008) *Rev. Scient. Instr.*, **79**, 083902.

47 Bosak, A., Krisch, M., Fischer, I., Huotari, S., and Monaco, G. (2007) *Phys. Rev. B*, **75**, 153408.

48 Minkiewicz, V.J., Shirane, G., and Nathans, R. (1967) *Phys. Rev.*, **162**, 528.

49 Reuss, A. (1929) *Z. Angew. Math. Mech.*, **9**, 49.

50 Voigt, W. (1928) *Lehrbuch der Kristallphysik*, Teubner Verlag, Leipzig.

51 Hill, R. (1952) *Proc. Phys. Soc. A*, **65**, 349.

52 Kröner, E. (1958) *Z. Phys.*, **151**, 504.

53 Kiewel, H. and Fritsche, L. (1994) *Phys. Rev. B*, **50**, 5.

54 Matthies, S. and Humbert, M. (1995) *J. Appl. Cryst.*, **28**, 254.

55 Hearmon, R.F.S. *Landolt-Börnstein New Series, Group III*, vols 11 and 18, (1984) Springer-Verlag.

56 (a) Mao, H.K., Shu, J., Shen, G., Hemley, R.J., Li, B., and Singh, A.K. (1998) *Nature*, **396**, 741; (b) Mao, H.K., Shu, J., Shen, G., Hemley, R.J., Li, B., and Singh, A.K. (1999) *Nature*, **399**, 280.

57 Steinle-Neumann, G., Stixrude, L., and Cohen, R.E. (1999) *Phys. Rev. B*, **60**, 791.

58 Bosak, A., Krisch, M., Mohr, M., Maultzsch, J., and Thomsen, C. (2007) *Phys. Rev. B*, **75**, 153408.

59 Zabel, H. (2001) *J. Phys.: Condens. Matter*, **13**, 7679.

60 Nicklow, R., Wakabayashi, N., and Smith, H.G. (1972) *Phys. Rev. B*, **5**, 4951.

61 Mohr, M., Maultzsch, J., Dobradzic, E., Reich, S., Milosevic, I., Damnjanovic, M., Bosak, A., Krisch, M., and Thomsen, C. (2007) *Phys. Rev. B*, **76**, 035439.

62 Seto, M., Yoda, Y., Kikuta, S., Zhang, X.W., and Ando, M. (1995) *Phys. Rev. Lett.*, **74**, 3828.

63 Sturhahn, W., Toellner, T.S., Alp, E.E., Zhang, X., Ando, M., Yoda, Y., Kikuta, S., Seto, M., Kimball, C.W., and Dabrowski, B. (1995) *Phys. Rev. Lett.*, **74**, 3832.

64 Needham, L.M., Cutroni, M., Dianoux, A.J., and Rosenberg, H.M. (1993) *J. Phys.: Condens. Matter*, **5**, 637.

65 de Wette, F.W. and Rahman, A. (1968) *Phys. Rev.*, **176**, 784.

66 Squires, G.L. (1978) *Introduction to the Theory of Thermal Neutron Scattering*, Cambridge University Press, Cambridge.

67 Bosak, A. and Krisch, M. (2005) *Phys. Rev. B*, **72**, 224305.

68 Kohn, V. and Chumakov, A. (2000) *Hyperfine Interact.*, **125**, 205.

69 Schütt, O., Pavone, P., Windl, W., Karch, K., and Strauch, D. (1994) *Phys. Rev. B*, **50**, 3746.

70 Ghose, S., Krisch, M., Oganov, A.R., Beraud, A., Bossak, A., Gulve, R., Seelaboyina, R., Yang, H., and Saxena, S.K. (2006) *Phys. Rev. Lett.*, **96**, 035507.

71 Drummond, N.D. and Ackland, G.J. (2002) *Phys. Rev. B*, **65**, 184104.

72 Parlinski, K., Zażewski, J., and Kawazoe, Y. (2000) *J. Phys. Chem. Solids*, **61**, 87.

73 Peckham, G. (1967) *Proc. Phys. Soc. Lond.*, **90**, 657.

74 Sangster, M.J.L., Peckham, G., and Saunderson, D.H. (1970) *J. Phys. C: Solid. State Phys.*, **3**, 1026.

75 Jackson, I. and Niesler, H. (1982) *High Pressure Research in Geophysics* (eds S. Akimoto and M.H. Manghnani), Center for Academic Publications, Tokyo, Japan, p. 93.

76 Fischer, I. (2008) PhD thesis. University of Grenoble.

77 Merkel, S., Wenk, H.R., Shu, J.F., Shen, G.Y., Gillet, P., Mao, H.K., and Hemley, R.J. (2002) *J. Geophys. Res.: Solid Earth*, **107**, 2271.

78 Bosak, A. and Krisch, M. (2006) *Radiat. Phys. Chem.*, **75**, 1661.

79 Werninghaus, T., Hahn, J., Richter, F., and Zahn, D.R.T. (1997) *Appl. Phys. Lett.*, **70**, 958.

80 Sanjurjo, J.A., López-Cruz, E., Vogl, P., and Cardona, M. (1983) *Phys. Rev. B*, **28**, 4579.

81 Serrano, J., Bosak, A., Arenal, R., Krisch, M., Watanabe, K., Taniguchi, T., Kanda, H., Rubio, A., and Wirtz, L. (2007) *Phys. Rev. Lett.*, **98**, 095503.

82 Karch, K. and Bechstedt, F. (1997) *Phys. Rev. B*, **56**, 7404.

83 Parlinski, K. (2001) *J. Alloys. Comp.*, **328**, 97.

84 Yu, W.J., Lau, W.M., Chan, S.P., Liu, Z.F., and Zheng, Q.Q. (2003) *Phys. Rev. B*, **67**, 014108.

85 Bechstedt, F., Grossner, U., and Furthmüller, J. (2000) *Phys. Rev. B*, **62**, 8003.

86 Kern, G., Kresse, G., and Hafner, J. (1999) *Phys. Rev. B*, **59**, 8551.

87 Grimsditch, M., Zouboulis, E.S., and Polian, A. (1994) *J. Appl. Phys.*, **76**, 832.

88 Solozhenko, V.L. (1994) *Properties of Group III Nitrides* (ed. J.H. Edgar), INSPEC, London.

89 Shimada, K., Sota, T., and Szuku, K. (1998) *J. Appl. Phys.*, **84**, 4951.

90 Lortz, R., Viennois, R., Petrovic, A., Wang, Y., Toulemonde, P., Meingast, C., Koza, M.M., Mutka, H., Bossak, A., and San Miguel, A. (2008) *Phys. Rev. B*, **77**, 224507.

91 Kawaji, H., Horie, H., Yamanaka, S., and Ishikawa, M. (1995) *Phys. Rev. Lett.*, **74**, 1427.

92 Yamanaka, S., Enishi, E., Fukuoka, H., and Yasukawa, M. (2000) *Inorg. Chem.*, **39**, 56.

93 Viennois, R., Toulemonde, P., Paulsen, C., and San Miguel, A. (2005) *J. Phys.: Condens. Matter*, **17**, L311.

94 Rachi, T., Yoshino, H., Kumashiro, R., Kitajima, M., Kobayashi, K., Yokogawa, K., Murata, K., Kimura, N., Aoki, H., Fukuoka, H., Yamanaka, S., Shimotani, H., Takenobu, T., Iwasa, Y., Sasaki, T., Kobayashi, N., Miyazaki, Y., Saito, K., Guo, F., Kobayashi, K., Osaka, K., Kato, K., Takata, M., and Tanigaki, K. (2005) *Phys. Rev. B*, **72**, 144504.

95 Kasper, J.S., Hagenmuller, P., Pouchard, M., and Cros, C. (1965) *Science*, **150**, 1713.

96 Cros, C., Pouchard, M., and Hagenmuller, P. (1970) *J. Solid State Chem.*, **2**, 570.

97 Connétable, D., Timoshevskii, V., Masenelli, B., Beille, J., Marcus, J., Barbara, B., Saitta, A.M., and Rignanese, G.-M. (2003) *Phys. Rev. Lett.*, **91**, 247001.

98 Tanigaki, K., Shimizu, T., Itoh, K.M., Teraokai, J., Moritomo, Y., and Yamanaka, S. (2003) *Nat. Mater.*, **2**, 653.

99 Tse, J.S., Iitaka, T., Kune, T., Shimizu, H., Parlinski, K., Fukuoka, H., and Yamana, S. (2005) *Phys. Rev. B*, **72**, 155441.

100 Dimitrov, D.A., Louca, D., and Roder, H. (1999) *Phys. Rev. B*, **60**, 6204.

101 Goodwin, A.L., Tucker, M.G., Dove, M.Y., and Keen, D.A. (2004) *Phys. Rev. Lett.*, **93**, 075502.

102 Fischer, I., Bosak, A., and Krisch, M. (2009) Phys. Rev. B, **79**, 134302.

103 Hannon, J.B., Mele, E.J., and Plummer, E.W. (1996) *Phys. Rev. B*, **53**, 2090.

104 Kwasniok, F. (1995) *Surf. Sci.*, **329**, 90.

105 Stedman, R., Amilius, Z., Pauli, R., and Sundin, F. (1976) *Met. Phys.*, **6**, 157.

5
Heat Capacity of Solids

Toshihide Tsuji

5.1
Introduction

Each atom in solids oscillates about its equilibrium position over a wide range of frequencies from zero up to a maximum value, and the conduction electrons in metals are freely mobile in solids. The oscillation of atoms and movement of electrons contribute to the internal energy of solids. When we heat a sample, solids absorb heat, and some of the phonons and electrons are excited thermally, so the internal energy of solids is expected to increase. The increase in internal energy due to lattice vibration of atoms and kinetic energy of free electrons in solids can attribute to the increase in heat capacity.

Heat capacity of solids is usually measured at constant pressure and defined as

$$C_p = (\mathrm{d}H/\mathrm{d}T)_p, \tag{5.1}$$

where H is the enthalpy, T is the absolute temperature, and P is the pressure. Heat capacity at constant pressure is one of the indispensable thermodynamic quantities to obtain the free energy function of solids. By using the thermodynamic quantities at the reference temperature θ, the free energy function (fef) of solids at temperature T is given as follows:

$$\begin{aligned}(\mathrm{fef})_{T(\theta)} &= (G_T^0 - H_\theta^0)/T \\ &= \left(\int_\theta^T C_p \mathrm{d}T\right)/T - \int_\theta^T C_p \mathrm{d}\ln T,\end{aligned} \tag{5.2}$$

where G_T^0 is the standard Gibbs energy at the absolute temperature T and H_θ^0 is the standard enthalpy at the reference temperature θ. From heat capacity data as a function of temperature in the temperature range from θ to T, the free energy function of solids can be determined by using Equation 5.2.

Equilibrium constants $K(T)$ of chemical reaction at temperature T are calculated from the standard Gibbs energy change of the reaction as follows:

$$\begin{aligned}\ln K(T) &= -\Delta G_T^0/(RT) \\ &= -\Delta H_\theta^0/(RT) - \Delta(\mathrm{fef})_{T(\theta)}/R,\end{aligned} \tag{5.3}$$

Thermodynamic Properties of Solids: Experiment and Modeling
Edited by S. L. Chaplot, R. Mittal, and N. Choudhury

ISBN: 928-3-527-40812-2

where ΔH_θ^0 is the standard enthalpy change at the reference temperature θ and R is the gas constant.

Thermodynamic equilibrium state is easily achieved at high temperatures, so the equilibrium constant of the chemical reaction, $K(T)$, can be calculated from Equation 5.3, and the precision and accuracy of thermodynamic quantities of the substances related to the chemical reaction depend on those of heat capacity data in the temperature range from θ to T and the standard enthalpies at the reference temperature.

In this chapter, the principles and experimental methods of calorimetry to measure heat capacity of solids will be described first. Then, the heat capacity at constant pressure (C_p) determined experimentally can be related to that at constant volume (C_v) calculated theoretically through the thermodynamic relation. The total heat capacity of C_v is represented by the sum of each contributed heat capacity, a theoretical calculation of which will be explained in some detail subsequently. Then, normal heat capacity data without phase transition will be compared with the experimental values. Finally, the phase transition mechanism of the second order–disorder phase transition will be discussed.

5.2 Principles and Experimental Methods of Calorimetry

Calorimetry is the measurement of the heat absorbed or generated in a solid under study, when the solid undergoes the physical and chemical changes from a well-defined initial state to a well-defined final state. The change of state in question may result from the physical changes such as melting, vaporization, and sublimation or its reverse processes, or chemical changes such as chemical reaction, dissolution, adsorption, or dilution. It may also result from the changes in temperature, pressure, magnetic field, and electric field. In the adiabatic calorimetry, the temperature increment (ΔT) rather than the heat is measured against a fixed amount of input energy (ΔE), so the heat capacity is obtained from the ratio of ΔE to ΔT. On the other hand, the enthalpy is measured for the temperature jump calorimetry. Heat capacity may then be calculated from the measured enthalpy as a function of temperature by Equation 5.1. An excellent textbook is available for the calorimetry and thermal analysis [1]. Principles and experimental methods of typical calorimetry will be briefly described in the following sections.

5.2.1 Adiabatic Heat Capacity Calorimetry

The absolute value of heat capacity is measured most accurately by an adiabatic calorimeter for the temperature region from liquid helium temperature to room temperature (low-temperature adiabatic calorimetry) and that for the region from room temperature up to about 1030 K (high-temperature adiabatic calorimetry).

Under adiabatic conditions, the heat capacity C_p is obtained as $C_p = \Delta E/\Delta T$ from the measurements of input energy ΔE to the sample vessel (or a sample and its vessel) and the resulting temperature increment ΔT. Before the energy input, the sample is in a thermodynamic equilibrium state at the initial temperature T_i. After the input energy of ΔE, the new equilibrium state at the final temperature T_f is achieved. As a small heat leak leads to a temperature drift of the calorimeter, the values of T_i and T_f are determined by extrapolating to the midpoint of the input energy, assuming adiabatic conditions during the energy input period. Thus, the heat capacity at the mean temperature $T_m = (T_i + T_f)/2$ is obtained as the average heat capacity between T_i and T_f. However, the curvature correction of the heat capacity should be made for a sharp anomaly due to a phase transition.

A typical cryostat of adiabatic heat capacity calorimeter in the temperature range from liquid helium temperature to room temperature (low-temperature adiabatic calorimetry) is shown in Figure 5.1 [1, 2]. The adiabatic condition is achieved in high vacuum of about 10^{-4} Pa, when the calorimeter vessel (F) is kept in the same temperature as the adiabatic shields (H). Liquid nitrogen and/or liquid helium are used as the refrigerant, depending on the temperature range involved. In order to cool the calorimeter, either a small amount of helium gas (about 1 Pa) is put into the vacuum space or mechanical heat switch is used. After cooling the calorimeter, either the helium gas is evacuated or the mechanical heat switch is turned off. When the adiabatic condition is achieved under high vacuum, the heat capacity measurement can be carried out. The heat capacity of a sample is calculated by subtracting the heat capacity of the empty calorimeter vessel from the measured total heat capacity of both sample and vessel.

The measuring principle of adiabatic calorimetry in the temperature range from room temperature up to about 1030 K (high-temperature adiabatic calorimetry) is the same as that of low-temperature adiabatic calorimetry, as described above. On constructing a calorimeter workable at high temperatures, one encounters a number of technical problems in calorimeter design and materials selection. The main problem intrinsic to high-temperature calorimeters is the heat exchange by radiation between the sample vessel and adiabatic shield especially at higher temperatures, because heat flow rate coefficient by radiation increases in proportion to T^3. The problem of heat exchange due to radiation is efficiently reduced by radiation shields. Other problems are heat leak due to gas and solid conduction through electric leads, thermocouple, heaters, and so on. Temperature measurement, the reaction between sample and its vessel, degradation of thermocouple as well as shielding and insulating materials, and so on should also be taken into consideration.

Figure 5.2 shows a high-temperature adiabatic calorimeter [1, 3]. The sample container of vitreous quartz is cylindrical having a central well for a resistance thermometer and Kanthal D heater. The container (a) fits into a silver calorimeter with removable bottom and end covers. This assembly is surrounded by top, side, and bottom shielded bodies (b) of silver. The temperature differences between the calorimeter and shields (d) are automatically controlled by shield heaters to maintain adiabatic conditions during input and drift periods. The calorimeter and shield systems are surrounded by guard heater silver bodies (c) and placed in a vertical tube

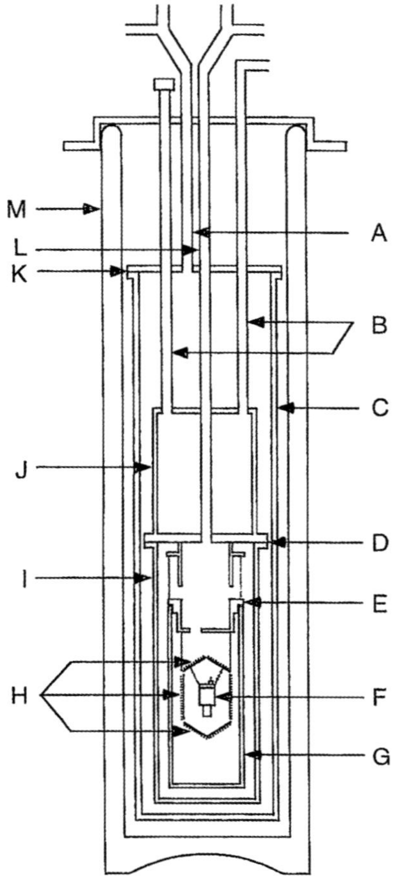

Figure 5.1 Cryostat of an adiabatic calorimeter [1, 2]. A, outer jacket evacuation tube; B, refrigerant transfer tubes; C, outer jacket; D, indium seal; E, thermal sink; F, calorimeter vessel; G, outer shield; H, adiabatic shields; I, inner jacket; J, refrigerant can; K, indium seal; L, inner jacket evacuation tube; M, liquid nitrogen Dewar vessel.

furnace. The energy inputs to the calorimeter are determined by measuring current through the Kanthal heater, the potential drop across it, and the input time. Heat capacity measurement on α-Al_2O_3 shows a standard deviation from the mean of 0.15% and a standard deviation from the NBS data of 0.25% over the temperature range from 300 to 1031 K.

5.2.2 Adiabatic Scanning Calorimetry

A double adiabatic scanning calorimeter (ASC) is shown in Figure 5.3 [1, 4]. The calorimeter is designed for the measurement of absolute enthalpy change of phase

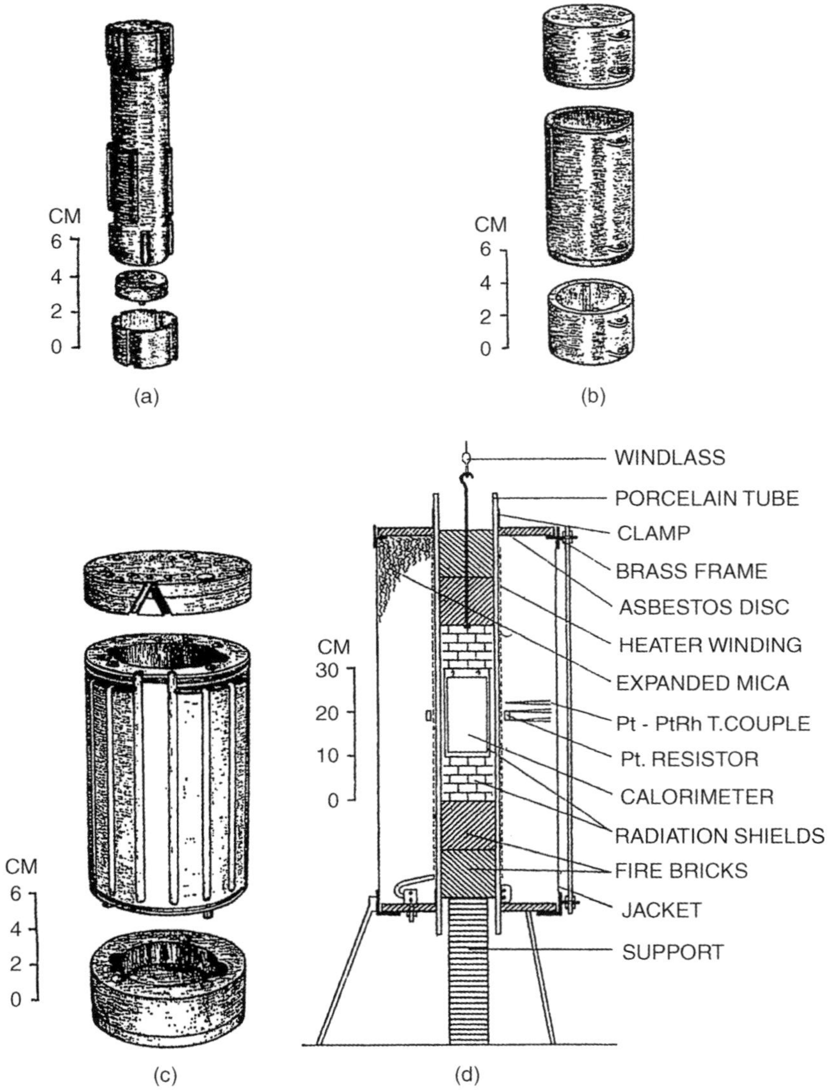

Figure 5.2 Adiabatic calorimeter at high temperatures [1, 3]. (a) Calorimeter proper with removal bottom and end covers. (b) Top, side, and bottom shielded bodies. (c) Top, side, and bottom guarded bodies. (d) Calorimeter thermostat.

transition at high temperatures, using a relatively small amount of sample (3–10 g). The necessary conditions for absolute heat capacity measurement are that the heat leak is minimal and constant for both measurements of vessel with and without sample.

The measuring principle of ASC is as follows. The temperature difference between the double platinum sphere and the first adiabatic shield and that between the first

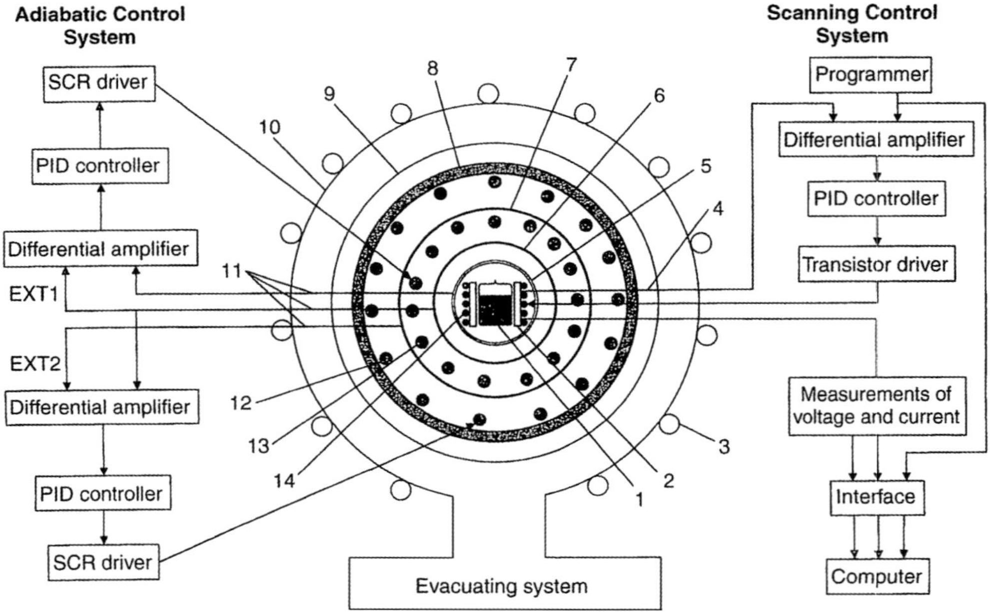

Figure 5.3 Double adiabatic scanning calorimeter at high temperatures [1, 4]. 1, Sample vessel of quartz; 2, alumina bobbin; 3, water cooling pipe; 4, thermocouple for programmer; 5, double platinum sphere for homogeneity of heat; 6, platinum sphere for the first adiabatic shield; 7, platinum sphere for the second adiabatic shield; 8, alumina sphere; 9, stainless steel sphere for thermal shield; 10, outermost sphere vessel; 11, thermocouple for adiabatic control; 12, second external heater; 13, first external heater; 14, internal heater. (Reprinted with permission from R.C. Chisholm and J.W. Stout, J. Chen. Phys. Vol. 36, 972, 1962. Copyright 1962, American Institute of Physics.)

and second shields are reduced to zero by controlling the power of the first and second external heaters, independently, using a PID thyristor-type controller. The sample is heated at a constant heating rate from 0.25 to 8 K min^{-1} by an electric programmer. The difference between the voltage of programmer and the electromotive force of the thermocouple in the sample vessel is brought to zero by controlling the power of an internal heater, using a PID transistor drive-type controller. The power of the internal heater is read by a wattmeter and recorder continuously. Heat capacity of a sample is calculated as follows:

$$C_p = M_{\mathrm{m}}^{-1} W_{\mathrm{e}} (\mathrm{d}T/\mathrm{d}t)^{-1},$$

where M_{m} is the number of moles of a sample, W_{e} is the energy input to the calorimeter, and $\mathrm{d}T/\mathrm{d}t$ is the heating rate.

ASC has been applied to measure the heat capacity of powder materials from 273 to 773 K. The heat capacity of α-Al_2O_3 standard sample is measured with an accuracy of 0.6%. The absolute enthalpy of transition is determined precisely, while the transition temperature is obtained by the extrapolation of the heating rate to zero.

5.2.3
Direct Pulse-Heating Calorimetry

The direct pulse-heating method for measuring heat capacity is very attractive, particularly for materials that are electrical conductors, because heat capacity, electrical resistivity, and hemispherical total emittance are measured simultaneously. The method involves rapid and continuous resistive heating and cooling of the sample by a single subsecond pulse at high temperatures (above 1900 K) with millisecond resolution.

In a high-temperature dynamic experiment of millisecond resolution, the major source of power loss is thermal radiation. Heat capacity of a sample may be expressed from the energy balance during heating as

$$C_p = [VI - \varepsilon\sigma A_s(T_s^4 - T_r^4)]/[n(dT/dt)_h], \tag{5.4}$$

where V is the terminal voltage between two knife edges, I is the current through the sample, ε is the spherical total emittance, σ is the Stefan–Boltzmann constant ($5.6697 \times 10^{-8}\,\mathrm{W\,m^{-2}\,K^{-4}}$), A_s is the effective surface area, T_s is the sample temperature and T_r is the room temperature, n is the effective amount of the sample, and $(dT/dt)_h$ is the heating rate. On the other hand, the energy balance during cooling can be written as

$$-C_p n(dT/dt)_c = \varepsilon\sigma A_s(T_s^4 - T_r^4], \tag{5.5}$$

where $(dT/dt)_c$ is the cooling rate. From combinations of Equations 5.4 and 5.5, ε is expressed as

$$\varepsilon = [VI(dT/dt)_c]/\{[\sigma A_s(T_s^4 - T_r^4)][(dT/dt)_c - (dT/dt)_h]\}. \tag{5.6}$$

Since ε is calculated from the experimental data on the right-hand side of Equation 5.6, heat capacity is calculated from Equation 5.4 or 5.5 by using the ε value.

Specific electrical resistivity ϱ is calculated with the aid of the following equation:

$$\varrho = R_s A_c / l, \tag{5.7}$$

where R_s is the resistance of effective sample, A_c is the effective cross-sectional area, and l is the effective length.

Schematic diagram of the arrangement of the sample, clamps, and potential probes is shown in Figure 5.4 [5]. The sample is heated from room temperature to close to its melting point in less than 1 s. The temperature of the rapidly heating sample is measured at blackbody radiation hole by means of a high-speed photoelectric pyrometer as a function of the measuring time. The current flow through the sample and the voltage drop between two knife edges consisting of the same materials as the measuring sample are obtained simultaneously.

5.2.4
Laser-Flash Calorimetry

The principle of the laser-flash calorimetry is as follows [1, 6]. After the sample is heated by a pulse from a ruby laser, the energy absorbed by the sample (E) and its

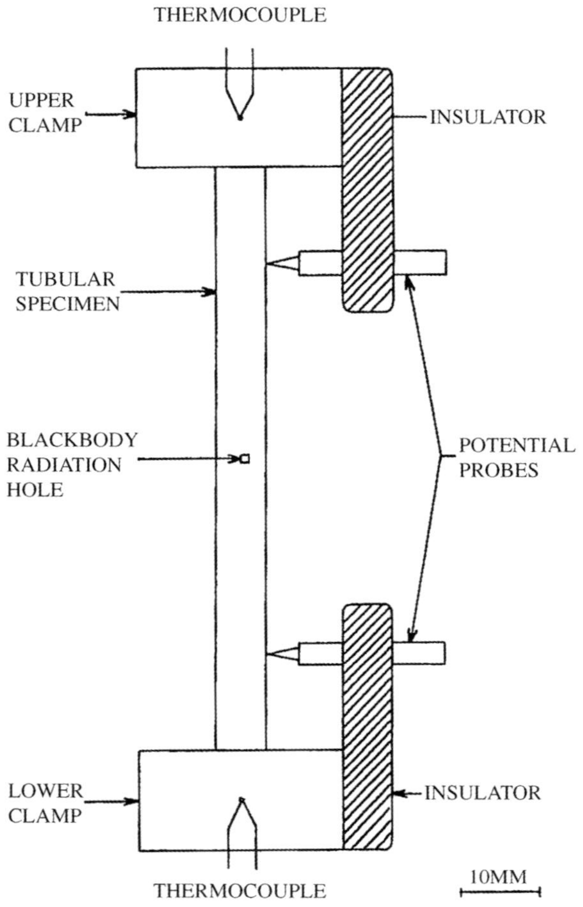

Figure 5.4 Schematic diagram of direct pulse-heating calorimeter. The arrangement of the sample, clamps, and potential proves is shown [5]. (Contribution of the National Institute of Standards and Technology).

resulting temperature increment (ΔT) are precisely measured. The sample is in the form of a small disk pellet having a diameter of 8–12 mm and a thickness of 0.5–5 mm. As shown in Figure 5.5 [1, 6], an absorbing disk and a thin glassy-carbon plate are attached on the front surface of the sample. For precise measurements of the sample temperature and its temperature increment, a thin thermocouple, attached to the back surface of the sample with silver paste, is used. The maximum temperature increment of a sample, $\Delta T_{\max}$, is determined after correction of the heat loss from the sample during the measurement. The amount of energy absorbed by the sample is measured by a Si-photoelectric cell and is corrected by measuring a standard material of known heat capacity of α-Al_2O_3 crystal. The heat capacity of the sample is then obtained by

$$C_p = \{(E/\Delta T_{\max}) - C)\}/M, \tag{5.8}$$

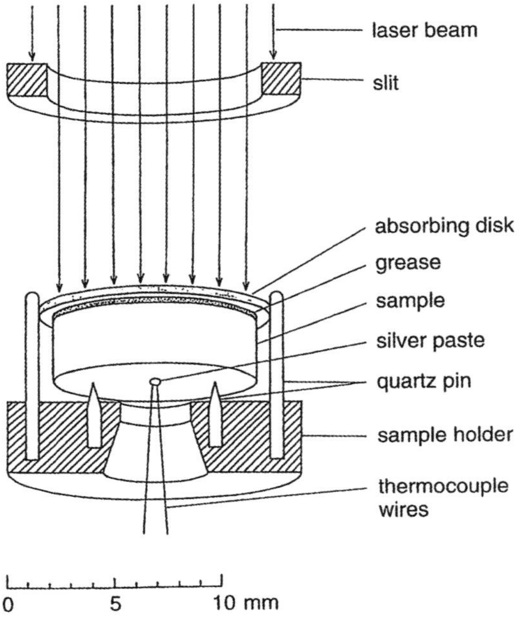

Figure 5.5 Sample holding assembly of laser-flash calorimeter [1, 6].

where M is the number of moles of a sample and C is the total heat capacity of the absorbing disk and the adhesive materials.

The experimental procedures for the heat capacity measurement are carried out in two steps. The first step is to determine the absolute heat capacity of the sample at room temperature by using a standard material of α-Al_2O_3 crystal. Next, the temperature dependence of the heat capacity is determined relative to the absolute value measured at room temperature. The estimated inaccuracies are within $\pm 0.8\%$ at 300 K and $\pm 2\%$ at 1100 K.

5.2.5
Temperature Jump Calorimetry

There exist two methods of temperature jump calorimetry: a sample at T is dropped into a calorimeter assembly at room temperature (room-temperature jump calorimetry) or, conversely, the sample at room temperature is dropped into the assembly at high temperatures (high-temperature jump calorimetry). In both methods of temperature jump calorimetry, the enthalpy change of the sample is measured in terms of the amount of heat absorbed by the calorimeter assembly in changing from an initial temperature to a final temperature. This measured value is then corrected to 298.15 K, so the tabulated enthalpy values of the sample are referred to 298.15 K, that is, $H_T - H_{298.15}$, where H_T and $H_{298.15}$ are the enthalpies at T and 298.15 K, respectively. When the reaction or transformation of the sample does not occur in the calorimeter, the heat capacity at constant pressure as a function of temperature may then be derived either from the smoothed enthalpy data obtained graphically or from the following equation:

$$C_p = \mathrm{d}(H_T - H_{298.15})_p/\mathrm{d}T. \tag{5.9}$$

On the other hand, enthalpy change by phase transformation can also be determined by a temperature jump calorimeter from the difference in the measured enthalpy change before and after the phase transformation.

A room-temperature jump calorimeter designed for measuring enthalpy of metals and alloys in the temperature range between 700 and 1800 K is schematically shown in Figure 5.6 [1, 7]. It consists of a heating furnace, a device for dropping a sample, and a calorimeter assembly. The calorimeter assembly, which consists of a Dewer vessel and a copper tube with 27 copper fins, contains distilled water. The temperature change of water in the calorimeter is measured with a precision of 0.001 K using a platinum resistance thermometer. The water equivalent of the calorimeter is determined to be 17.7 kJ K^{-1} using sapphire as a standard material. The temperature of the sample in the furnace is measured with a Pt/Pt-13%Rh thermocouple situated beside the platinum holder.

5.3
Thermodynamic Relation Between C_p and C_v

The heat capacity of solids is ordinarily measured at constant pressure C_p, as described in the previous section. On the other hand, the heat capacity at constant volume C_v is calculated theoretically if the interatomic distance is kept constant, independent of the temperature changes, and is defined as $C_v = (\mathrm{d}E/\mathrm{d}T)_v$, where E is the internal energy of solids and V is the volume. The dilatometric term C_d is the difference between C_p and C_v and is derived from the classical thermodynamic relation

$$C_\mathrm{d} = C_p - C_v = -T\,(\partial V/\partial T)_p^2/(\partial V/\partial p)_T. \tag{5.10}$$

From the definition of the isothermal compressibility $\varkappa_T = -(\partial V/\partial p)_T/V$ and the coefficient of the volume thermal expansion $\beta_p = (\partial V/\partial T)_p/V$, Equation 5.10 may be expressed as

$$C_\mathrm{d} = C_p - C_v = (V\beta_p^2/\varkappa_T)T. \tag{5.11}$$

Since the Grüneisen constant γ_e is defined by $\gamma_\mathrm{e} = \beta_p V/(\varkappa_T C_v)$, Equation 5.11 is rewritten as

$$C_p = C_v + C_v\gamma_\mathrm{e}\beta_p T. \tag{5.12}$$

For isotropic substances, the coefficient of the volume thermal expansion may be expressed in terms of the coefficient of linear expansion α_p as

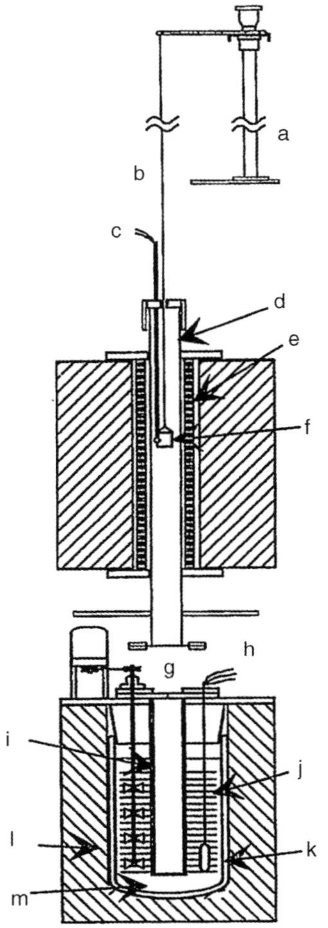

Figure 5.6 Construction of a temperature jump calorimeter [1, 7]. a, Dropping mechanism; b, Pt-Rh wire; c, Pt/Pt-Rh thermocouple; d, alumina tube; e, molybdenum silicide heater; f, Pt crucible; g, shutter; h, Pt resistance thermometer; i, copper tube; j, copper fins; k, Dewar vessel; l, insulating material; m, distilled water.

$$\beta_p = (\partial V/\partial T)_p/V = 3\{(\partial L/\partial T)_p/L\} = 3\alpha_p. \tag{5.13}$$

Hence, from Equation 5.11,

$$C_p - C_v = (9V\alpha_p^2/\varkappa_T)T. \tag{5.14}$$

Dividing Equation 5.12 by C_v, the following equation can be derived:

$$C_p/C_v = 1 + \gamma_e \beta_p T. \tag{5.15}$$

If we assume $\gamma_e = 2$ and $\beta_p = 10 \times 10^{-6}$ (K^{-1}) at $T = 300$ K, the second term in Equation 5.15 is estimated to be 6×10^{-3} (0.6%). It means that the heat capacity at constant volume may be assumed to be approximately equal to the heat capacity at constant pressure at room temperature. C_v may be equal to C_p at cryogenic temperatures, because β_p is zero at the absolute zero temperature. However, $C_p > C_v$ holds at higher temperature, because values of both $\gamma_e\beta_p$ and T are positive.

For a lattice heat capacity contributed to lattice vibrations, an approximation called the Nernst–Lindemann formula is often used:

$$C_d = C_p - C_v = AC_p^2 T,$$

where A is a system-dependent constant. The values of C_d at the temperature T are estimated from the C_p values measured experimentally.

5.4 Data Analysis of Heat Capacity at Constant Volume (C_v)

When a sample is heated under adiabatic conditions, the increment of total internal energy (ΔE_v) of the solid is attributed as the sum of each contributed internal energy (ΔE_i) from the conservation of energy as follows:

$$\Delta E_v = \Sigma \Delta E_i = \Delta E_l + \Delta E_{l,a} + \Delta E_{e,c} + \Delta E_{e,sh} + \Delta E_m + \Delta E_f + \cdots. \tag{5.16}$$

As the derivative of the internal energy by the temperature is the heat capacity at constant volume, the heat capacity at constant volume C_v can be expressed as

$$C_v = C_l + C_{l,a} + C_{e,c} + C_{e,sh} + C_m + C_f + \cdots, \tag{5.17}$$

where C_l is the lattice heat capacity, $C_{l,a}$ is the heat capacity due to anharmonic lattice vibration, $C_{e,c}$ is the electronic heat capacity due to conduction electron, $C_{e,sh}$ is the electronic Schottky-type heat capacity due to electronic excitation to higher energy levels, C_m is the magnon heat capacity due to the excitation of the spin system in magnetically ordered substances, and C_f is the heat capacity due to the formation of vacancies. Main contribution of heat capacity at constant volume is the lattice heat capacity (C_l) due to lattice vibrations and will be discussed in the next section.

5.4.1 Lattice Heat Capacity (C_l)

5.4.1.1 Classical Theory of Lattice Heat Capacity

Dulong and Petit observed from the heat capacity measurement at constant pressure that the heat capacity of many elemental solids is about 6 cal K^{-1} mol^{-1} at room temperature. The theoretical justification of the experimental law of Dulong and Petit was demonstrated by the equipartition of energy theorem by Boltzmann as follows.

The total energy of a linear harmonic oscillator, E, consists of kinetic and potential energies, that is,

$$E = (1/2)mv^2 + (1/2)Kx^2, \tag{5.18}$$

where m is the mass, v is the velocity, K is the force constant, and x is the distance from the equilibrium position. From the theorem of equipartition of energy, the energy of a particle per degree of freedom in equilibrium is $(k_BT/2)$, where k_B is the Boltzmann constant. A three-dimensional oscillator that has three degrees of freedom for kinetic energy as well as potential energy will therefore have an internal energy of $3k_BT$ at thermal equilibrium. One mole of an elemental solid has N_A atoms, where N_A is the Avogadro constant, and thus its internal energy is $3N_Ak_BT$. The lattice heat capacity at constant volume, C_l, is obtained by differentiating the internal energy with respect to temperature at constant volume, that is,

$$(dE/dT)_v = C_l = 3N_Ak_B.$$

The product of Avogadro constant and Boltzmann constant is equal to the gas constant R. Therefore,

$$C_l = 3R = 5.96\ \text{cal K}^{-1}\ \text{mol}^{-1} = 24.94\ \text{J K}^{-1}\ \text{mol}^{-1}.$$

Hence, the law of Dulong and Petit, where heat capacity of the elemental solids is about 6 cal K^{-1} mol^{-1} at room temperature, can be explained on the basis of classical statistical mechanics.

For chemical compounds, Kopp and Neumann extended the experimental law of Dulong and Petit and suggested that $C_v = 3nR$, where n is the number of atoms per formula unit (or "molecular").

Table 5.1 shows the molar heat capacity at constant pressure of solids at room temperature. The C_p values of most of metallic elements are in nearly good agreement with $3R = 24.94\ \text{J K}^{-1}\text{mol}^{-1}$, although the atomic heat capacities of C (diamond) and Si are considerably lower than the values predicted by the experimental law of Dulong and Petit and its explanation will be discussed below. This means that the main contribution of heat capacity at constant pressure is the lattice heat capacity. It is also seen in the table that the experimental values at room temperature of compounds except SiC and SiO_2 (quartz) agree well with $6R$ ($= 49.88\ \text{J K}^{-1}\text{mol}^{-1}$) and $9R$ ($= 74.82\ \text{J K}^{-1}\text{mol}^{-1}$) for two and three atoms per formula unit, respectively.

5.4.1.2 **Einstein's Model of Lattice Heat Capacity**

Einstein proposed a simple harmonic vibration model of crystal lattice to explain the increase in lattice heat capacity at constant volume from zero at the absolute

Table 5.1 Molar heat capacity at constant pressure of solids at 298.15 K.

Solids	C_p (J K^{-1} mol^{-1})	Solids	C_p (J K^{-1} mol^{-1})
Ag	25.4	NaCl	49.69
Al	24.35	SiC	26.65
C (diamond)	6.12	FeS	50.54
Cu	24.43	SiO_2 (quartz)	44.43
Fe	25.0	BaF_2	71.21
Si	20.0	$CaCl_2$	72.61

temperature to $3R$ value per mole at high-temperature limit. His physical model considers the lattice vibrations of N_A atoms as a set of $3N_A$ independent harmonic oscillators in one dimension, having the same frequency ω. A harmonic oscillator does not have a continuous energy spectrum in a classical model, but can accept energy values equal to an integer times $\hbar\omega$, where ω is the frequency of oscillators and $\hbar$ is equal to $h/2\pi$ (h is the Planck constant). The possible nth energy level, ε_n, of an oscillator may be given by

$$\varepsilon_n = \{n + (1/2)\}\hbar\omega, \quad n = 0, 1, 2, 3, \ldots. \tag{5.19}$$

The probability existing in the nth energy level, P_n, is represented by the Boltzmann factor as

$$\begin{aligned} P_n &= \exp\{-\varepsilon_n/(k_B T)\} \Big/ \sum_{n=0}^{\infty} \exp\{-\varepsilon_n/(k_B T)\} \\ &= \exp\{-n\hbar\omega/(k_B T)\} \Big/ \sum_{n=0}^{\infty} \exp\{-n\hbar\omega/(k_B T)\}. \end{aligned}$$

The average energy of an oscillator at temperature T is thus expressed as

$$\begin{aligned} \langle\varepsilon\rangle &= \sum_{n=0}^{\infty} P_n \varepsilon_n \\ &= (\hbar\omega/2) + \left\{(\hbar\omega)\left(\sum_{n=0}^{\infty} n e^{-nx}\right)\right\} \Big/ \left(\sum_{n=0}^{\infty} e^{-nx}\right), \end{aligned} \tag{5.20}$$

where $x = \hbar\omega/(k_B T)$.

Since the second term in Equation 5.20 is calculated by mathematical technique as

$$\begin{aligned} \left(\sum_{n=0}^{\infty} n e^{-nx}\right) \Big/ \left(\sum_{n=0}^{\infty} e^{-nx}\right) &= -\left\{\mathrm{d}\ln\left(\sum_{n=0}^{\infty} e^{-nx}\right) \Big/ \mathrm{d}x\right\} \\ &= -\mathrm{d}\ln\{1/(1-e^{-x})\}/\mathrm{d}x = 1/(e^{x}-1), \end{aligned}$$

the average energy of an oscillator is thus given by

$$\langle\varepsilon\rangle = (\hbar\omega/2) + (\hbar\omega)/[\exp\{\hbar\omega/(k_B T)\} - 1]. \tag{5.21}$$

The vibrational energy of an elemental solid, $\langle E\rangle$, having $3N_A$ independent harmonic oscillators is expressed as

$$\langle E\rangle = 3N_A\langle\varepsilon\rangle = 3N_A(\hbar\omega/2) + 3N_A(\hbar\omega)/[\exp\{\hbar\omega/(k_B T)\} - 1]. \tag{5.22}$$

The heat capacity of an elemental solid is thus obtained by differentiating Equation 5.22 with respect to temperature:

$$C_l = (\partial\langle E\rangle/\partial T)_v = 3N_A k_B\{\hbar\omega/(k_B T)\}^2 \exp\{\hbar\omega/(k_B T)\}/[\exp\{\hbar\omega/(k_B T)\} - 1]^2. \tag{5.23}$$

For convenience, the characteristic Einstein temperature defined by $\theta_E = \hbar\omega/k_B$ may be introduced in Equation 5.23 to obtain

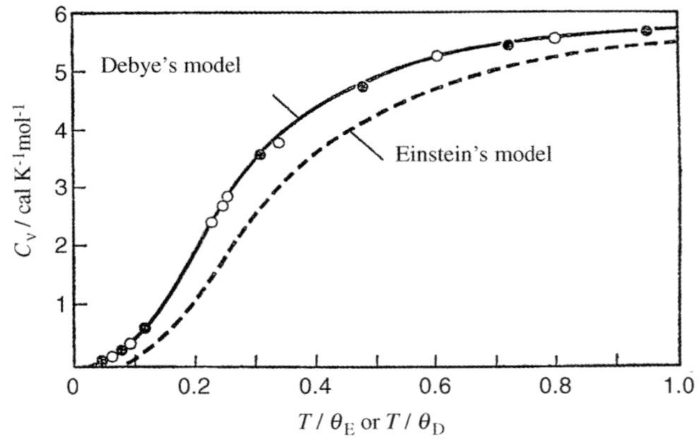

Figure 5.7 Theoretical heat capacity data calculated by Einstein's and Debye's models are compared with the experimental values of Al (○) and Cu (●) metals.

$$C_l = \{3R(\theta_E/T)^2\exp(\theta_E/T)\}/\{\exp(\theta_E/T)-1\}^2. \tag{5.24}$$

In the high-temperature limit, $T \gg \theta_E$, Equation 5.24 upon expansion in power series becomes

$$C_l = 3R = 5.96\ \text{cal}\ K^{-1}\ \text{mol}^{-1} = 24.94\ \text{J}\ K^{-1}\ \text{mol}^{-1},$$

where the result of Einstein's theory agrees with that of the classical Boltzmann's theory.

In the low-temperature region, $T \ll \theta_E$, Equation 5.24 may be written approximately as

$$C_l \approx 3R(\theta_E/T)^2\exp(-\theta_E/T). \tag{5.25}$$

According to Equation 5.25, the low-temperature heat capacity of solids should approach zero exponentially. As seen in Figure 5.7, the experimental heat capacities of Al and Cu approach to zero more slowly than the theoretical values predicted by Einstein's model. The reason for the discrepancy between Einstein's theoretical prediction and the experimental results may be explained on the basis of the assumption made in the theory that each atom in a solid vibrates independently of the other atoms with the same frequency.

5.4.1.3 **Debye's Model of Lattice Heat Capacity**

Debye's model of lattice heat capacity assumes the continuum model for all possible vibrational modes of the solid, where the wavelength is larger compared with the interatomic distances, and a solid may appear like a continuous elastic medium. Debye has also given a limit to the total number of vibrational modes equal to $3N_A$, and the frequency spectrum to an ideal continuum is cut off in order to comply with a total of $3N_A$ modes. This procedure thus provides a maximum frequency ω_D (Debye frequency) that is common to both longitudinal and transverse modes.

Since each vibrational mode associates with a harmonic oscillator of the same frequency, the internal energy of an elementary solid is the sum of vibration modes expressed in Equation 5.21 and is represented as

$$\langle E\rangle = \sum_{i=1}^{3}\sum_{q}\{\hbar\omega_i(q)\}/[\exp\{\hbar\omega_i(q)/(k_B T)\}-1], \tag{5.26}$$

where the first sum is that of one longitudinal and two transverse waves and the second sum is that of wave number vector, q, from 0 to n. The zero-point energy in Equation 5.21 is neglected for Equation 5.26, because this value is independent of temperature and has no effect on the final result. Since q is a continuous parameter and density of state is expressed as $V/(2\pi)^3$, Equation 5.26 may be written as

$$\langle E\rangle = \{V/(2\pi)^3\}\int_{i=1}^{3}\iint dq_x\, dq_y\, dq_z\{\hbar\omega_i(q)\}/[\exp\{\hbar\omega_i(q)/(k_B T)\}-1], \tag{5.27}$$

where V is the volume.

The dispersion relation between $\omega(q)$ and q may be written as

$$\omega(q) = vq.$$

The total number of vibrational modes is equal to $3N_A$, and the frequency spectrum to an ideal continuum is cut off in order to comply with a total of $3N_A$ modes as follows:

$$3(4\pi/3)q_m^3\{V/(2\pi)^3\} = 3N_A \quad \text{or} \quad q_m = (6\pi^2 N_A/V)^{1/3}$$

$$\iiiint dq_x\, dq_y\, dq_z = 4\pi\int_0^{q_m} q^2 dq, \tag{5.28}$$

where q_m is the maximum of wave number vector.

Thus, Equation 5.27 is expressed as

$$\begin{aligned}\langle E\rangle &= \{3V/(2\pi^2)\}\int_0^{q_m} dq\cdot q^2(\hbar vq)/[\exp\{\hbar vq/(k_B T)\}-1]\\ &= \{3V/(2\pi^2)\}\{(k_B T)^4/(\hbar v)^3\}\int_0^{x_m} x^3 dx/(e^x-1),\end{aligned} \tag{5.29}$$

where $x = \hbar\omega/(k_B T)$ and $x_m = \hbar\omega_D/(k_B T)$.

From Equation 5.28 and definition of the Debye temperature $\theta_D = \hbar\omega_D/k_B$, Equation 5.29 is written as

$$\langle E\rangle = 9RT(T/\theta_D)^3\int_0^{\theta_D/T}\{x^3 dx/(e^x-1)\}, \tag{5.30}$$

where $x_m = \hbar\omega_D/(k_B T) = \theta_D/T$.

Lattice heat capacity is calculated by differentiating Equation 5.30 with respect to the temperature as

$$C_l = 9R(T/\theta_D)^3\int_0^{\theta_D/T}\{e^x x^4 dx/(e^x-1)^2\} = 3RD(\theta_D/T), \tag{5.31}$$

where $D(\theta_D/T)$ is the tabulated Debye function, $x = \theta_D/T$.

In the high-temperature limit ($T \gg \theta_D$), x_m ($= \theta_D/T$) is small compared with unity for the whole integration range, and Equation 5.31 could easily be integrated to obtain the expression

$$C_l = 3R = 5.96\ \mathrm{cal}\ K^{-1}\ \mathrm{mol}^{-1} = 24.94\ \mathrm{J}\ K^{-1}\ \mathrm{mol}^{-1}.$$

This result explains the law of Dulong and Petit derived from the classical theory.

At very low temperatures, $T \ll \theta_D$, the upper limit of integration in Equation 5.31 may be replaced by infinity, since $\hbar\omega/(k_B T) \to \infty$ as $T \to 0$. It is now possible to integrate Equation 5.31 as follows:

$$\int_0^{\theta_D/T} \{e^x x^4 dx/(e^x - 1)^2\} = 4\pi^4/15.$$

Hence,

$$C_l = (12/5)\pi^4 R(T/\theta_D)^3 \qquad T < (\theta_D/50). \tag{5.32}$$

Debye's theory predicts a cube law dependence of the heat capacity of the elemental solids for temperatures $T < (\theta_D/50)$. As seen in Figure 5.7, the prediction of Debye's theory agrees quite well with the experimental heat capacity data of Al and Cu and improves Einstein's theory definitely.

Lindemann derives the relation between Debye's temperature θ_D and melting point T_m of solids as follows:

$$\theta_D = (3\hbar/x_m)(4\pi/3)^{1/3} k_B^{-1/2} \{T_m/(MV^{2/3})\}^{1/2}, \tag{5.33}$$

where M is the mass of solid, V is the molar volume of solid, and x_m is the atomic displacement of lattice vibration at melting point. Debye temperatures θ_D are plotted as a function of $\{T_m/(MV^{2/3})\}^{1/2}$ for various materials in Figure 5.8 [8]. A good linear relation holds for the same crystal structure materials such as alkali halide, covalent materials having diamond or zinc blende structure, and metals having face-centered cubic structure (FCC) or body-centered cubic structure (BCC).

According to the prediction of Debye's theory, the Debye temperature θ_D of a solid is constant, independent of temperature, but in fact varies with temperature. The deficiency of the Debye theory may be explained on the basis of the approximation made in treating solids as continuous elastic media and of neglecting the discreteness of the atoms.

5.4.1.4 **Anharmonic Term of Lattice Heat Capacity**

In Debye's continuum model, the harmonic lattice vibration of solid is assumed. The anharmonic term of lattice heat capacity, $C_{l,a}$, is usually described as a term proportional to the temperature T, which is proposed by Peierls [9]:

$$C_{l,a} = \mathrm{const.}\ T.$$

Anharmonic term of lattice heat capacity for many metal and UO_2 is analyzed by the above equation.

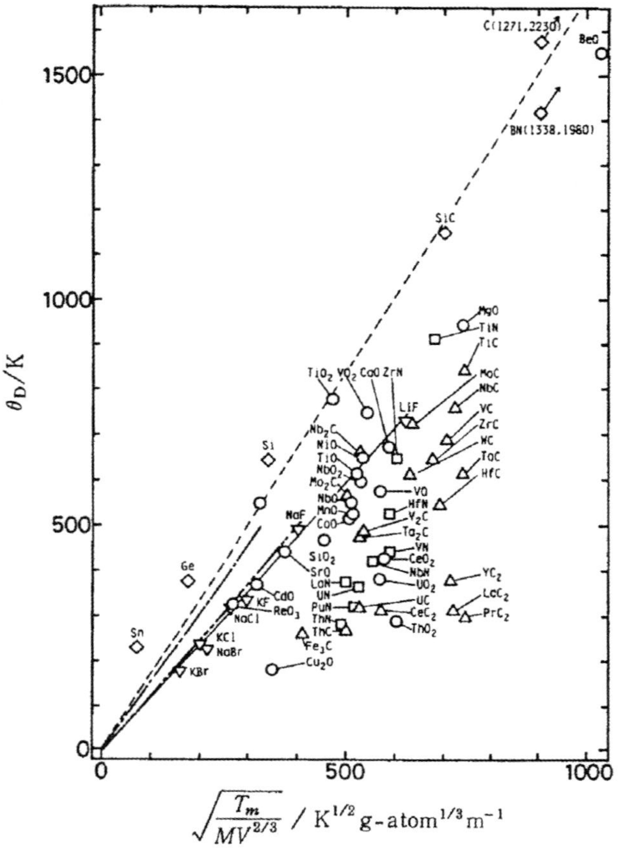

Figure 5.8 Debye temperatures θ_D are plotted as a function of $\{T_m/(MV^{2/3})\}^{1/2}$ for various materials [8]. ○, oxide; △, carbide; □, nitride; –◇–, covalent materials with diamond or zinc blende structure; –▽–, alkali halide, – – –, face-centered cubic metal; ——- - - -——, body-centered cubic metal.

5.4.2
Other Terms Contributed to Heat Capacity at Constant Volume

5.4.2.1 **Electronic Heat Capacity ($C_{e,c}$)**

When heat is supplied to a metal, the energies of both lattice vibrations and free electrons are increased. However, a significant contribution to the heat capacity comes from the free electrons present in the material especially at low temperatures.

As expected from classical theory for the atoms of a monoatomic gas (see Section 5.4.1.1), the electronic contribution to heat capacity should be (3/2) $N_A k_B = (3/2)R$ from a free-electron model for metals in thermal equilibrium with the atoms of the solid, if a metal gives one valence electron and all electrons are freely mobile. But the observed electron contribution at room temperature is usually less than 0.01 of this expected value.

A qualitative solution to the problem of heat capacity for the conduction electron gas is explained as follows. If N_A is the total number of electrons, only a fraction of electrons having the order of $E/E_F = k_BT/E_F$, where $E_F = k_BT_F$ is the Fermi energy and T_F is the Fermi temperature, can be excited thermally at temperature T, because only these electrons lie within an energy range of the order of k_BT near the top of the Fermi–Dirac energy distribution. Since each of these $N_A(T/T_F)$ electrons has a thermal energy of the order of k_BT, the total electronic thermal kinetic energy is of the order of $E_K \approx N_A(T/T_F)k_BT$ and the electronic heat capacity due to conduction electron, $C_{e,c}$, is given by

$$C_{e,c} = (dE_K/dT)_v \approx 2N_Ak_B(T/T_F)$$

and is directly proportional to T, in good agreement with the experimental results. At room temperature, $C_{e,c}$ is smaller than the classical value $(3/2)N_Ak_B = (3/2)R$ by a factor of the order of 0.01 or less, for $T_F \sim 5 \times 10^4$ K.

Using the Fermi–Dirac distribution function and the density of state, which is the number of orbitals per unit energy range, the following expression for the electronic heat capacity due to conduction electrons may be obtained at low temperatures:

$$C_{e,c} = (1/2)\pi^2 N_Ak_B(T/T_F) = \gamma T, \tag{5.34}$$

where γ is the proportionality constant.

5.4.2.2 Schottky-type Heat Capacity ($C_{e,sh}$)

The electronic heat capacity due to electronic excitation from the ground state to higher energy levels, which is usually called Schottky-type heat capacity, $C_{e,sh}$, is described as follows. The partition function Q of the system is generally given as

$$Q = \sum_{i=1}^{n} g_i \exp\{-E_i/(RT)\},$$

where E_i is the energy difference between the ground state and the ith excited state and g_i is the degeneracy of ith excited state. When we put $E_0 = 0$ for the ground state, the average energy $\langle E \rangle$ is expressed as

$$\langle E \rangle = Q^{-1} \sum_{i=1}^{n} g_i E_i \exp\{-E_i/(RT)\}. \tag{5.35}$$

Schottky-type heat capacity is obtained by differentiating the average energy of Equation 5.35 with respect to the temperature:

$$\begin{aligned} C_{e,sh} &= (\partial\langle E \rangle/\partial T)_v \\ &= Q^{-0}R^{-1}T^{-2}\left[Q\sum_{i=1}^{n} g_i E_i^2 \exp\{-E_i/(RT)\} - \left\{\sum_{i=1}^{n} g_i E_i \exp(-E_i/(RT))\right\}^2\right] \\ &= R^{-1}T^{-2}\{\langle E^2 \rangle - (\langle E \rangle)^2\}. \end{aligned}$$

For a two-level system consisting of the ground state and the first excited state, $C_{e,sh}$ may be written as

$$C_{e,sh} = \{E_i^2/(RT^2)\}(g_1/g_0)\exp\{-E_i/(RT)\}/[1+(g_1/g_0)\exp\{-E_i/(RT)\}]^2, \tag{5.36}$$

where E_1 is the energy difference between the ground state and the first excited state.

At very low temperatures, $T \ll E_1/R$, Equation 5.36 may be written as

$$C_{e,sh} = (g_1/g_0)\{E_1^2/(RT^2)\}\exp\{-E_1/(RT)\}. \tag{5.37}$$

On the other hand, at high temperatures, $T \gg E_1/R$, Equation 5.36 becomes

$$C_{e,sh} = g_0 g_1 (g_0+g_1)^{-2}\{E_1^2/(RT^2)\}. \tag{5.38}$$

5.4.2.3 Magnetic Heat Capacity (C_m)

There are two types of materials that exhibit a magnetic contribution to the total heat capacity, namely, the ferromagnetic and the ferrimagnetic materials.

A ferromagnetic material has a spontaneous magnetic moment even in the absence of an external magnetic field and shows a magnetic ordering with parallel alignment of adjacent spins. A ferromagnetic material shows a phase transition from the ordered ferromagnetic phase at low temperatures to the disordered paramagnetic phase at high temperatures above the Curie temperature, T_c, which is defined as the temperature above which magnetization disappears.

On the other hand, ferrimagnetic materials are similar to the ferromagnetic materials, but in the former the adjacent spins are unequal and antiparallel. The Nèel temperature of a ferrimagnetic material is defined as the phase transition temperature from the ordered ferrimagnetic phase at low temperatures to the disordered paramagnetic phase at high temperatures.

For ferri- and ferr-omagnetic materials, the mean internal energy $\langle E\rangle$ is given by the following equation:

$$\langle E\rangle = 4\pi V(2\alpha_f J s_p a^2)\{(k_B T)/(2\alpha_f J s_p a^2)^{5/2}\}\int_0^x x^4 dx/\{\exp(x^2)-1\}, \tag{5.39}$$

where α_f is a constant depending upon crystal structure, a is the lattice constant, J is the quantum mechanical exchange constant, s_p is the magnitude of the spin vector for ferri- and ferr-omagnetic materials, and V is the volume of the material.

At low temperatures, the upper limit for x may be taken equal to infinity and hence the integral may be easily determined. Differentiating Equation 5.39 with respect to temperature gives the magnetic heat capacity $C_{m,f}$,

$$C_{m,f} = d\langle E\rangle/dT = C_a N_A k_B\{k_B T/(2J s_p)\}^{3/2} = C_a R\{k_B T/(2J s_p)\}^{3/2}, \tag{5.40}$$

where C_a is the constant depending upon crystal structure.

For an antiferromagnetic material, spins are ordered in an antiparallel arrangement, but there is no net magnetic moment below the Nèel temperature. The expression for the mean internal energy of antiferromagnetic material may be written as

$$E = 4\pi V(2\alpha_a J' s_p a^2)\{(k_B T)/(2\alpha_a J' s_p a^2)\}^4 \int x^3 dx/(e^x - 1), \quad (5.41)$$

where α_a is a constant depending upon crystal structure and J' is the magnitude of the quantum mechanical exchange constant for the antiferromagnetic material.

The upper limit for integration may be taken as equal to infinity at low temperatures, so differentiation of Equation 5.41 with respect to temperature gives the magnetic heat capacity:

$$C_{m,af} = C_{af} N_A k_B \{k_B T/(2J' s_p)\}^3 \quad (5.42)$$

where C_{af} is a constant depending upon the type of lattice.

5.4.2.4 Heat Capacity due to Activation Process

Formation of vacancies as well as the formation of the electron and hole pairs may occur at high temperatures through the activation process. Heat capacity due to the vacancy formation may be given as

$$C_f = \{\xi_f \Delta H_f^2/(RT^2)\}\exp\{-\Delta H_f/(RT)\}, \quad (5.43)$$

where ξ_f is the entropy term of the vacancy formation given by exp $(\Delta S_f/R)$ and ΔH_f is the energy of vacancy formation.

On the other hand, heat capacity due to the electron and hole pairs is expressed as

$$C_{eh} = \{\xi_{eh} \Delta E_{eh}^2/(RT^2)\}\exp\{-\Delta E_{eh}/(RT)\}, \quad (5.44)$$

where ξ_{eh} is the entropy term of the electron and hole pairs represented by exp$(\Delta S_{eh}/R)$ and ΔE_{eh} is the formation energy of the electron and hole pairs.

5.5 Estimation of Normal Heat Capacity

5.5.1 Analysis of Heat Capacity Data

5.5.1.1 Heat Capacity Data at Low Temperatures

When heat is supplied to a metal or an alloy, the energies of both lattice vibrations and free electrons are increased. In the absence of contributions from magnetic heat capacity, the heat capacity of metals or alloys below both the Debye temperature and the Fermi temperature may be expressed as the sum of electron and phonon contributions:

$$C_v = C_{e,c} + C_l = \gamma T + \beta T^3, \quad (5.45)$$

where γT is the electronic contribution and βT^3 is the phonon contribution of the lattice vibrations. At sufficiently low temperatures ($T < 1$ K), the electronic heat

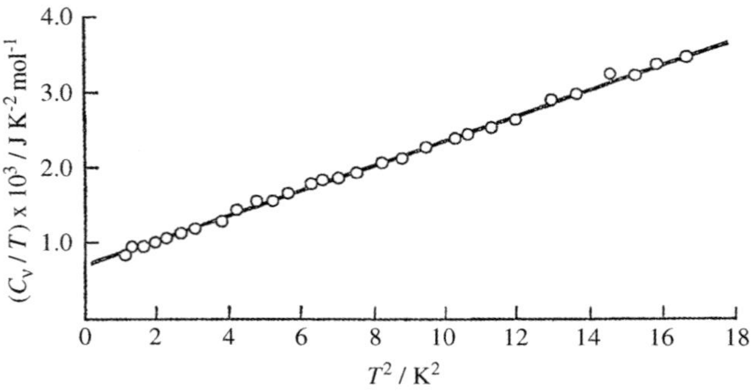

Figure 5.9 Heat capacity of Ag at low temperatures plotted by Equation 5.46.

capacity is dominant, while at high temperatures the phonon contribution is predominant. Dividing Equation 5.45 by T gives the following equation:

$$C_v/T = \gamma + \beta T^2. \tag{5.46}$$

Figure 5.9 shows a good linear relation of Equation 5.46 in the plot of C_v/T for Ag metal as a function of T^2. The values of β and γ are calculated from the slope of a straight line and intercept ($\gamma = 0.646\ \mathrm{m\,J\,K^{-2}\,mol^{-1}}$) of the plot, respectively.

For nonmetals, the electronic contribution may be very small compared with the lattice term, so

$$C_v = C_l = \beta T^3. \tag{5.47}$$

5.5.1.2 **Heat Capacity Data of Metal Oxides with Fluorite-Type Crystal Structure**

Heat capacities of the stoichiometric CeO_2, ThO_2, UO_2, NpO_2, and PuO_2 having a fluorite-type crystal structure are shown in Figure 5.10 [10] as a function of

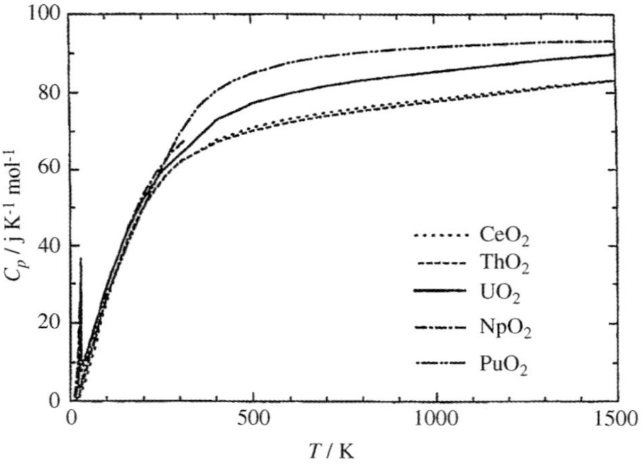

Figure 5.10 Dependence of heat capacity (C_p) on temperature for the stoichiometric CeO_2, ThO_2, UO_2, NpO_2, and PuO_2 [10].

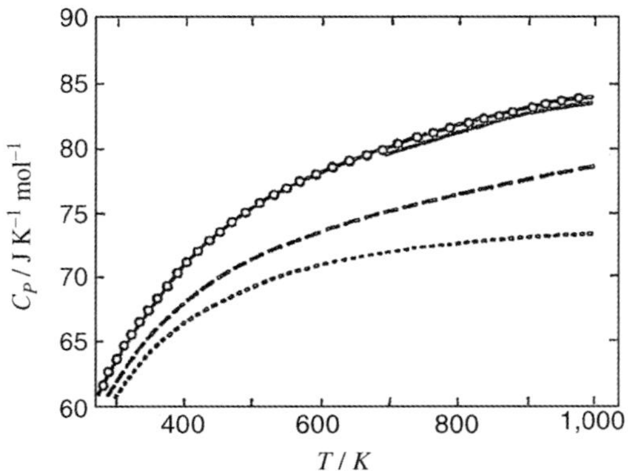

Figure 5.11 Dependence of heat capacity (C_p) on temperature for the stoichiometric UO_2 [11]. o–o, corrected experimental results; ······, C_l; - - - - - -, $C_l + C_d$; —, $C_l + C_d + C_{e,sh}$.

temperature. The magnetic phase transitions from the ordered antiferromagnetic phase to the paramagnetic phase are seen at 30.4 and 25.2 K for the stoichiometric UO_2 and NpO_2, respectively. Heat capacities of CeO_2 (Ce^{4+}) without 4f electrons agree well with those of ThO_2 (Th^{4+}) without 5f electrons, and both heat capacity data are thus near to the Dulong–Petit value of $9R$ (=74.82 J K^{-1} mol^{-1}) at high temperatures. The increase in heat capacity of UO_2 (U^{4+} : 5f^2), NpO_2 (Np^{4+} : 5f^3), and PuO_2 (Pu^{4+} : 5f^4) with increasing 5f electrons may be explained by Schottky-type heat capacity due to excitation of 5f electrons.

The experimental heat capacity data of the stoichiometric UO_2 are shown as a function of the temperature in Figure 5.11 [11], where the sum of lattice heat capacity (C_l), dilatometric heat capacity (C_d), Schottky-type heat capacity ($C_{e,sh}$), and small amount of heat capacity due to vacancy formation (C_f) is plotted. The contribution to lattice heat capacity C_l is large at low temperatures, but the contributions of C_d and $C_{e,sh}$ increase with increasing temperature.

5.5.1.3 Heat Capacity Data of Negative Thermal Expansion Materials ZrW_2O_8

Most of the materials expand with increasing temperature because of the increase in the distance between the constituent particles that vibrate in an anharmonic potential. However, some of the materials contract with increasing temperature, the so-called negative thermal expansion materials. Isomorphous ZrW_2O_8 and HfW_2O_8 contract isotropically over a very wide temperature range over 1000 K. Their crystal structures are very scarce in nature and consist of ZrO_6 (or HfO_6) octahedra and WO_4 tetrahedra, sharing the oxygen atoms at vertexes of these polyhedra. The polyhedra form some low-energy librational and translational modes, which have been proved to be closely related to the negative thermal expansion.

Corrected heat capacity at constant volume (C_v) of ZrW_2O_8 obtained from the experimental heat capacity data is shown as a function of temperature in Figure 5.12a and b [12], where heat capacity contribution of five functions (C_D, C_{E1}, C_{E2}, C_{R1}, and

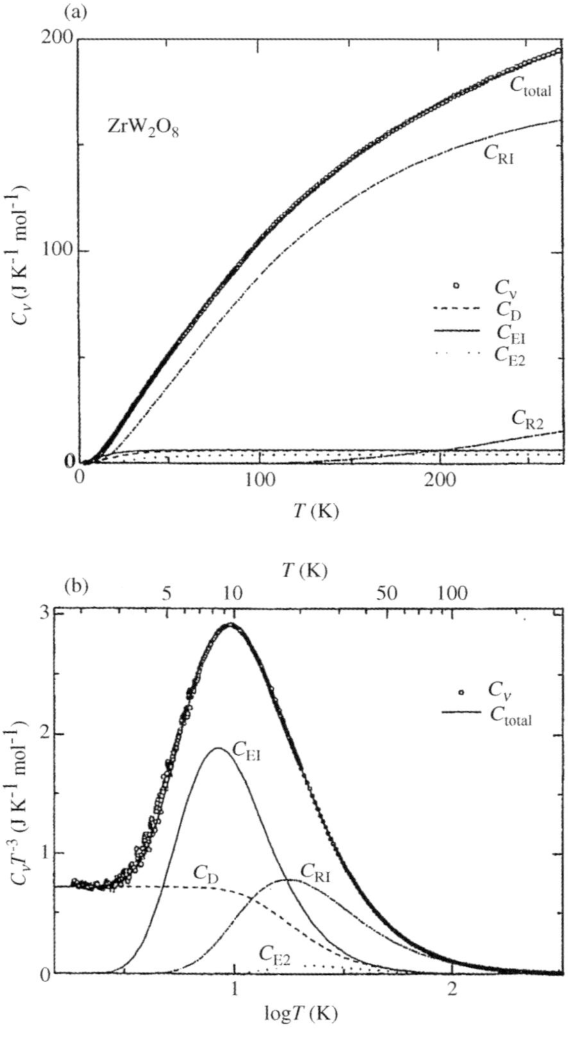

Figure 5.12 Heat capacity contributions of five functions (C_D, C_{E1}, C_{E2}, C_{R1}, and C_{R2}), their sum ($C_{total} = C_D + C_{E1} + C_{E2} + C_{R1} + C_{R2}$), and C_v obtained from the experimental heat capacity data for ZrW_2O_8 in the C_v–T plot (a) and C_vT^3 versus log T plot (b) [12]. In (b), C_{R2} is excluded due to the small contribution to heat capacity. (Reprinted with permission from Y. Yamamura, N. Nakajima, *et al.*, Phys. Rev. B, Vol. 66, 014301, 2002. Copyright 2002 by the American Physical Society.)

C_{R2}) and their sum ($C_{total} = C_D + C_{E1} + C_{E2} + C_{R1} + C_{R2}$) are plotted. Measured C_p data are converted approximately to heat capacity at constant volume (C_v) between 80 and 210 K by using Equation 5.12, although the C_p value was identified as the C_v value (about 0.6% of C_p at 100 K) between 1.8 and 80 K. In the figure, the total heat capacity C_{total} reproduces C_v very well. Most parts of the heat capacity below 300 K are attributed to C_{R1}. C_D, C_{E1}, C_{E2}, C_{R1}, and C_{total} are plotted as C_vT^{-3} versus log T in Figure 5.12b. It is clear that the maximum peak at about 9.5 K mainly consists of $C_{E1}T^{-3}$ and $C_{R1}T^{-3}$. The maximum of $C_{E2}T^{-3}$ is located at about 22 K and overlaps with $C_{R1}T^{-3}$. Although the fit seems to be carried out without the E_2 mode at first glance, the good fit could not be obtained without the E_2 mode. To describe a broad distribution over a rather wide frequency range, a rectangular distribution was introduced. The rectangular distribution enables us to describe complex phonon density of states (DOSs) and to find some characteristic vibration modes. The Debye temperature ($\theta_D = 311$ K) was calculated from the Debye T^3 law in the narrow temperature range 1.8–2.6 K by using Equation 5.32. The detailed analysis indicates that two characteristic Einstein modes (E_1, E_2) have negative mode Grüneisen parameters in ZrW_2O_8. The E_1 mode originates from the vibrations including the translational motion of Zr (or Hf) atoms or undistorted ZrO_6 octahedron, and E_2 mode originates from the vibrations including the librational motion of the octahedron.

Figure 5.13a and b [12] shows the effective DOS of ZrW_2O_8, together with the experimental DOS, $g_n(\omega)$, obtained from inelastic neutron scattering. It is not easy to

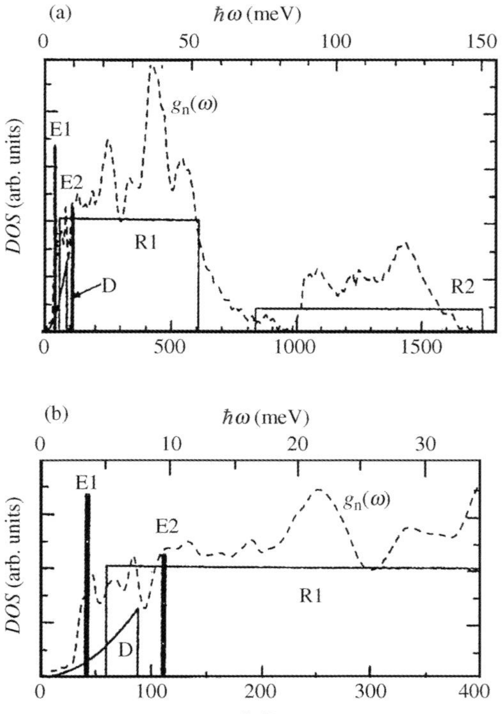

Figure 5.13 Calculated effective phonon DOS of ZrW_2O_8 compared with the experimental phonon DOS, $g_n(\omega)$, obtained from inelastic neutron scattering [12]. (b) is an enlarged plot of (a). (Reprinted with permission from Y. Yamamura, N. Nakajima, *et al.*, Phys. Rev. B, Vol. 66, 014301, 2002. Copyright 2002 by the American Physical Society.)

compare the effective DOS with the experimental $g_n(\omega)$, because $g_n(\omega)$ is a neutron cross section weighted DOS. It is, however, still possible to compare the experimental data to the effective DOS at some points. Any phonon mode above 150 meV was not found by the neutron study. The upper limit value agrees very well with the highest characteristic temperature of R_2 ($\theta_{R2H} = 1747$ K, 150.5 meV). On the other hand, the lowest limit of the characteristic temperature of all the modes except the Debye mode is E_1 ($\theta_{E1} = 41.8$ K, 3.60 meV), which is the same as the peak at the lowest frequency of $g_n(\omega)$.

5.5.2
Kopp–Neumann Law

Dulong and Petit's law was extended to predict the molar heat capacity of alloys by Kopp and Neumann. According to the Kopp–Neumann law, if an alloy consists of elements 1, 2, 3, ..., n having atomic fraction X_1, X_2, X_3, ..., X_n and atomic heat capacity C_{p1}, C_{p2}, C_{p3}, ..., C_{pn} at constant pressure, the heat capacity of the alloy at constant pressure is given by

$$C_p = \sum_{i=1}^{n} X_i C_{pi}. \tag{5.48}$$

However, Equation 5.48 should be applied with caution for alloys especially near magnetic and phase transitions. As will be shown in Figure 5.18, the experimental heat capacity data at constant pressure of β-brass agree closely with the calculated values between 0 and 150 °C. However, the Kopp–Neumann law cannot be applied to a compound (UO_2) whose physical and chemical properties of one element (U) are different from those of other element (O) by using the equation $C_p(UO_2) = C_p(U) + 2C_p(O)$.

More important extension of the Kopp–Neumann law was also applied to chemical compounds, that is, the molar specific heat capacity of a compound is equal to the sum of the molar fraction and heat capacity of its constituent compounds having similar physical and chemical properties. For example, the heat capacities of the nonstoichiometric UO_{2+x} and the $U_{1-x}Th_xO_2$ solid solution are estimated as

$$\begin{aligned} C_p(UO_{2+x}) &= \{(3+x)/3\}C_p(UO_2) \\ C_p(U_{1-x}Th_xO_2) &= (1-x)C_p(UO_2) + xC_p(ThO_2). \end{aligned} \tag{5.49}$$

Heat capacity data of UO_2, $UO_{2.25}$ (U_4O_9), and $UO_{2.667}$ (U_3O_8) per mole of uranium compounds at constant pressure are calculated by using Equation 5.49, and the results are shown as a function of temperature in Figure 5.14 [13], where the broken lines are normal heat capacities of U_4O_9 and U_3O_8 estimated by using the Kopp-Neumann law. Experimental heat capacity data of $UO_{2.25}$ (U_4O_9) at temperatures between 500 and 850 K except phase transition are in good agreement with normal heat capacity data, whereas the experimental heat capacity data of $UO_{2.667}$ (U_3O_8) are

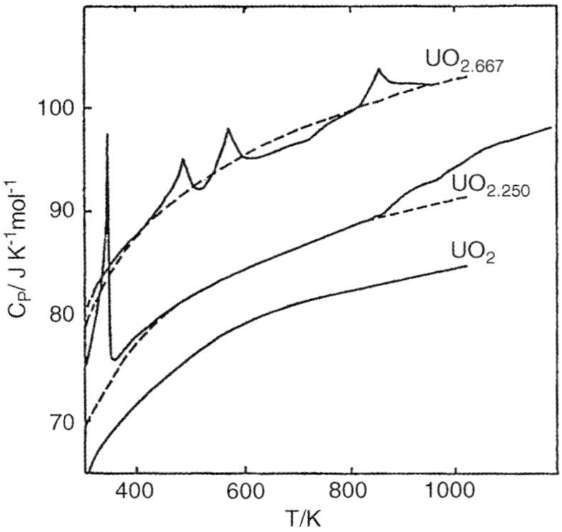

Figure 5.14 Heat capacities of UO_2, $UO_{2.250}$, and $UO_{2.667}$. The broken lines show the estimated heat capacities of $UO_{2.250}$ and $UO_{2.667}$ by Equation 5.49 [13].

lower than the normal values. The difference of the latter may be caused by the overestimation of Schottky-type heat capacity due to U^{4+} in UO_2, as shown in Figure 5.11, in spite of the small contribution of U^{4+} to $UO_{2.667}$ (U_3O_8).

5.5.3
Estimation of Heat Capacity Data from Thermal Expansion Coefficient

Heat capacity at constant pressure of isotropic crystals is expressed from the thermodynamic relation as

$$C_p = \beta_p V_m/(K_a\gamma_a) = 3\alpha_p V_m/(K_a\gamma_a), \qquad (5.50)$$

where K_a is the adiabatic compressibility, γ_a is the adiabatic compressibility, V_m is the molar volume, β_p is the coefficient of the volume thermal expansion, and α_p is the coefficient of the linear thermal expansion. The values of K_a and γ_a in Equation 5.50 are assumed to be independent of temperature, and K_a and γ_a of CeO_2 at 298.15 K are used in the subsequent calculations. Then, C_p at temperatures from 150 to 800 K is calculated from the measured $3\alpha_p$ and V_m by using Equation 5.50. Figure 5.15 shows the temperature dependence of the calculated C_p for CeO_2, together with those of two compiled databases [14]. The calculated values agree reasonably with the reported ones in a wide temperature range, although the calculated values are slightly higher than the reported data at low and high temperatures. The small deviation may be caused by the assumption that the values of K_a and γ_a are independent of temperature.

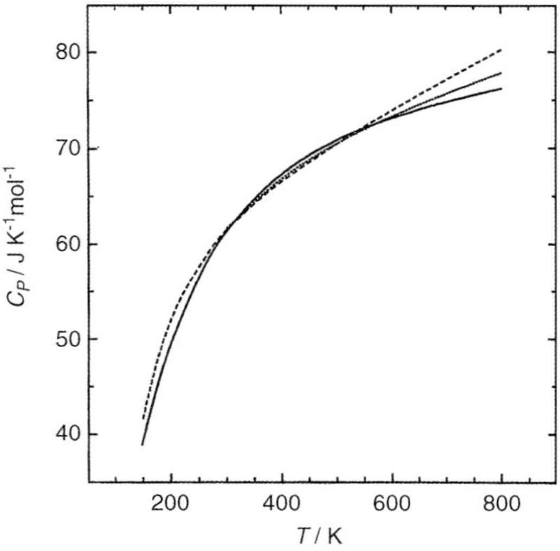

Figure 5.15 Heat capacities of CeO_2 as a function of temperature. The estimated heat capacity data (···) are compared with those of the compiled two databases (---, –) [14].

5.5.4
Corresponding States Method

Corresponding states method is applied to the system in which heat capacities of a series of compounds having the same crystal structure are measured [15, 16]. In this method, it is assumed that the heat capacity of a sample at temperature T, $C_{p,\mathrm{normal}}(\text{sample}, T)$, can be represented by the heat capacity of the standard material at temperature rT, $C_{p,\mathrm{normal}}(\text{standard material}, rT)$, as follows:

$$C_{p,\mathrm{normal}}(\text{sample}, T) = C_{p,\mathrm{normal}}(\text{standard material}, rT). \tag{5.51}$$

Materials without any anomaly in the temperature range of phase transition are selected as standard materials. For example, at the magnetic phase transition, an antiferromagnetic material is used as a standard material, which will be discussed in Section 5.6.2

5.5.5
Volumetric Interpolation Schemes

The volumetric interpolation scheme proposed by Westrum [17] involves the linear interpolation on the basis of the fractional molar volume increment of the compounds in question. Since the $Ln(OH)_3$ (Ln = Pr, Eu, Tb) and $LnCl_3$ (Pr, Eu) compounds are part of isoanionic series having relatively small lattice contributions and the lower lying energy levels have been spectroscopically determined for these compounds, this lanthanide series is the most nearly ideal system to resolve Schottky contributions of the heat capacity in the temperature range from 5 to 350 K. Normal heat capacities of $Ln(OH)_3$ (Ln = Pr, Eu, Tb) and $LnCl_3$ (Pr, Eu) compounds at constant pressure may be calculated from antiferromagnetic lanthanum compounds, not having 4f electron, and gadolinium compounds, having half-filled 7 electrons, as follows:

$$\begin{aligned} C_{p,\text{normal}}(\text{LnL}_3) &= xC_{p,\text{normal}}(\text{LaL}_3)+(1-x)C_{p,\text{normal}}(\text{GdL}_3) \\ x &= \{V_\text{m}(\text{LnL}_3)-V_\text{m}(\text{LaL}_3)\}/\{V_\text{m}(\text{GdL}_3)-V_\text{m}(\text{LaL}_3)\}, \end{aligned} \tag{5.52}$$

where Ln is the lanthanide ion, V_m is the molar volume, and L is an anion of OH^- or Cl^-. Figure 5.16 shows the calorimetrically and spectroscopically determined Schottky-type contributions of heat capacity of the $Ln(OH)_3$ (Ln = Pr, Eu, Tb) and $LnCl_3$ (Pr, Eu) compounds [17]. Excess heat capacities determined from the experimental heat capacity data at constant pressure and normal heat capacity obtained by

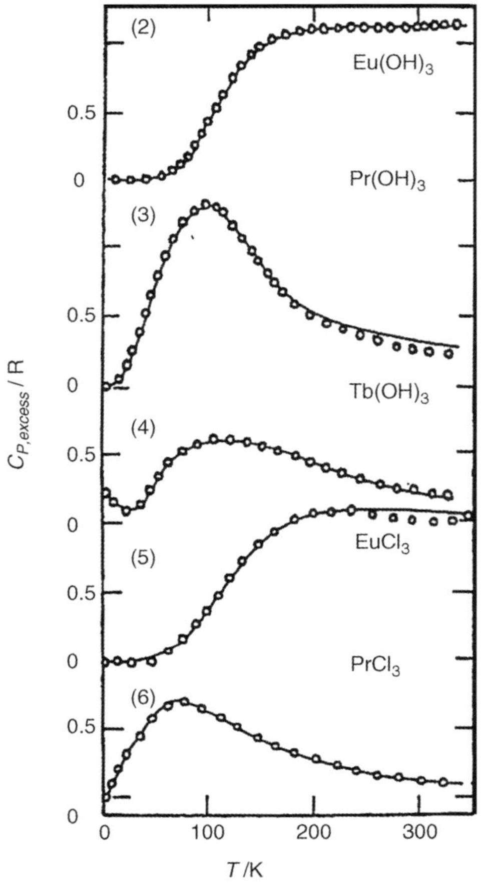

Figure 5.16 Calorimetrically (–) and spectroscopically (O) determined Schottky contributions for $Eu(OH)_3$, $Pr(OH)_3$, $Tb(OH)_3$, $EuCl_3$, and $PrCl_3$. The successive curves are displaced by 1 unit of C_p/R [17].

using Equation 5.52 are in good agreement with heat capacity data calculated from the spectroscopic data by distribution function.

5.6
Phase Transition

5.6.1
Second-Order Phase Transition

5.6.1.1 Order–Disorder Phase Transition due to Atomic Configuration

The Bragg–Williams theory [10] of the order–disorder phase transition has been developed in binary alloys AB, where the number of A atoms is equal to the number of B atoms. The alloy has two crystallographic sites α and β. In the low-temperature phase, when all the A atoms occupy α-sites and all the B atoms occupy β-sites, the alloy is completely ordered. On the other hand, when the distribution of A and B atoms is completely random over α- and β-sites in the high-temperature phase, the alloy is completely disordered. If N_R and N_W are, respectively, the number of the right and wrong sites that the atoms occupy, the order parameter s is defined as

$$s = (N_R - N_W)/(N_R + N_W).$$

Therefore, $s = 1$ for the complete order state and $s = 0$ for the complete disorder state. Internal energy of the system, $E(s)$, is expressed as a function of the long-range order parameter s by

$$E(s) = E(0) - (1/4)NZVs^2$$
$$V = (1/2)\,(V_{AA} + V_{BB}) - V_{AB},$$

where $E(0)$ is the internal energy at completely disordered state ($s = 0$), N is the total number of crystallographic sites, Z is the coordination number, and V_{AA}, V_{BB}, and V_{AB} are the interaction energies of AA, BB, and AB pairs, respectively.

The heat capacity for the long-range order–disorder phase transition of binary alloy is described in terms of the long-range order parameter:

$$\Delta C_v = RT_c^2\{1 - \tanh(2T_c s/T)\}/[T^2 - TT_c\{1 - \tanh(2T_c s/T)\}], \quad (5.53)$$

where ΔC_v is the heat capacity due to the order–disorder rearrangement and T_c is the critical temperature.

Partition function of configuration, $P(s)$, is written as

$$P(s) = [(N/2)!/\{(N/4)(1+s)\}!\{(N/4)(1-s)\}!]^2 \times \exp[-\{E(0)/(k_B T)\} + (1/4)NZVs^2/(k_B T)]. \quad (5.54)$$

In the case of $dP(s)/ds = 0$, the relation between s and T is expressed as

$$s = \tanh\{ZVs/(2k_B T)\}, \quad (5.55)$$

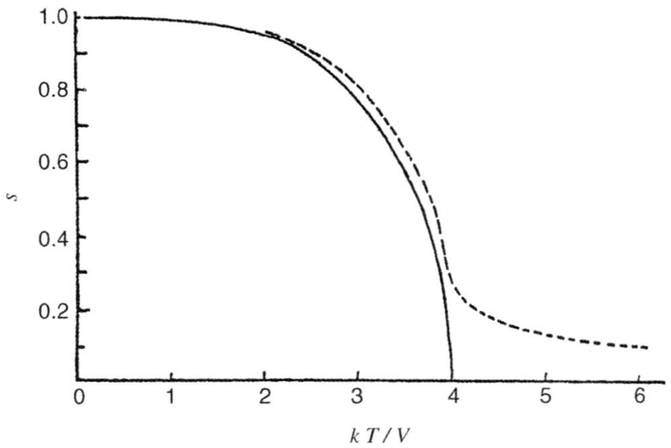

Figure 5.17 Long-range order parameter calculated by the Bragg–Williams model [18] and a short-range order parameter for β-brass [19] as a function of temperature. The solid and broken lines are the long- and short-range order parameters, respectively. (Reprinted with permission from J.M. Cowley, Phys. Rev., 120, 1648, 1960. Copyright 1960 by the American Physical Society.)

and the following relation holds at transition temperature T_c:

$$ZV = 2k_B T_c.$$

Figure 5.17 shows the long-range order parameter calculated by the Bragg–Williams model [18] and an example of a short-range parameter [19] for β-brass (CuZn) as a function of temperature. The solid and broken lines are the long- and short-range order parameters, respectively. The order parameter s decreases continuously up to T_c with increasing temperature, which is characteristic of a typical second-order phase transition. The experimental heat capacity data of the CuZn alloy [20] and that calculated by Equation 5.53 are shown in Figure 5.18 [18]. A sharp heat capacity anomaly is observed experimentally, and the excess heat capacity remains above phase transition temperatures, compared to the calculated heat capacity data, because the Bragg–Williams approximation takes into consideration only long-range ordering and not short-range ordering.

5.6.1.2 Order–Disorder Phase Transition due to Orientation in ZrW_2O_8

The negative thermal expansion of ZrW_2O_8 originates in the rare crystal structure having the large open space and the low-energy vibrations of WO_4 tetrahedra and ZrO_6 octahedra. This low-energy vibrational motion of WO_4 units brings about another interesting physical property of ZrW_2O_8, which undergoes a structural phase transition around 440 K from an acentric to centric structure with increasing temperature. The structural phase transition is related to the orientation of unshared vertex of the WO_4 unit. In the low-temperature phase ($P2_13$), two neighboring WO_4 tetrahedra on the [111] body diagonal in the unit cell point to their unshared vertexes in the [111] direction, whereas they are randomly oriented in the [111] or −[111] directions in the high-temperature phase ($Pa\bar{3}$). Two disordering mechanisms have been proposed so far. One mechanism is that the two WO_4 tetrahedra on the [111] body diagonal change their orientation concertedly. If this is the case, the entropy of

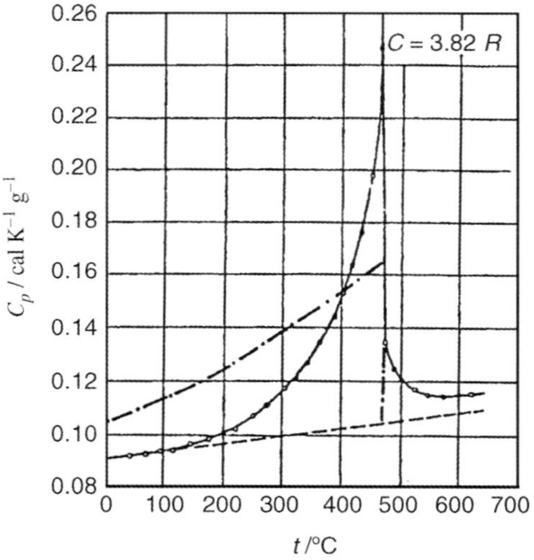

Figure 5.18 Heat capacity (•) of β-brass as a function of temperature [20]. The broken line is the normal heat capacity estimated from the Kopp–Neumann law by using Equation 5.48, and the chain line is the heat capacity calculated by the Bragg–Williams model [18].

transition is $R \ln 2$ ($=5.8\,\mathrm{J\,K^{-1}\,mol^{-1}}$) like a magnetic transition in a spin 1/2. On the other hand, if two tetrahedra can independently take two orientations, the expected entropy is $R \ln 4$ ($=11.5\,\mathrm{J\,K^{-1}\,mol^{-1}}$).

In order to know phase transition mechanism, heat capacities of ZrW_2O_8 are measured in the temperature range from 1.8 to 483 K by using two adiabatic calorimeters and a commercial relaxation calorimeter, and the results are shown in Figure 5.19 [21]. A large λ-type heat capacity anomaly due to an order–disorder

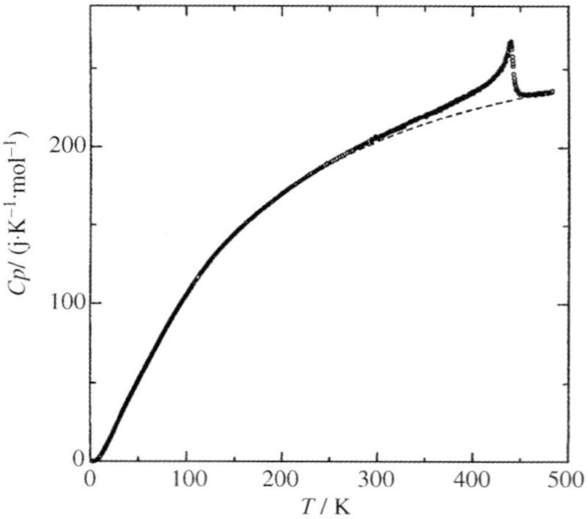

Figure 5.19 Heat capacity of ZrW_2O_8 plotted against temperature. The broken line is the normal portion of the heat capacity to separate the excess heat capacity from the observed values [21].

phase transition is observed at 440 K. The heat capacity data below 180 K and above 470 K are fitted with five functions by using a linear least-squares method, as described in Section 5.5.1.3. The estimated normal heat capacity is shown as the broken curves. The enthalpy and entropy of phase transition are determined to be 1.56 kJ mol^{-1} and 4.09 J K^{-1} mol^{-1}, respectively. The entropy of transition is about 70% of R ln 2 (=5.8 J K^{-1} mol^{-1}), and this result is expected for a typical order–disorder phase transition in which the two WO_4 tetrahedra on the [111] body diagonal change their orientation concertedly.

5.6.2
Magnetic Order–Disorder Phase Transition

Equation 5.40 shows that the magnetic contribution to the heat capacity at low temperatures for the ferromagnetic and ferrimagnetic materials is proportional to the three-halves power of the absolute temperature. When heat is supplied to a metal, the energies of free electrons, lattice vibrations, and spins are increased. For ferromagnetic metals, the total heat capacity at constant volume is equal to the sum of the electronic, lattice, and magnetic terms, that is,

$$C_v = C_{e,c} + C_l + C_m = \gamma T + \beta T^3 + \delta T^{3/2}. \tag{5.56}$$

For ferrimagnets, which are electrical insulators, the electronic term in Equation 5.56 is negligibly small, compared with the other terms, so the total heat capacity at constant volume may be given by the following expression:

$$C_v = C_l + C_m = \beta T^3 + \delta T^{3/2}. \tag{5.57}$$

Both sides of Equation 5.57 may be divided by $T^{3/2}$ to give

$$C_v/T^{3/2} = \beta T^{3/2} + \delta. \tag{5.58}$$

A plot of $C_v/T^{3/2}$ versus $T^{3/2}$ should give a straight line with slope β and intercept δ.

The magnetic contribution to the heat capacity by ferromagnetic and ferrimagnetic materials is the $T^{3/2}$ dependence, whereas that by antiferromagnetic materials is T^3 dependence, as described in Section 5.4.2.3. Hence, for antiferromagnetic materials, the temperature dependence is of the same form as the Debye's T^3 law, so the separation of the spin wave contribution from the heat capacity in antiferromagnetic materials is very difficult.

Heat capacities of $CoCl_2$, $CuCl_2$, and $CrCl_2$ minus the heat capacity of $MnCl_2$ are shown in the temperature range from 10 to 300 K in Figure 5.20 [15]. Between 35 and 90 K, the total heat capacity of $CoCl_2$ is less than the heat capacity of $MnCl_2$. To avoid this negative heat capacity, the heat capacity of $CoCl_2$ is approximated by the corresponding states method. The heat capacity of $CoCl_2$ at temperature T is represented by the lattice heat capacity of $MnCl_2$ at temperature rT, and the lattice entropy of $CoCl_2$ at T is that of the $MnCl_2$ lattice at temperature rT.

Figure 5.21 [15] shows the magnetic entropy and heat capacity of $CoCl_2$ calculated for $r = 0.90$. The magnetic entropy of $CoCl_2$ reaches R ln 2 at near 75 K, and it is

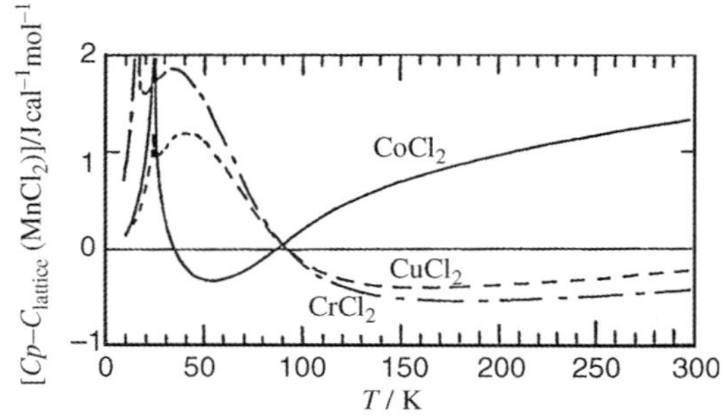

Figure 5.20 Heat capacities of $CoCl_2$, $CuCl_2$, and $CrCl_2$ minus the lattice heat capacity of $MnCl_2$ versus temperature [15].

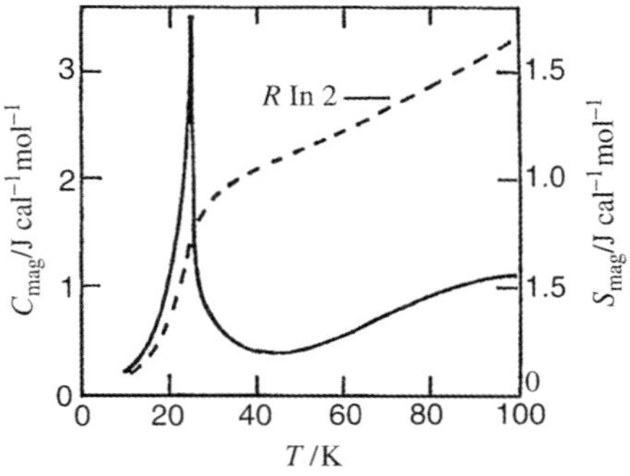

Figure 5.21 Magnetic contributions to the heat capacity and entropy of $CoCl_2$ versus temperature [15]. The lattice heat capacity and entropy of $MnCl_2$ at temperature 0.90T are the estimates of the lattice heat capacity and entropy of $CoCl_2$ at temperature T.

apparent that there are only two states associated with the magnetic ordering corresponding to the heat capacity anomaly with a peak at 24.91 K.

5.7 Summary

5.7.1 Heat Capacity Measurement

There exist two typical heat capacity measurements: an adiabatic heat capacity calorimetry and a temperature jump calorimetry.

5.7.1.1 Adiabatic Heat Capacity Calorimetry

The absolute value of heat capacity, C_p, is obtained most accurately as $C_p = \Delta E/\Delta T$ from the measurements of input energy ΔE to the sample vessel (or a sample and its vessel) and the resulting temperature increment ΔT.

5.7.1.2 Temperature Jump Calorimetry

The enthalpy change of the sample is measured as the amount of heat absorbed by the calorimeter in changing from an initial temperature to a final temperature. The C_p value as a function of temperature may then be derived from the smoothed tabulated enthalpy data as $C_p = \mathrm{d}(H_T - H_{298.15})_p/\mathrm{d}T$. The enthalpy change by phase transformation can also be determined from the difference in the measured enthalpy change before and after the phase transformation.

5.7.2 Thermodynamic Relation Between C_p and C_v

The heat capacity of solids is ordinarily measured at constant pressure, $C_p = (\mathrm{d}H/\mathrm{d}T)_p$, whereas the heat capacity at constant volume, $C_v = (\mathrm{d}E/\mathrm{d}T)_v$, is calculated theoretically if the interatomic distance is kept constant, independent of the temperature changes. The dilatometric term C_d is the difference between C_p and C_v and is obtained from the classical thermodynamic relations as

$$\begin{aligned} C_\mathrm{d} &= C_p - C_v = (V\beta_p^2/\varkappa_T)T \\ &= C_v \gamma_\mathrm{e} \beta_p T \end{aligned}$$

where $\varkappa_T$ is the isothermal compressibility, β_p is the volume thermal expansion, and γ_e is the Grüneisen constant.

For a lattice heat capacity contributed to lattice vibrations, an approximation called the Nernst–Lindemann formula is often used:

$$C_\mathrm{d} = C_p - C_v = AC_p^2 T,$$

where A is a system-dependent constant. The values of C_d at the temperature T are estimated from the values of C_p measured experimentally.

5.7.3 Estimation of Normal Heat Capacity

The heat capacity at constant volume C_v can be expressed as the sum of each contributed heat capacity:

$$C_v = C_\mathrm{l} + C_\mathrm{l,a} + C_\mathrm{e,c} + C_\mathrm{e,sh} + C_\mathrm{m} + C_\mathrm{f} + \cdots,$$

where C_l is the lattice heat capacity, $C_\mathrm{l,a}$ is the heat capacity due to anharmonic lattice vibration, $C_\mathrm{e,c}$ is the electronic heat capacity due to conduction electron, $C_\mathrm{e,sh}$ is the electronic Schottky-type heat capacity due to electronic excitation to higher energy levels, C_m is the magnon heat capacity due to the excitation of the spin system in magnetically ordered substances, and C_f is the heat capacity due to the formation of vacancies.

The main contribution of heat capacity at constant volume is the lattice heat capacity (C_l) due to lattice vibrations. Debye's theory predicts a cube law dependence of the heat capacity of the elemental solids for temperatures $T < (\theta_D/50)$:

$$C_l = (12/5)\pi^4 R(T/\theta_D)^3.$$

5.7.3.1 Nonmagnetic Metals and Alloys at Low Temperatures

The heat capacity of nonmagnetic metals and alloys below both the Debye temperature and the Fermi temperature may be expressed as the sum of phonon and electron contributions:

$$C_v = C_l + C_{e,c} = \beta T^3 + \gamma T.$$

At sufficiently low temperatures ($T < 1$ K) the electronic heat capacity is dominant, while at high temperatures the phonon contribution is predominant.

5.7.3.2 Nonmetals and Non-Alloys Without Magnetic Transition at Low Temperatures

For nonmetals and non-alloys, the electronic contribution may be very small compared with the lattice term, so

$$C_v = C_l = \beta T^3.$$

5.7.3.3 Ferromagnetic and Ferrimagnetic Materials at Low Temperatures

For ferromagnetic metals and alloys, the total heat capacity at constant volume is equal to the sum of the lattice, electronic, and magnetic terms, that is,

$$C_v = C_l + C_{e,c} + C_m = \beta T^3 + \gamma T + \delta T^{3/2}.$$

For ferrimagnets, which are electrical insulators, the electronic term in equation is negligibly small, compared with the other terms, so the total heat capacity at constant volume may be given by the following expression:

$$C_v = C_l + C_m = \beta T^3 + \delta T^{3/2}.$$

5.7.3.4 Antiferromagnetic Materials at Low Temperatures

For antiferromagnetic materials, the magnon heat capacity shows T^3 dependence:

$$C_v = C_l + C_{e,c} + C_m = \beta T^3 + \gamma T + \delta T^3.$$

The temperature dependence is of the same form as the Debye's T^3 law, so the separation of the spin wave contribution from the heat capacity in antiferromagnetic materials is very difficult.

5.7.3.5 Metal Oxides with Fluorite-Type Crystal Structure at High Temperatures

The heat capacity data of metal oxide with fluorite-type crystal structure at high temperatures are expressed as the sum of the lattice, the electronic Schottky-type, and the vacancy formation terms:

$$\begin{aligned} C_v &= C_l + C_{e,sh} + C_f \\ &= 3RD(T/\Theta_D) + (g_1/g_0)\{E_1^2/(RT^2)\}\exp\{-E_1/(RT)\} \\ &\quad + \{\xi_f \Delta H_f^2/(RT^2)\}\exp\{-\Delta H_f/(RT)\}. \end{aligned}$$

5.7.4
Second-Order Phase Transition

5.7.4.1 Order–Disorder Phase Transition due to Atomic Configuration

According to the Bragg–Williams theory of the order–disorder phase transition, the heat capacity for the long-range order–disorder phase transition of binary alloy is described in terms of the long-range order parameter *s*:

$$\Delta C_v = RT_c^2\{1-\tanh(2T_c s/T)\}/[T^2 - TT_c\{1-\tanh(2T_c s/T)\}],$$

where ΔC_v is the heat capacity due to the order–disorder rearrangement and T_c is the critical temperature. The following relation holds at transition temperature T_c:

$$ZV = 2k_B T_c,$$

where Z is the coordination number.

A sharp heat capacity anomaly of CuZn alloy is observed experimentally, and the excess heat capacity remains above phase transition temperatures, compared to the calculated heat capacity data, because the Bragg–Williams approximation takes into consideration only long-range ordering and not short-range ordering.

5.7.4.2 Order–Disorder Phase Transition due to Orientation in ZrW_2O_8

The negative thermal expansion material of ZrW_2O_8 undergoes a structural phase transition around 440 K from an acentric to centric structure with increasing temperature. In the low-temperature phase, two neighboring WO_4 tetrahedra on the [111] body diagonal point to their unshared vertexes in the [111] direction, whereas they are randomly oriented in the [111] or −[111] directions in the high-temperature phase. Two disordering mechanisms have been proposed so far. One mechanism is that the two WO_4 tetrahedra on the [111] body diagonal change their orientation concertedly, where the entropy of transition is $R \ln 2$ (=5.8 J K^{-1} mol^{-1}) like a magnetic transition in a spin 1/2. On the other hand, if two tetrahedra can independently take two orientations, the expected entropy is $R \ln 4$ (=11.5 J K^{-1} mol^{-1}).

The entropy of phase transition is determined experimentally to be 4.09 J K^{-1} mol^{-1}, which is about 70% of $R \ln 2$ (=5.8 J K^{-1} mol^{-1}), supporting the transition mechanism that the two WO_4 tetrahedra on the [111] body diagonal change their orientation concertedly.

References

1 Sorai, M. (ed.) (2004) *Comprehensive Handbook of Calorimetry and Thermal Analysis*, John Wiley & Sons, Ltd.
2 Tanaka, T., Atake, T., Nakayama, N., Eguchi, T., Saito, K., and Ikemoto, I. (1994) *J. Chem. Thermodyn.*, **26**, 1231.
3 Grϕnvold, F. (1967) *Acta Chem. Scand.*, **21**, 1965.
4 Naito, K., Inaba, H., Ishida, M., Saito, Y., and Arima, H. (1974) *J. Phys. E*, **7**, 464.
5 Cezairliyan, A., Morse, M.S., Berman, H.A., and Beckett, C.W. (1970) *J. Res. Natl. Bur. Stand.*, **74A**, 65.
6 Takahashi, Y., Yokokawa, H., Kadokura, H., Sekine, Y., and Mukaibo, T. (1979) *J. Chem. Thermodyn.*, **11**, 379.
7 Yamaguchi, K. and Itagaki, K. (2002) *J. Therm. Anal. Calor.*, **69**, 1059.
8 Inaba, H. and Yamamoto, T. (1983) *Netsu Sokutei*, **10**, 132.
9 Peierls, R.E. (1955) *Quantum Theory of Solids*, Oxford.
10 Tsuji, T. (2000) *Netsu Sokutei*, **27**, 88.
11 Grϕnvold, F., Kveseth, N.J., Sveen, A., and Tichy, J. (1970) *J. Chem. Thermodyn.*, **2**, 665.
12 Yamamura, Y., Nakajima, N., Tsuji, T., Koyano, M., Iwasa, Y., and Katayama, S. (2002) *Phy. Rev.*, **B66**, 014301.
13 Inaba, H. (1985) *Netsu Sokutei*, **12**, 30.
14 Nakajima, N., Mitani, H., Yamamura, Y., and Tsuji, T. (2001) *J. Nucl. Mater.*, **294**, 188.
15 Chisholm, R.C. and Stout, J.W. (1962) *J. Chem. Phys.*, **36**, 972.
16 Stout, J.W. and Chisholm, R.C. (1962) *J. Chem. Phys.*, **36**, 979.
17 Westrum, E.F. Jr. (1981) *Netsu Sokutei*, **8**, 106.
18 Bragg, W.L. and Williams, E.J. (1935) *Proc. R. Soc. Lond.*, **A151**, 540.
19 Cowley, J.M. (1960) *Phys. Rev.*, **120**, 1648.
20 Moser, H. (1936) *Phys. Z.*, **37**, 737.
21 Yamamura, Y., Tsuji, T., Saito, K., and Sorai, M. (2004) *J. Chem. Thermodyn.*, **36**, 525.

6
Diffraction and Thermal Expansion of Solids

Avesh Kumar Tyagi and Srungarpu Nagabhusan Achary

6.1
Introduction

Thermal expansion of materials is a topic of age-old interest to mankind in view of the chronological development of several devices and technologies [1, 2]. The concept of thermal expansion and knowledge of thermal expansion coefficient of materials remain important for any structural materials experiencing a temperature gradient. The structural materials include any material used in technology to build a mechanically or electronically integrated physical entity. In particular, the thermal properties are of interest when any structural assembly faces a temperature variation. Thermal expansion effects have been well considered right from the metals and ceramic parts of cookwares to the highly sophisticated mechanical structures, such as buildings, bridges, air/spacecraft, vessels, kilns, furnaces, and so on. In particular, metals such as steel, copper, and so on that are important structural parts show a significant expansion with temperature. Thus, in all these materials, thermal expansion was exploited either to enhance or to nullify any temperature-induced dimensional instability, for example, the utilization of thermal expansion in automatic cut-off switch using bimetals is well known. The discovery of Invar (an alloy of 35% Ni and 65% Fe) and subsequent modified compositions as Elinvar, Kovar, Alnico, and so on have found immense applications in modern technologies, such as design of high-accuracy clocks, shadow mask in television screen, and so on [3]. In addition, low thermal expansion in fused silica and controllable thermal expansion in materials with glass and glass–ceramic compositions have been discovered and used in precise optical instruments, mirror substrates, glass–metal junctions [4, 5]. A large number of crystalline materials with low thermal expansion are also discovered [6]. Thermal expansion data of ceramics have been a prime consideration in the designing of electrolyte and electrodes of solid oxide fuel cells [7]. Similarly, thermal expansion data of nuclear fuel materials are significant in preventing detrimental effects of fuel–clad interaction and unwanted swelling of fuel pins [8]. Nuclear reactor performance is mainly restricted by the thermal expansion and thermal conductivity of the fuel pellets under irradiation conditions. In these

Thermodynamic Properties of Solids: Experiment and Modeling
Edited by S. L. Chaplot, R. Mittal, and N. Choudhury

ISBN: 928-3-527-40812-2

applications, the material property is more significant than the structural assembly and hence the structural materials are tuned as per their thermal expansion behavior. However, the thermal expansion behavior of electronic materials causing temperature-induced dimensional instability leading to circuit failure also needs to be dealt with. Considering such aspects, several Invar-type alloys and metal matrix composites are developed in recent years [9]. Thus, the thermal expansion data of any material are the first requisite for the development of most of the technologies.

In general, it is known that all materials expand or contract with the rise or fall of temperature. The thermal expansion of any material is explained as a relative change in dimension with a unit rise in temperature. The bulk thermal expansion coefficient is thus presented as (α_l) and defined as

$$\alpha_l = \frac{l_T - l_0}{l_0(T - T_0)}, \tag{6.1}$$

where l_T is the length of the material at temperature T; l_0 is the reference length, that is, the length of the material at reference temperature T_0; and α_l is the coefficient of thermal expansion in $^\circ C^{-1}$ or K^{-1}.

The α_l determined by Equation 6.1 represents the function of two extremes of temperature, thus giving the average thermal expansion over the range. However, nonlinear variation of the dimension is very commonly observed in practice. Hence, often α_l is represented as a function defined in a range of temperature. This variation of dimension with temperature is expressed as polynomial function:

$$L_T = L_0 + a(T - T_0) + b(T - T_0)^2 + c(T - T_0)^3 + d(T - T_0)^4 + \cdots, \tag{6.2}$$

where T_0 is the reference temperature with respect to which the thermal expansion coefficients are defined. If the reference temperature is assumed at absolute zero, the above equation reduces to

$$L_T = L_0 + a(T) + b(T)^2 + c(T)^3 + d(T)^4 + \cdots.$$

Hence, the coefficient of thermal expansion is defined as

$$\alpha_l = \frac{1}{L_0}\left(\frac{dL_T}{dT}\right). \tag{6.3}$$

With the development of technology, interest in the thermal expansion grew only from the utility point of view. However, the basic understanding of the phenomena greatly advanced after a close correlation between coefficients of thermal expansion and the internal structure and the effect of temperature on vibrations of atoms were revealed [10]. Thus, the lattice structure and dynamics have been extensively studied to explain the thermal expansion behavior of materials. The concept of internal structure leads to the definition of two different coefficients of thermal expansions, namely, bulk (average) and lattice (unit cell) thermal expansion coefficients.

The intrinsic thermal expansion of the material is the actual lattice expansion. The measurements of the bulk and lattice thermal expansions are respectively based on macroscopic and microscopic observation of the effects of temperature. However, contributions from the grain boundary, micro/macrocracks, and voids are the additional important factors, which contribute to the bulk thermal expansion of a material [11]. The average thermal expansion and the intrinsic unit cell expansion agree well only in the direct measurement on a single crystal. In general, the unit cell expansion is anisotropic in nature unless the crystal is isotropic. Thus, the average thermal expansion obtained in the bulk measurement is an average of this anisotropy, and is significant in the utilization of a material. In addition, the bulk thermal expansion is easier to tune than the lattice thermal expansion [9, 12]. Usually, the presence of micro/macrocracks and grain boundary and voids significantly mask expansion and thus withstand considerable thermal shock, even though they have degraded mechanical strength compared to single crystals. Therefore, measurements of the bulk or average thermal expansion as well as unit cell expansion govern the response of any material under heating.

Mostly, the lattice thermal expansions measured from diffraction data are given in terms of volumetric thermal expansion coefficients (α_V in $^\circ C^{-1}$ or K^{-1}) and similar to Equation 6.3, they are defined as a relative change of unit cell volume at infinitesimal change in temperature:

$$\alpha_V = \frac{1}{V_0}\left(\frac{dV}{dT}\right). \tag{6.4}$$

Similarly, the average volume thermal expansion of a material is defined in terms of the volumes at initial and final temperatures and thus

$$\alpha_v = \frac{V_T - V_0}{V_0(T - T_0)}, \tag{6.5}$$

where V_T is the unit cell volume of the materials at the temperature T and V_0 is the reference volume, that is, volume of the material at reference temperature T_0.

The coefficients of the thermal expansion can also be expressed as a variation of density with temperature. For isotropic materials, the linear thermal expansion as well as the bulk thermal expansion can be related easily as $\alpha_V = 3 \times \alpha_l$. However, the situation is not such trivial where the thermal expansion is anisotropic, as in the case of noncubic systems. In such cases, the axial thermal expansion (α_l) measured along some reference axes (often preferred along unit cell directions) is equated to the volume thermal expansion by appropriate symmetry considerations.

6.2 Strain Analysis

In order to relate the thermal expansion to the elastic constants of a solid and hence to the other related properties, thermal expansion coefficients are preferentially given in strain notation [13–18]. Usually the strain in lattice is determined with respect to an unstrained lattice at constant stress, that is, normally at zero pressure. The temperature derivative of strain represents the thermal expansion.

In a simple orthogonal systems, the deformation in lattice due to the expansion along the three orthogonal axes a_0, b_0, and c_0 results in a new set of orthogonal axes a, b, and c. The new orthogonal axes due to rise in temperature ΔT can thus be given as

$$\left.\begin{aligned} a &= a_0 + \Delta a \\ b &= b_0 + \Delta b \\ c &= c_0 + \Delta c \\ \text{and hence} \quad V &= V_0 + \Delta V = (a_0 + \Delta a)(b_0 + \Delta b)(c_0 + \Delta c) \end{aligned}\right\}. \tag{6.6}$$

From the definition of strain, the linear strain along a, b, and c axes and the volume strain, given in Equation 6.7, are called Lagrangian strain:

$$\Delta a/a_0,\ \Delta b/b_0,\ \Delta c/c_0 \text{ and } \Delta V/V_0. \tag{6.7}$$

The temperature derivative of these Lagrange strain components give the axial or volume thermal expansion coefficients, which can be written as

$$\begin{aligned} &\alpha_a = (\Delta a/a_0)/\Delta T,\ \alpha_b = (\Delta b/b_0)/\Delta T,\ \alpha_c = (\Delta c/c_0)/\Delta T \text{ and} \\ &\alpha_V = (\Delta V/V_0)/\Delta T. \end{aligned} \tag{6.8}$$

Ignoring the product of two or more strain components, the volume strain can be obtained from Equation 6.6 as

$$\Delta V/V_0 = \Delta a/a_0 + \Delta b/b_0 + \Delta c/c_0 \tag{6.9}$$

and similarly the coefficient of volume thermal expansion can be written as

$$\alpha_V = \alpha_a + \alpha_b + \alpha_c. \tag{6.10}$$

In particular, the expansions of single crystals are measured usually along the crystallographic habits or along the directional vectors of the suitably cut planes. The linear thermal expansion in any arbitrary direction with direction cosines l, m, and n is given by

$$\alpha_{lmn} = \alpha_a \times l^2 + \alpha_b \times m^2 + \alpha_c \times n^2. \tag{6.11}$$

In the case of lower symmetric (viz., triclinic) systems, the lattice strains are defined with six independent Lagrange strain components (ε_{ij}) tensor expressed as

$$\begin{bmatrix} \varepsilon_{11} & \varepsilon_{12} & \varepsilon_{13} \\ \varepsilon_{21} & \varepsilon_{22} & \varepsilon_{23} \\ \varepsilon_{31} & \varepsilon_{32} & \varepsilon_{33} \end{bmatrix}, \tag{6.12}$$

where the components ε_{11}, ε_{22}, and ε_{33} (i.e., ε_{ij} for $i = j$) are strains along the principal axes and ε_{ij} (for $i \neq j$) is defined with an angle between two axes, that is, with respect to two principal axes. The ε_{ij} terms are defined as $\varepsilon_{12} = a \times b - \delta_{ab}$, $\varepsilon_{23} = b \times c - \delta_{bc}$, and $\varepsilon_{13} = a \times c - \delta_{ac}$. Thus, the coefficients ε_{11}, ε_{22}, and ε_{33} are strains along the normal crystallographic axes and are called principal strain coefficients. Similarly, the coefficients ε_{12}, ε_{23}, ε_{31}, and so on are called shear strains.

From Equation 6.11, the temperature derivative terms that represent thermal expansion are as follows:

$$\begin{bmatrix} \alpha_{11} & \alpha_{12} & \alpha_{13} \\ \alpha_{21} & \alpha_{22} & \alpha_{23} \\ \alpha_{31} & \alpha_{32} & \alpha_{33} \end{bmatrix}. \tag{6.13}$$

Hence, the volume thermal expansion coefficient in terms of the unit cell parameters is defined as

$$\alpha_V = \alpha_a + \alpha_b + \alpha_c + \frac{1}{X}\left[\phi_\alpha \frac{\partial \cos\alpha}{\partial T} + \phi_\beta \frac{\partial \cos\beta}{\partial T} + \phi_\gamma \frac{\partial \cos\gamma}{\partial T}\right], \tag{6.14}$$

where

$$X = \frac{V}{abc} = (1 - \cos^2\alpha - \cos^2\beta - \cos^2\gamma + 2\cos\alpha \times \cos\beta \times \cos\gamma)^{1/2},$$

$$\begin{aligned} \phi_\alpha &= \cos\beta \times \cos\gamma - \cos\alpha \\ \phi_\beta &= \cos\alpha \times \cos\gamma - \cos\beta\,. \\ \phi_\gamma &= \cos\alpha \times \cos\beta - \cos\gamma \end{aligned}$$

Besides, no shear components exist in the case of isotropic orthogonal crystals, that is, with $\alpha = \beta = \gamma = 90\,^\circ\mathrm{C}$, and thus the three principal strain are only used for lattice expansion behavior. In such cases, Equation 6.13 reduces to Equation 6.10, and thus the sum of the axial thermal expansion coefficients represents the volume thermal expansion coefficient.

6.3 Thermodynamics of Thermal Expansion

Using the definition of thermal expansion coefficients along with the Maxwell equations, the thermal expansion can be explained with other thermodynamic parameters. The relations between thermophysical properties of crystals have been extensively explained in several reports [19–23]. In this section, the relations of the thermal expansion coefficients with various thermodynamic and elastic parameters are briefly explained.

According to the definition (Equation 6.9), the volume thermal expansion coefficient (α_V) at constant pressure is given by

$$\alpha_V = \frac{1}{V}\left(\frac{\mathrm{d}V}{\mathrm{d}T}\right)_p.$$

Using the Maxwell relation $(\mathrm{d}V/\mathrm{d}T)_p = -(\mathrm{d}S/\mathrm{d}P)_T$, the above equation can be written as

$$\alpha_V = -\frac{1}{V}\left(\frac{\mathrm{d}S}{\mathrm{d}P}\right)_T = -\frac{1}{V}\left(\frac{\mathrm{d}S}{\mathrm{d}V}\right)_T\left(\frac{\mathrm{d}V}{\mathrm{d}P}\right)_T = \chi\left(\frac{\mathrm{d}S}{\mathrm{d}V}\right)_T, \tag{6.15}$$

where χ, the isothermal compressibility, is defined as Equation 6.16 and its reciprocal represents bulk modulus:

$$\chi = -\frac{1}{V}\left(\frac{dV}{dP}\right)_T. \tag{6.16}$$

The isothermal compressibility is related to the elastic constants (S_{ij}) of the lattices as

$$\chi = \sum_{i,j=1-3} S_{ij}. \tag{6.17}$$

Further, from the alternative Maxwell relation, α_V can be defined as

$$\alpha_V = \frac{1}{B_T}\left(\frac{dP}{dT}\right)_p, \tag{6.18}$$

where B_T is the isothermal bulk modulus.

In order to explain the thermal expansion behavior of solids, Grüneisen [24] had introduced the Grüneisen parameter (γ), defined as

$$\gamma = \frac{V}{C_V}\left(\frac{dP}{dT}\right)_p, \tag{6.19}$$

where V is the volume of the unit cell and C_V is the heat capacity at constant volume. Hence, the thermal expansion coefficient can be related to the thermodynamic parameters as

$$\alpha_V = \frac{C_V}{B_T V}\gamma. \tag{6.20}$$

Similarly, using the heat capacity at constant pressure, an equivalent relation for α_V can be derived as

$$\alpha_V = \frac{C_P}{B_S V}\gamma, \tag{6.21}$$

where B_s is the adiabatic bulk modulus.

Thus, it can be inferred from Equations 6.20 and 6.21 that the coefficient of thermal expansion has a direct relation with the heat capacity and inverse relation with the bulk modulus of a solid. Since the bulk modulus is weakly temperature dependent, the specific heat and thermal expansion behave similarly, but with a constant difference [23–25].

In principle, the thermal properties of a solid are the consequence of temperature variation of the Helmholtz free energy (F), which has been defined as $U - TS$, where U is internal energy and S is the entropy. Also, the pressure can be defined as $P = -(dF/dV)_T$. The volume of the material at pressure P and temperature T is decided by the minimum value of the $F(V, T)$. The increase in temperature can alter the minima of the free energy variation with the volume and the shift in the minima represents the change in volume with the temperature [26, 27].

6.4 Origin of Thermal Expansion

The atomistic analysis of thermal expansion shows that the thermal expansion originates from the anharmonic nature of the interatomic potential energy curve represented by potential well model [18, 28]. Several theoretical studies on thermal expansion of solids using various interatomic potential models have been discussed in detail in the literature [29–33]. The typical interatomic potential energy can be written as

$$U(x) = ax^2 - bx^3 - cx^4 \cdots , \tag{6.22}$$

where x is the displacement of atoms from the equilibrium position.

The coefficients a and b represent symmetric and asymmetric components of the vibrations, respectively, and c represents the damping of vibration at large vibration amplitude [18]. The typical potential energy variation with the interatomic separation is shown in Figure 6.1. An increase in the width of the asymmetry with temperature due to the longitudinal vibration leads to an increase in the equilibrium separation of the bonded pair by δr value, as shown in Figure 6.1. The mean separation (r) and vibration amplitude depend on the force constant of the chemical bond of the paired atoms. In this chapter, we have qualitatively mentioned that the thermal expansion of a solid is a consequence of the expansion of chemical bonds due to the asymmetric nature of the interatomic potential energy curve.

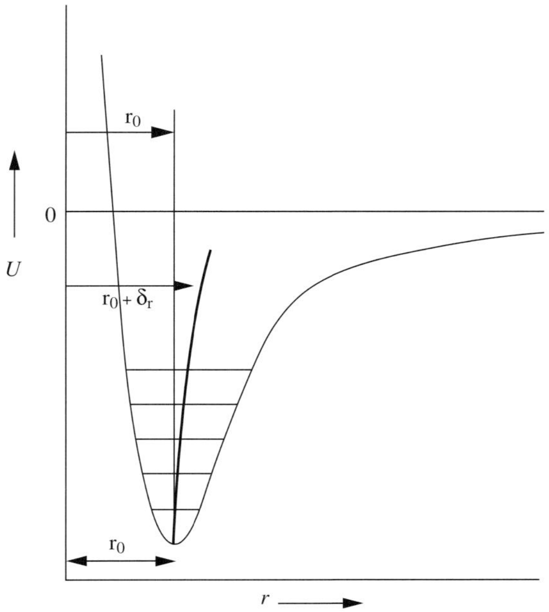

Figure 6.1 Variation of interatomic potential (U) with a separation of a pair of atoms (r).

Considering the vibration energy, the free energy function can be written in simpler relation [34–37]:

$$F = E + kT\sum_j \ln\left(\frac{h\omega_j}{kT}\right) \tag{6.23}$$

where k is the Boltzmann's constant, ω_j is the vibrational frequency, and E is the internal energy of the lattice.

Assuming the phonon frequencies as temperature independent, from the thermodynamic definition of pressure $P = -(\mathrm{d}F/\mathrm{d}V)_T$ and Equation 6.18, the coefficients of volume thermal expansion α_V is

$$\alpha_V = -k\chi\sum_j \frac{1}{\omega_j}\frac{\mathrm{d}\omega_j}{\mathrm{d}V}, \tag{6.24}$$

where V is the volume of the unit cell and χ is the compressibility.

The Grüneisen parameter represents the strength of anharmonic forces in the crystal and can be defined in terms of vibrational frequency (Equation 6.25):

$$\gamma_j = -\frac{\mathrm{d}\ln\omega_j}{\mathrm{d}\ln V}. \tag{6.25}$$

From Equations 6.24 and 6.25, the coefficients of volume thermal expansion can be related to the Grüneisen parameters:

$$\alpha_V = -\frac{k\chi}{V}\sum_j \frac{V}{\omega_j}\frac{\mathrm{d}\omega_j}{\mathrm{d}V} = \frac{k\chi}{V}\sum_j \gamma_j. \tag{6.26}$$

The overall Grüneisen parameter (γ) can be defined as a sum of all the mode Grüneisen parameters (γ_j). Including the temperature dependencies of vibrational frequencies, Equation 6.26 can be transformed to Equation 6.20. Thus, from Equations 6.20 and 6.26, it can be concluded that the magnitude and sign of γ are directly related to the coefficients of thermal expansion. For most materials, the positive values of γ are observed and they show normal positive thermal expansion behaviors. The variation of vibration energies with volume generally governs the sign of γ and thus the positive or negative thermal expansion coefficients.

In summary, the thermal expansion of crystals is related to their bonding strength and in turn to the lattice energy and crystal structure. Thus, the thermal expansion of a solid depends on the nature and strength of the chemical bond, mass of the vibrating atoms, melting point, crystal structure, and so on.

The above relations are based on the general assumption that thermal expansion or specific heat originating from the vibrational contribution of the lattice approach to zero following T^3 relation as the temperature approaches to 0 K. Similarly, they attain constant values at temperature greater than the characteristic temperature defined as the Debye temperature (θ_D). However, these expressions show appreciable deviation in the case of metals. It can be mentioned here that in the case of conductors or metals, a significant contribution to thermophysical properties arises from the electrons [28, 38, 39]. Hence, it is appropriate to consider the contribution of electrons along with the phonon or their interactions to the thermal expansion. It can also be mentioned here that most of the electron states have very little effect on specific heat and thermal expansion. However, electrons close to the Fermi surface significantly contribute to these properties. Besides, it can also be emphasized here that the Grüneisen parameter has a temperature dependency and thus cannot be assumed as constant over all temperatures in practice. It has been observed that the Grüneisen parameter approaches to a constant value as the temperature approaches to 0 K and also to another constant as the temperature exceeds θ_D [18, 38]. Considering the contribution of electrons, the thermal expansion coefficient can be by

$$\alpha_V = \frac{1}{BV}\left(\gamma_\omega^{\text{phonon}} C_V^{\text{phonon}} + \gamma^{\text{electron}} C_V^{\text{electron}}\right) \tag{6.27}$$

where γ^{phonon} is the Grüneisen parameter for the phonons, γ^{electron} is the Grüneisen parameter for electrons, C_V^{phonon} is the lattice contribution to specific heat, and C_V^{electron} is the electronic specific heat.

Assuming the electron gas model, γ^{electron} can be approximated to 2/3 [28] and hence the above equation can be expressed as

$$\alpha_V = \frac{1}{BV}\left(\gamma_\omega^{\text{phonon}} C_V^{\text{phonon}} + \frac{2}{3} C_V^{\text{electron}}\right). \tag{6.28}$$

The thermal expansions of metals have been extensively studied in order to decipher the contribution of both electron and phonon [28, 38, 39]. It has been observed that at very low temperatures, the contribution from electron are appreciable and thus the thermal expansion falls linearly with T instead with T^3 as in the insulators. The electronic contribution to thermal expansion can thus show a negative thermal expansion (NTE) behavior at low temperatures [19, 34]. Besides these electronic effects, magnetic interaction and correlated electrons also significantly contribute to the nature of thermal expansion [40–48], which has been discussed subsequently.

6.5 Techniques for Measurement of Thermal Expansion

The measurement of thermal expansion developed chronologically from crude marker comparison methods to the presently used more sophisticated and automated measurement devices such as dilatometer and interferometers. The principles and procedures of measurement of thermal expansion of solids have been discussed in detail in several literatures [49–55]. On the basis of the measured data, the measurements can be grouped as bulk and lattice thermal expansion measurements. The bulk thermal expansion of a material is measured by a direct observation of the change of dimension with temperature. In practice, the change in length is measured by a dilatometer or an interferometer. However, the lattice thermal expansion is generally determined by the diffraction methods like variable-temperature X-ray or neutron diffraction. In these methods the delineation of exact structural information at various temperatures provides plausible reason of thermal expansion.

The measurement of thermal expansion needs a mode to quantify the dilation of a material at a particular temperature with respect to a reference temperature. Thus, the sample is usually heated to a particular temperature and dilation or its effect is measured with various modes, such as physical measurement of length or measurement of other effects such as electrical or optical properties. Highly sensitive attachments to the direct measurement methods enhance the accuracy and shorten the data acquisition time. Some of the techniques are briefly touched upon here.

6.5.1
Dilatometer

The pushrod-type dilatometer is the most widely used technique for thermal expansion measurement, due to its simplest and efficient features [2, 3, 34, 56, 57]. In this method, the length of the sample is measured either continuously or periodically while heating or cooling it. The sample is placed in a furnace and a pushrod is placed touching the sample. The expansion in the sample is caused by movement of the pushrod. The magnitude of the expansion is measured by a transducer via mechanical, optical, or electrical means. The sensitivity of the transducer decides the accuracy of the expansion data. The vertical and horizontal designs of the dilatometer are the most commonly used geometries. Usually the horizontal geometry dilatometers have less thermal gradient compared to the vertical ones due to less convection loss of heat. Appropriate furnace assembly minimizing the convection loss can make the vertical design more effective. Besides, the shrinkage and negative thermal expansions are more suitably studied with the vertical dilatometers as sample position is intact without sliding from the end plate due to the self-weights of the sample and pushrod. The limitation to maximum temperature of measurements usually arises from the nature of pushrod. Fused silica rods are frequently used as pushrods due to their low thermal expansion ($\alpha_l = 0.5 \times 10^{-6}\,°C^{-1}$). The fused silica rods can be safely operated below the devitrification temperature of about 1000 °C. High-purity sintered alumina rods ($\alpha_l = 8\text{–}8.5 \times 10^{-6}\,°C^{-1}$) are widely used for high temperatures (up to 1700 °C) due to their high melting temperature and mechanical strength. In the case of alumina pushrod-based dilatometer, a blank run data are needed for correcting the values of overall expansion. For measurements at very high temperatures, metals such as Mo, Ta, and so on (up to 2000 °C) and graphite rods (up to 3000 °C) are often used, but only in vacuum or inert atmosphere.

The usual dilatometers have normal dial gauges based on mechanical strains to more sensitive linear variable differential transformers (LVDT) and resistive or capacitive or squid-based electrical transducer to quantify the thermal dilation. All these transducers are extensively used for the measurement of the bulk thermal expansion [2, 3, 56]. In the mechanical dial gauge, the dilation from the sample is transmitted to the strained coil attached to a pointer. The pointer deflects from its equilibrium position with the external stress and the extent of deflection indicates the dilation of the sample. More sensitive measurements of the dilation are done with the electrical transducers such as LVDT [58], where the pushrod is connected to the core placed inside the coils of a transformer. The motion of the core inside the coils generates *emf* signal in secondary coil and the signal is measured by potentiometers. In variable transformer, the null position is created with equilibrium position of the core in two oppositely connected transformers, and the development of the signal with positive or negative direction due to the sample expansion or contraction of the sample is measured directly. Also, the pushrod can be connected to a slider over a coil of a variac assembly. The measurement of resistance can give the position of the slider and hence the extent of expansion. Similarly, the pushrod can also be connected to one of the plates of a parallel plate capacitor and the measured capacitance governed by the separation of the plates is a measure of the dilation in the sample. The dilation in length (Δl) is measured with an increase in the temperature (ΔT) and thus the thermal expansion is calculated.

6.5.2 Interferometer

Interferometers have been extensively used for the measurement of thermal expansion from the basic concept of interference of light [59]. Usually these are sensitive even up to 0.025 μm in displacement over a wide range of temperatures. The design details and application of interferometer for the thermal expansion of materials have been explained in several reports [60–64]. In general, the fringes are formed by two optically flat transparent plates separated by the spacer (sample) due to interference when illuminated with a parallel monochromatic beam of light. The expansion or contraction of the sample due to thermal treatment changes the separation of the two reflecting plates and hence the variation of fringe patterns, which is observed from the fringe motion. The number of fringes passed through a reference point due to the variation of spacing of plates is counted. The variation of this spacing due to the increase in temperature is calculated from the following relation:

$$\Delta l = \Delta N \frac{\lambda}{2}, \tag{6.29}$$

where ΔN is the number of fringes passed through the reference points, λ is the wavelength of the illuminating light, and Δl is the change in the separation of the two plates and hence the change in the dimension of the sample.

Several designs, namely, Fizeau [65] and Fabry–Perrot [66] type interferometers are commonly used for thermal expansion measurements. In an interferometer, the fringes are counted by a telescope monitoring constantly. However, later developments of automated interferometers use photographic plates and photoelectric recorders to overcome this constraint. The changes in intensity at a reference point placed on the top plate are detected by photocell and the current pulse produced after each fringe movement is recorded as light intensity with time.

6.5.3
Telescope Methods

Telescopic methods are the most direct observation of thermal expansion where the dilation of the sample is measured by a microtelescope [67]. Often, a twin telescope is used to see two marked points of the sample placed in a heater and illuminated with a light source. The absolute expansion measurement can be very accurate, even up to few ppm (i.e., 10^{-6}) levels [68]. However, similar accuracy can be easily obtained by pushrod-type dilatometers. Thus, these methods are rarely used for measurements of thermal expansion.

6.5.4
Diffraction Methods

Diffraction (in particular X-ray diffraction) has been extensively used for fast and accurate measurement of the thermal expansion data. There are several monographs that explain the methods of determination of thermal expansion coefficients using X-ray diffraction [69–72]. In such measurements, the sample holder of a normal diffractometer is heated either directly or indirectly without any other modifications except the electrical and thermal protection of the other parts, such as goniometer or camera and detectors. Similarly, for the measurement of diffraction data at low temperatures, the sample is cooled either by a cryostat or by blowing cold gas. The diffraction methods are advantageous over other techniques for thermal expansion measurements as it requires small amount of sample and there is no need to have large single crystals or densely packed samples. The problem of temperature gradient is also minimized due to the small sample size. In fact, data obtained from the diffraction methods are the intrinsic thermal expansion data, that is, thermal expansion of the unit cell, in contrast to the bulk (overall) data in other methods. Thus, the anisotropy of the thermal expansion can be determined by diffraction methods. Also, the thermal expansion of a sample of interest in a mixture or in the absence of a phase-pure specimen can also be determined. The effects like phase transition, decomposition as well as effect of crystallization, strain, and so on can also be delineated by the diffraction methods. However, it should be mentioned here that unlike dilatometry, diffraction methods cannot be used in studying the thermal expansion behavior of amorphous materials.

More commonly, the diffraction experiments are meant for the structural studies of the crystalline materials. It is well known that crystal structure can be explained with three-dimensional arrays of parallel rows of atoms, which are equivalent to parallel planes of a normal grating. Thus, they produce bright and dark fringes of diffraction pattern when illuminated with a radiation having wavelength of the order of separation of planes. The Bragg law (Equation 6.30) of diffraction relates the possibility of constructive interference to the interplanar separations.

$$\frac{1}{{d_{hkl}}^2} = \frac{n^2\lambda^2}{4 \times \sin^2\theta_{hkl}} \quad \text{(Bragg's law)}, \tag{6.30}$$

where λ is the wavelength of radiation (Å), θ is the glancing angle (called as the Bragg angle) (°), d is the interplanar separations (Å), and n is the order of diffraction.

For measuring the axial thermal expansion, the unit cell parameters at various temperatures are determined by the observed reflections of the diffraction pattern recorded at different temperatures. It is known that the unit cell parameters are related to the position of the reflections (θ) and in turn to d (interplanar spacing) values as below.

The relation between the interplanar spacing and the unit cell parameters is given by

$$\frac{1}{d_{hkl}^2} = \frac{\frac{h^2}{a^2}\sin^2\alpha + \frac{k^2}{b^2}\sin^2\beta + \frac{l^2}{c^2}\sin^2\gamma + \frac{2hk}{ab}(\cos\alpha\cdot\cos\beta - \cos\gamma) + \frac{2kl}{bc}(\cos\beta\cdot\cos\gamma - \cos\alpha) + \frac{2lh}{ca}(\cos\gamma\cdot\cos\alpha - \cos\beta)}{1 - \cos^2\alpha - \cos^2\beta - \cos^2\gamma + 2\cos\alpha\cdot\cos\beta\cdot\cos\gamma}, \tag{6.31}$$

where d is the interplanar spacing of hkl planes; θ is the Bragg angle; λ is the wavelength of the X-ray used; and a, b, c, α, β, and γ are the unit cell parameters.

The diffraction method can use radiation sources, such as X-ray, neutron, or electron. However, for most practical studies, the X-ray and neutron diffractions are used. There are several pros and cons of neutron and X-ray diffractometers and hence they are often used as complementary techniques for the thermal expansion measurements. Since the neutron sources for the diffraction experiment are central facilities and are not as common as laboratory X-ray diffraction facilities, variable-temperature X-ray diffractometers are very commonly used for thermal expansion studies. The details of the procedures for thermal expansion measurements by X-ray diffraction are explained in the following section.

6.6
X-Ray Diffraction in Thermal Expansion

X-ray diffraction experimental setup requires an X-ray source, the sample under investigation, and a detector to count the diffracted X-rays. The radiation source can be monochromatic or polychromatic. The X-ray diffraction experiments can be carried out either with single-crystal or polycrystalline samples. The former type of samples are studied by the Laue method, the Weissenberg photograph method, or most commonly with the automated four-circle diffractometers [73, 74]. The latter types of samples are studied either by the Debye–Scherrer photographic methods or automated powder diffractometers. Similarly, the detection of scattered radiation can be made with a photographic film or radiation counter. The earliest methods of detection of X-rays are based on film methods that are also used even today. Modern single-crystal and polycrystalline diffractometers use image plate for detection of X-rays, which are fast and efficient. In a powder X-ray diffractometer, the powdered sample either in the form of a smeared layer or compacted flat pack or filled in a capillary is exposed to monochromatic beam of X-ray and intensity of the diffracted beam is collected over a range of angles (2θ with respect to the incident beam). The

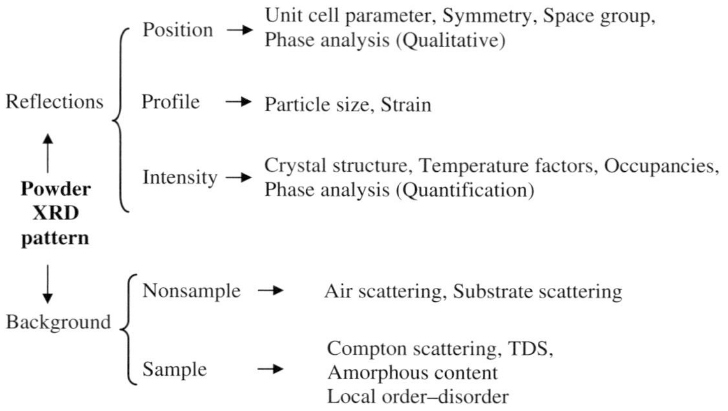

Figure 6.2 Schematic of typical information present in a powder X-ray diffraction pattern.

intensity corresponding to a constructive interference of the diffracted beam from a crystallographic plane is observed as a peak, corresponding to the Bragg angle (θ), while background is observed at all other angles. Besides, the symmetry in a lattice can also extinct some peaks in diffraction pattern. Qualitatively, the formal information present in a diffraction pattern can be depicted as shown in Figure 6.2.

The design of the diffractometer and heating assembly are based on the requirements [69–72]. In the case of variable-temperature X-ray diffractometer, the sample holder is placed in a heater or cryostat. The heater usually is a resistive heater that acts as sample holder in most commonly used high-temperature (HT) X-ray diffractometer with focusing geometry. In diffractometer with the Debye–Scherrer geometry, radiation heater surrounding the sample or reflection heaters are more commonly used than the direct resistive heaters. The utility of such heater is limited by its melting temperature and mechanical strength as well as the compatibility with sample and atmosphere [71]. Platinum or Pt-Rh strip is very commonly used for temperatures up to 1600 °C, while for higher temperatures W or Ta heaters are used but in inert atmospheres or vacuum.

The sample preparation and data collection (step width and step time) are determined by the sample character. For a good peak shape, appreciable intensity is essential, which is useful for getting the position of maxima accurately. The most important source of error to the intensity of the diffraction patterns arises from the preferred orientation, sample displacement, or transparency. The latter two also affect the position of the Bragg peaks. In addition, the sample heating rate and the temperature stability are also important parameters for generating good diffraction pattern for thermal expansion measurements. Selected reflections required for the determination of unit cell parameters are scanned for a time and width suitable for determining accurate peak position and hence the interplanar spacing. For the determination of unit cell parameters, Equation 6.31 can be rewritten in reciprocal lattice parameters as

$$\frac{1}{d^2} = h^2(a^*)^2 + k^2(b^*)^2 + l^2(c^*)^2 + 2hka^*b^*\cos\gamma^* + 2klb^*c^*\cos\alpha^* + 2lhc^*a^*\cos\beta^*, \tag{6.32}$$

where a^*, b^*, c^*, α^*, β^*, and $\gamma^*$are reciprocal lattice parameters and they are given by

$$\left.\begin{aligned} a^* &= \frac{bc\sin\alpha}{V}, \quad b^* = \frac{ca\sin\beta}{V}, \quad c^* = \frac{ab\sin\gamma}{V} \\ \cos\alpha^* &= \frac{\cos\beta\cdot\cos\gamma - \cos\alpha}{\sin\beta\cdot\sin\gamma}, \quad \cos\beta^* = \frac{\cos\gamma\cdot\cos\alpha - \cos\beta}{\sin\gamma\cdot\sin\alpha}, \\ \cos\gamma^* &= \frac{\cos\alpha\cdot\cos\beta - \cos\gamma}{\sin\alpha\sin\beta} \\ V^2 &= a^2b^2c^2(1 - \cos^2\alpha - \cos^2\beta - \cos^2\gamma + 2\cos\alpha\cdot\cos\beta\cdot\cos\gamma) \end{aligned}\right\}. \tag{6.33}$$

In the above equations, a^*, b^*, c^*, α^*, β^*, and γ^* are the unknown quantities. In a simplified manner, this equation is written as

$$Q^*_{hkl} = h^2 \times A + k^2 \times B + l^2 \times C + 2hk \times D + 2kl \times E + 2hl \times F, \tag{6.34}$$

where Q is equal to $1/d^2$.

The terms *A–F* in Equation 6.34 are the unknown parameters to be determined by the observed Q^* values by substituting appropriate integers for the Miller indices *h*, *k*, and *l* by trial and errors. In most of the trial and error methods, the crystal system is assumed to be cubic initially and thus reducing the above equation to simpler forms [69, 70]. The trials for indexing all the observed reflections are made by lowering the lattice symmetry to tetragonal, orthorhombic, and so on. In the Ito's method of indexing, a reverse approach is followed [69, 70, 75], where the reciprocal lattice parameters are calculated from appropriately selected Q_{hkl} using the following relations:

$$\left.\begin{aligned} &Q_{h00} = h^2 Q_{100} \quad Q_{0k0} = k^2 Q_{010} \quad Q_{001} = l^2 Q_{001} \\ &\cos\alpha^* = \frac{Q_{0kl} - Q_{0k\bar{l}}}{4klb^*c^*}, \quad \cos\beta^* = \frac{Q_{h0l} - Q_{h0\bar{l}}}{4hla^*c^*}, \quad \text{and} \quad \cos\gamma^* = \frac{Q_{hk0} - Q_{h\bar{k}0}}{4hka^*b^*} \end{aligned}\right\}. \tag{6.35}$$

The reciprocal lattice vectors are further refined by least-squares method to get their best values and then direct unit cell parameters are obtained using Equation 6.32. Several computer programs, namely, TREOR, VISER, ITO, CELL, UNITCELL, POWDER, INDEXING, and so on are used for automatic determination of the unit cell parameters from a set of observed *d*-values. The solution with the highest figure of merits with smallest unit cell volume and higher symmetry is usually preferred. The typical figures of merit of indexing are given either by the de-Wolff's M20 or M_N [76] or by the Louer's F_N [77] defined in Equation 6.36.

$$M_N = \frac{Q_N}{2|\overline{\Delta Q}|}\frac{N_{\text{obs}}}{N_{\text{pos}}}, \quad F_N = \frac{1}{|\overline{\Delta\theta}|}\frac{N_{\text{obs}}}{N_{\text{pos}}}, \tag{6.36}$$

where $Q_N = (1/d_N^2)$ for Nth reflection, $\overline{\Delta Q}$ and $\overline{\Delta \theta}$ are average errors in Q_s and $2\theta_s$, and N_{pos} is the number of possible lines within the last observed reflections.

The systematic absence in the indices can give the lattice type and symmetry of the unit cell. The determination of the space group by the powder data is mostly susceptible to mistake owing to the high degree of peak overlapping. However, it is also possible to get the unit cell and structural parameters accurately at various temperatures from the full powder X-ray diffraction pattern. Rietveld refinements [78, 79] have been extensively used to derive the structural parameters. So the general practice is the stepwise collection of full diffraction pattern and subsequent refinement of the structural parameters to determine exact crystal structure [80]. The basic principle of the Rietveld refinement is based on the calculation of intensity of an unknown compound assuming a model with all the structure and profile parameters. The structural model assumes space group, position coordinates of all the atoms, and fairly close unit cell parameters. The profile is defined with specific function, namely, Gaussian, Lorentzian, or their combination as pseudo-Voigt function. A smooth varying function or linear interpolation of the selected background points is used for modeling of the background:

$$Y_{ci} = y_{bi} + s\sum_{hkl} L \times P \times n \times |F_{hkl}|^2 \phi(2\vartheta_i - 2\vartheta_{hkl}) \times P_{hkl} \times A, \tag{6.37}$$

where Y_{ci} is the calculated intensity at the ith step, y_{bi} is the background intensity at ith step, L is the Lorentz factor, P is the polarization factor, $|F_{hkl}|^2$ is the structure factor for hkl reflections, $\varphi(2\theta_i - 2\theta_{hkl})$ is the profile function, P_{hkl} is the preferred orientation function, A is the absorption correction, and S is the scalar factor.

The difference between the calculated and observed intensity at each step is minimized by least-squares refinements. Thus, the minimizing factor is given by

$$D = \sum_{i=1}^{n} w_i (Y_{io} - Y_{ic})^2, \tag{6.38}$$

where w_i is the weighting factor and usually $1/Y_{oi}$.

Several computer packages, namely, DBWS, Fullprof, GSAS, Rietan, and so on, are used for the Rietveld refinements. The progress of the refinement and goodness of the refinement are judged by the residual indicators defined as R-values and the difference plot of the observed and calculated diffraction profiles.

In order to determine the thermal expansion behavior, the diffraction data are collected at various temperatures and subjected to similar analysis procedures to get the unit cell and other structural parameters at those temperatures. The temperature factors (B, defined as $B = 8\pi^2 u^2$, where u is the root mean square amplitude of vibration of atom from the mean position) are also refined at various temperatures. This is very important to understand the internal expansion of the crystal structure and thus the intricacy of the thermal expansion [81]. But it can be mentioned here that the refinement of thermal parameters and in particular anisotropic refinement is limited by the quality and strategy of collected diffraction data. Often, neutron diffraction data prove to be superior for the refinement of the anisotropic temperature factors.

Thus, it can be inferred that the bulk thermal expansion of materials is measured by the dilatometry or interferrometric techniques, while the lattice thermal expansion has been studied by the diffraction techniques. As mentioned earlier, the greatest source of errors in the measurement of the bulk thermal expansions arise from the microcrack and voids that can also be caused by the phase transitions with significant change in molar volumes. Thus, they often show lower values of coefficients of thermal expansion compared to true values. Thus, the observation of negative thermal expansion behavior in the bulk thermal expansion measurement is often an artifact. Suitable sized single crystals for such materials are also not easy to prepare. Diffraction methods provide the most reliable data for low or negative thermal expansion materials. Hence, all such materials are essentially studied by diffraction methods only.

6.7
Positive and Negative Thermal Expansions

A large number of thermal expansion data of diversified materials, in the interest of either mineralogy or technology or fundamental interests, have been generated by varieties of techniques. The thermal expansion data of these materials have been compiled in several databases. An extensive thermal expansion data of various metallic and nonmetallic solids have been compiled in Refs [82–84]. Similar compilations of a large number of compounds, in particular interest to geology or mineralogy, are also available in other literatures [85–89]. The most reliable data on thermal expansion and the structural explanation of thermal expansion behavior have been generated only after the development of high-temperature camera for diffraction studies [87, 89–92]. More commonly, the thermal expansion data were collected at ambient pressure, that is, 1 atm. Later with the theoretical and experimental development of equation of states with pressure, volume, and temperature, attention has been extended to the measurement of thermal expansion data under nonambient pressures. Initially such studies were focused on the materials of geological interest and later on technologically relevant materials [93, 94]. Based on the thermal expansion behaviors, most of the materials exhibit moderate ($\alpha_l \sim 10–30 \times 10^{-6}\,°C^{-1}$) to high positive ($\alpha_l \sim 50–60 \times 10^{-6}\,°C^{-1}$) coefficients. Certain materials such as low melting metals and ionic solids exhibit thermal expansion coefficients of the order of $80–100 \times 10^{-6}\,°C^{-1}$, while strongly bonded covalent solids such as Si, diamond, and so on exhibit low thermal expansion coefficients ($\alpha_l \sim 1–5 \times 10^{-6}\,°C^{-1}$). Most of the ceramic oxides exhibit significant positive thermal expansion. Data for a large number of fluorite and fluorite-related compounds having relevance to nuclear or fuel cell technology have also been extensively reported in the literature [7, 95–98]. Similarly, the studies in perovskite-type compounds bear fundamental interest in understanding the phase transition and structure-dependent thermal expansion behavior as well as interesting physical properties [99–104]. A large number of thermal expansion studies are aimed at technologically important optical and electronic materials for their practical application in the form and growth of crystals.

Extensive studies on the thermal expansion of ABO_4-type compounds with the scheelite-and zircon-type materials show that zircon-type silicates, vanadates, and phosphates show lower thermal expansion compared to the scheelite analogues [105–110]. Several examples of discerning the thermal expansion data from the high-temperature XRD and dilatometer are explained in the subsequent section.

In addition to these common phenomena, certain materials exhibit negative values for thermal expansion coefficients and are known as negative thermal expansion materials [5, 6]. The low and negative thermal expansion materials have attracted a lot of attention due to their technological importance as well as for fundamental understanding [111–123]. Also, the negative thermal expansion materials are very promising for tailoring thermal expansion behavior. Several materials, though relatively scarce, namely, AX_2O_8, AY_2O_7 (A = Zr, Hf; X = W, Mo; Y = P and V) [111, 113, 125–128], $A_2(XO_4)_3$ (A = trivalent cations, such as Sc^{3+}, Y^{3+}, Er^{3+}, Yb^{3+} etc. and X = W^{6+} or Mo^{6+}) [129–136], A_2O (A = Cu^+, Ag^+) [137, 138], AXO_2 (A = Cu^+, Ag^+ and X = Sc^{3+}, Al^{3+} and Ln^{3+}) [139, 140], various phosphates, such as $NbOPO_4$, VPO_5 [141, 142], and $Zr_2O(PO_4)_2$ [143], show contraction behavior with increase in temperature. Various silica polymorphs and related compounds are also known to exhibit low thermal expansions [144]. Anomalous or low thermal expansion in the β-eucryptite [145], cordierite ($Mg_2Al_2Si_5O_{18}$) [146], β-spodumene [147], and NZP family compounds [148–153] due to compensation of expansion and contraction along the different crystallographic axes are also reported. Various types of zeolite-type materials such as fauzasite and MCM are also reported to exhibit low or negative thermal expansion behavior [154, 155]. Several tetrahedrally bonded elements, such as Si, Ge, and so on, and amorphous silica also exhibit negative thermal expansion at low temperatures [156]. The negative thermal expansion of ice near the melting point is also an important example to reveal the bonding and structural effects [157]. Negative thermal expansion coefficients in polymeric structures such as elastomers and crystalline polyethylene are also known. Negative thermal expansions in f-block elements such as Pu and Ce have also been reported [158, 159]. Several alloys such as YbGaGe [160], $CeAl_3$ [161], and so on exhibiting negative and controllable thermal expansions along with the electrical or magnetic properties are also known. Alloys such as Invar, Covar, and so on also show such anomalous expansion behavior with very low thermal expansion coefficients [3]. Glass and glass ceramic compositions with low controllable thermal expansion behavior have been used extensively in various applications [162]. Anomalous or negative thermal expansion behaviors in some perovskite are also reported in the literature. In particular, some of the ferroelectric perovskite-type solid solutions also exhibit negative thermal expansion over a small temperature range [163–169]. The negative thermal expansion in ZrO_2 near the phase transition is also known [170]. A large number of compounds with cyanide groups are also known to exhibit negative or low thermal expansion behavior. $Zn(CN)_2$ [171] and $Cd(CN)_2$ [172] have negative thermal expansion coefficients that are almost double of that of ZrW_2O_8. Several complex cyanides exhibiting negative thermal expansion have been recently discovered [173, 174]. A large anisotropic thermal expansion coefficients of about $+140 \times 10^{-6}\,°C^{-1}$ (for α_a) and $-125 \times 10^{-6}\,°C^{-1}$ (for α_c) have been reported for trigonal $Ag_3Co(CN)_6$ [175]. Systematic temperature-dependent structural studies of these compounds revealed the structural intricacies relating to the origin of anomalous thermal expansion behavior. These aspects have been explained in subsequent sections.

6.8 Factors Affecting the Thermal Expansion Coefficients

From the origin of thermal expansion behaviors of materials and experimental data of large number of diversified materials, namely, weak van der Waals to strong covalently bonded network to closely packed structures, metals to insulators, non-magnetic to strongly ferromagnetic, and so on, it is observed that several of their physical properties have strong correlation with coefficients of thermal expansion. Thus, thermal expansion behavior can be qualitatively predicted from the melting point, compressibility, and strength of the chemical bonds of a material. The correlation of the thermal expansion coefficients with these properties are briefly explained below.

6.8.1 Melting Points

Melting temperature of a solid can be directly related to the thermal expansion as the melting process is related to the chemical bond expansion followed by dissociation. A general trend of the melting temperature with the thermal expansion coefficients has been observed [176, 177]. A comparison of melting temperature with thermal expansion coefficients of several isostructural materials is given in Table 6.1. The

Table 6.1 Comparison of melting temperature (mp) and thermal expansion coefficient (α) of some elements and compounds.

Name	mp (°C)	$\alpha_l \times 10^6$ °C^{-1}	Name	mp (°C)	$\alpha_l \times 10^6$ °C^{-1}	Name	mp (°C)	$\alpha_a \times 10^6$ °C^{-1} (Temperature range)
Li	179	47	$CoAl_2O_4$	1955	8.5	ThO_2[a]	2375	9.6 (20–1200 °C)
Na	98	70	$ZnAl_2O_4$	1952	8.7	CeO_2[a]	2600	11.6 (20–1200 °C)
K	64	83	$MgAl_2O_4$	2130	7.8	YSZ[a]	2700	10.6
Rb	40	90	$NiAl_2O_4$	2020	8.4	UO_2[b]	2840	9.38 (20–1425 °C)
Cs	29	97						
BCC			Spinel			Fluorite		
References	c)	d)			e)			

a) M.D. Mathews, B.R. Ambekar, and A.K. Tyagi, *J. Nucl. Mater.*, 341, 19, 2005.
b) A.C. Momin and M.D. Mathews, *Indian J. Chem.*, 15a, 1096, 1977.
c) *Handbook of Chemistry and Physics*, 50th edn, 1969.
d) R.K. Kirby, Thermal expansion, *American Institute of Physics Handbook*, 3rd edn, 1972.
e) L.M. Foster and H.C. Stumph, *J. Am. Ceram. Soc.*, 73, 1590, 1951.

thermal expansion coefficient can be approximated to the melting temperature by a simple empirical relation [176]:

$$\alpha_V = \frac{1}{16.6T_m}. \tag{6.39}$$

Owing to the inverse relation with melting temperature, the low melting elements such as alkali metals have very high thermal expansion coefficients compared to the transition metal elements, which have higher melting temperature. Similar trend is observed in the case of compounds also (Table 6.1). It can be mentioned here that only about 8–10% dilation is observed in materials up to melting temperature. Thus, this general trend of melting points does not hold over all the temperature range but to some extent within half of the melting point.

6.8.2
Bond Strengths

A close correlation of the bond valence with the expansion or compression of bond has been used to explain thermal expansion and compressibility of the crystal [178–180]. The stronger force of attraction leads to a rigid structure and thus inhibits the expansion or compression. From the Born potential model, Megaw [178, 179] has generalized the thermal expansion of chemical bond in terms of Pauling bond valence (S) as an empirical relation given below:

$$\alpha_{\text{bond}} \propto \frac{1}{S^2}, \tag{6.40}$$

where S is the bond valence defined as S is equal to z/n, z being the charge of the cation and n the coordination number around the cation.

Using the bond valences, Cameron *et al.* [181] and Hazen and Prewitt [180] generalized the thermal expansion of the bonds with empirical relations as

$$\alpha_{\text{bond}} \approx 32.9(0.75 - S_i) \times 10^{-6\circ}\text{C}^{-1}. \tag{6.41}$$

The definition of Pauling's bond valence has later been modified by Brown as Equation 6.42 [182, 183] to explain the thermal expansion of irregular polyhedra. In calculating the bond valences, the parameter B is usually fixed at a constant value of 0.37 and the equilibrium bond length (r_0) is defined empirically from bond lengths of regular polyhedron. Extensive compilations for the values of r_0 are reported in Refs [182–184]:

$$S_i = \exp\left(\frac{r - r_0}{B}\right), \tag{6.42}$$

where S_i is the bond valence of the ith bond.

The rigid nature of some chemical bonds, such as Si—O, Ge—O, P—O, V—O, and so on, as suggested from the above relations, has been validated in several temperature variation crystal structure studies [88, 92, 178–185]. In addition, the low or negative thermal expansion behaviors of several examples are also explained from such considerations.

6.8.3 Compressibility and Packing Density

As it has been mentioned earlier, the thermal expansion coefficient has a direct relation with the compressibilities. In general, more compressible materials have poorly packed structure and show higher thermal expansion. Thus, the molecular solids show higher values for thermal expansion coefficients [186]. The compressibility to thermal expansion can be related empirically as $\alpha_l \sim A \times \chi$, where χ is the compressibility and A is a constant. Hazen and Prewitt [180] have empirically related the thermal expansion coefficients and compressibilites as

$$\frac{\alpha}{\chi} = (32.9(0.75 - S_i)/37.0(d^3/z), \tag{6.43}$$

where S_i is equal to z/n, z being the charge of the central cation and n the coordination number, and d is the equilibrium bond length.

These empirical relations generally represent the trend of thermal expansion behavior. However, deviations from the normal behaviors are evident in several cases.

6.8.4 Defects and Impurities or Alloy Formation

Thermal expansion of a material is also affected significantly by the presence of defects in the lattice. The normal point defects such as vacancies (Schottky) or interstitial (Frenkel) types have significant role in the thermal expansion. The point defect concentrations are estimated from the difference in the bulk and lattice thermal expansion coefficients determined by dilatometry and high-temperature XRD [11]. In general, the linear thermal expansion measured by normal dilatometry shows higher thermal expansion coefficients compared to that determined by the X-ray diffraction methods due to the presence of vacancies. The difference and variation of the thermal expansion difference is used to determine the vacancy concentration as well as their formation energies [11]. Usually the thermal expansion data obtained by dilatometry and HT-XRD are similar at low temperature region but differ appreciably at higher temperature region.

Similarly, the extrinsic impurities also affect the thermal expansion behavior of a material. The alloying effect can show the averaging effect of the chemical bond strengths unless there is no significant alteration in the interaction of the atoms. Thus, most of the alloys show the weighted average thermal expansion coefficients of the component elements. Often deviations are observed in the case of additional segregated phase or clusters formation in the sample. In such cases, the thermal expansion measurement from the diffraction studies are also affected due to the additional strain in the lattice. Besides, the additional component in the solid solution can also alter the electronic structures and hence the drastic difference in the normal averaging behaviors.

6.8.5
Phase Transitions (Magnetic and Electronic Transitions)

Crystallographic phase transitions show discontinuity in the variation of the unit cell parameters with temperature due to the difference in the unit cell and bonding patterns. The two phases can have difference in the thermophysical properties due to the intrinsic internal arrangement of atoms. Also, the phase transition can cause either increase or decrease in the specific volume. In both the cases, the materials develop stress leading to a damaged mechanical integrity. In such cases, the thermal expansion measured by the bulk techniques lead to erroneous results. The diffraction experiments can be successfully used to study the difference in the thermal expansion behavior of these phases. The thermal expansion in several ferroelectric or relaxor compounds near the transition temperature shows discontinuity in the variation of unit cell parameters with temperature [163–169]. It has already been mentioned that the electrons also contribute to the thermal expansion in addition to the phonons. Thus, the materials exhibiting electronic or magnetic transitions often show anomaly in thermal expansion behaviors. The variation of the magnetic interaction of the ions is reflected as discontinuity or anomaly in the thermal expansion behaviors. The influence of magnetic interactions in several Fe-based alloys (Invar and related alloys) on thermal expansions is also known [3]. Similarly, the influence of electronic transitions on the thermal expansion in governing the thermal expansion coefficient can also be significant. The low or negative thermal expansion in several metals such as Pu, U, and Ce are explained with this effect [158, 159]. The negative thermal expansion of YbGaGe and $CeAl_3$ [160, 161] are also the consequence of such electronic transitions.

6.9
Structure and Thermal Expansion

Recent studies on anomalous thermal expansion materials, that is, materials with negative and anisotropic thermal expansion behavior indicate that the crystal structure of the materials plays a vital role in governing the thermal expansion behavior. High-temperature crystal chemistry studies on such materials revealed that the nature and type of bonding, polyhedra around the cations and packing of polyhedra in the unit cell, amplitude and anisotropy of vibration parameters of atoms, and so on are the key features in governing the magnitude and anisotropy of thermal expansion behavior.

In order to reduce the repulsive forces between the central cations, the polyhedra with the higher charge cations prefer to share only corners to form a lattice. Thus, an appreciably lower density structures arise in these arrangements. More implicitly, polyhedra formed with central ions with higher charges favor such framework

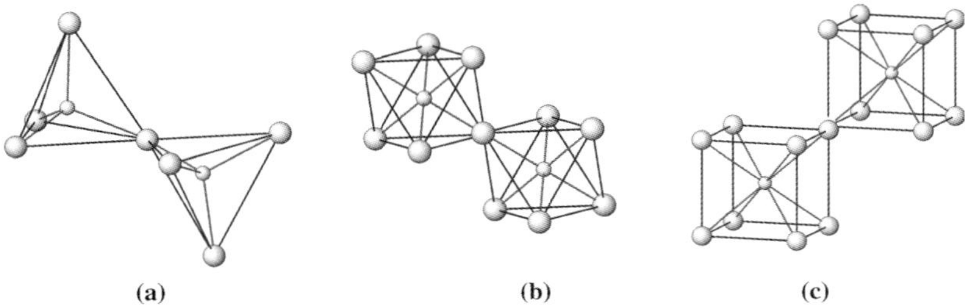

Figure 6.3 Typical connection of two tetrahedra (a), octahedra (b), and cubes (c) framework-structured compounds.

structure. The polyhedra around the central atoms can be tetrahedra, octahedra, or cube depending on the ionic radii. The typical polyhedral connections in such structures are shown in Figure 6.3. In an unstrained lattice, the maximum separation of the central ions is possible in linear linkage, that is, M–X–M = 180°, where M is the central ion and X is the ligand of a polyhedra. In such cases, the separation between the central atoms can be given by 2 × M–X.

However, in real crystals, the effective packing may restrict not only the nature of the polyhedra but also their angles. Such angled features lead to the tilting of the polyhedra, which are commonly observed in real crystals. As an example of framework structure, silica structure is formed with SiO_4 tetrahedra linked by sharing the corner oxygen atoms. The typical Si–O–Si bond angles (147–152°) are significantly lower than the expected 180°. The lower values of Si–O–Si bond angles indicate that the tetrahedra are tilted to accommodate the space with minimum strain. However, sufficient strains show appreciable expansion in the lattice to relieve it to an equilibrium values. These polyhedral tilting effects are observed in the perovskite lattices leading to phase transition as well as anisotropic thermal expansion behavior.

Besides, such framework materials do not show appreciable expansion due to transverse vibration of the bridging atoms of polyhedral units. The typical schematic figure showing the effects of transverse and longitudinal vibrations of the framework solid is depicted in Figure 6.4.

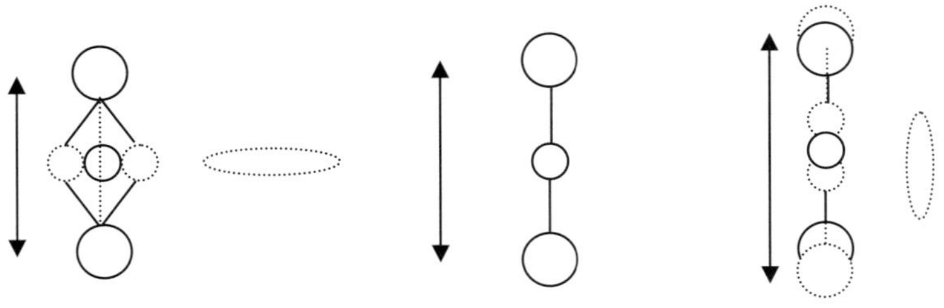

Figure 6.4 Schematic for representing the vibration of shared atoms leading to difference in separation of nonbonded atoms.

As observed from Figure 6.4, with the increase in temperature, the increasing amplitude of transverse vibration of a bridging atom can effectively lower the separation between the next nearest atoms, while the increase in the vibration amplitude in longitudinal direction, that is, along the length of the bond, leads to increase the same. Such phenomena have been observed in the materials with the two coordinated atoms. The anisotropy of the thermal vibration depends on the coordination as well as arrangement of atoms in the unit cell. The negative thermal expansion behavior of cubic ZrW_2O_8 or HfW_2O_8 has been explained by such effects [111]. This observation has been generalized in a wide varieties of framework materials showing negative thermal expansion behavior [6, 111, 126, 128, 129, 135, 141, 142, 167, 169, 187–197]. Some typical examples are explained in the subsequent section of this chapter.

Besides the tilting or rotation of polyhedral units, the distortion within the polyhedra also significantly affects the thermal expansion behavior of materials. In general, in real lattices, the polyhedra are often irregular and exhibit significant distortion. The polyhedral distortion arising from the variation of the various central to apex distances are crucial for the expansion of polyhedra, in addition to the interligand repulsion. The typical variations of bond valence with interatomic separation (r) are shown in Figure 6.5. A small change in r for $r < r_0$, where r_0 is the equilibrium distance, causes a significant change in bond valences compared to an identical change in r for $r > r_0$. Thus, a small increase in the shorter bonds needs a large decrease in the longer side. In the homogenization process of distorted polyhedra, the longer bonds decrease significantly and thereby reduction of polyhedral volume is observed.

The polyhedral distortions have been quantified in various relations explained in several reports. The most commonly used polyhedral distortion parameters are

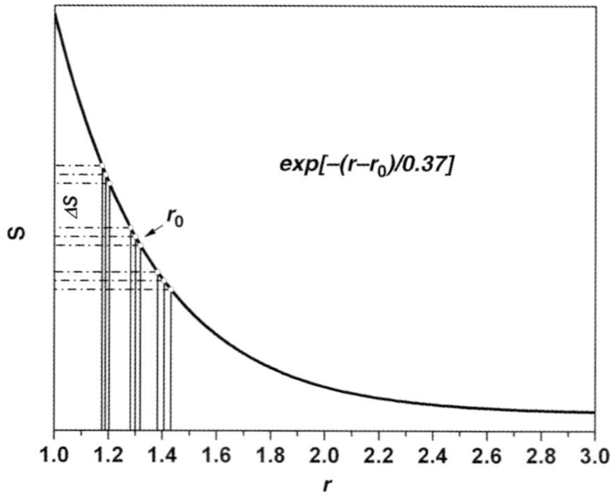

Figure 6.5 Variation of bond valence parameters (S) with interatomic distances (r).

defined by bond angle variance (σ^2) and quadratic elongation $\langle\lambda\rangle$ terms defined by Robinson *et al.* [197] as

$$\langle\lambda\rangle = \sum_{i=1}^{n} \frac{\left(\frac{l_i}{l_0}\right)^2}{n} \quad (\sigma^2) = \sum_{i=1}^{n} \frac{(\theta_i - \theta_0)^2}{n-1}, \tag{6.44}$$

where l_i and θ_i are the observed bond lengths and bond angles, l_0 are the bond lengths of corresponding regular polyhedra, θ_0 are the regular polyhedral angles, and n is the coordination number.

With an increase in temperature, more distorted polyhedra homogenize the bonds and that is reflected in its expansion behaviors. However, exceptions with increasing distortion at higher temperatures are also known. These are basically controlled by other structural parameters, such as interligand, interpolyhedral, or counterions repulsion in the lattice, and electronic characters of the ions [46].

Thus, in conclusion, the thermal expansion can be treated as a structure-sensitive property. In view of this, an extensive study of temperature variation of crystal structure provides the plausible reason for their behavior with temperature.

6.10 Examples

The salient results of thermal expansion behavior obtained for different systems have been grouped and explained on the basis of their structure type in this section. The similarities and the differences observed in the bulk thermal expansion from dilatometry and lattice thermal expansion from X-ray diffraction are explained. Also, the relation of crystal chemistry with the nature and magnitude of thermal expansion will be discussed.

6.10.1 Fluorite-Type AO_2 Compounds

Thermal expansion data on fluorite and related compounds have significant utilities in several areas, such as crystal growth of optical materials, design of fuel cells, gas sensors, and nuclear reactors [7, 99]. Also, rich crystallography of fluorite lattice due to the formation of anion or cation excess or deficient lattice is of fundamental interest [197b]. In particular, the fluorite-type oxides have a direct relevance to the nuclear industry as most efficient nuclear fuel materials are of fluorite types. Due to the higher thermal stability and melting points as well as the better irradiation and postirradiation managements [95–98], the fluorite-type nuclear fuels are often preferred. The empty crystallographic sites of the fluorite lattice can accommodate a large amount of fission products formed during the burning process. Some of the materials relevant to the nuclear industry are UO_2, ThO_2, PuO_2, AmO_2, CeO_2, and so on, which crystallize in the fluorite-type lattice. In particular interest of the Indian

Table 6.2 Thermal expansion coefficients of some of the fluorite-type (AO_2) compounds.

Name	$\alpha_a \times 10^6$ °C^{-1}	Temperature range (°C)	References	Name	$\alpha_a \times 10^6$ °C^{-1}	Temperature range (°C)	References
UO_2	10.8	20–2300	[96]	ThO_2	9.04	20–900[a)]	[198]
	10.0	20–1000	[97]		9.54	20–900	[198]
					10.41	20–1027	[199]
					10.24	20–927	[83]
					11.00	25–900	[200]
					10.4	20–927	[202]
					9.67	20–1000	[97]
					9.58	20–1200	[203]
PuO_2	11.61	20–927	[199]	CeO_2	10.58	20–900[a)]	[198]
	12.00	20–927	[83]		11.76	20–900	[198]
	12.14	25–900	[200]		12.06	20–1200	[198]
	11.4	20–1000	[97]				
NpO_2	10.80	20–1027	[199]	AmO_2	9.3	20–927	[200]
	10.99	25–900	[200]				
	11.14	20–927	[202]				
	9.5	20–700	[201]				
	11.0	20–1000	[201]				
BkO_2	13.2	25–900	[200]	CmO_2	10.10	20–1000	[200]

a) Dilatometric data.

Nuclear Energy Programme, ThO_2 and its substituted materials have been extensively studied in the recent years for their application as fuel in Advanced Heavy Water Reactor (AHWR) [204]. In this aspect, the detailed phase diagrams as well as thermal expansion behaviors have been investigated afresh with various simulated or actual fuel matrices [205]. The typical thermal expansion data on some of the fluorite-type materials relevant to nuclear technology are summarized in Table 6.2.

Similarly, thermal expansion behaviors of fluorite-type lattices containing rare-earth or alkaline-earth elements have also been studied by high-temperature X-ray diffraction. Some of such studies are explained in this section. In such cases, high-temperature diffraction studies have been extensively used to determine the lattice thermal expansion and to understand the internal scenario during the expansion of fuels under irradiation conditions.

6.10.1.1 **Isovalent Substituted AO_2 Lattices**

Isovalent cation substitution in the ThO_2 lattice form solid solution without any cation or anion vacancies. In such cases, the solubility limits are often restricted due to the ionic size mismatch of the host and guest cations. The formation of fluorite-type solid solutions have been reported in CeO_2–ZrO_2 [206, 207], CeO_2–HfO_2 [208], ThO_2–CeO_2 [98, 198, 209], CeO_2–ThO_2–ZrO_2 [207, 211], CeO_2–YSZ [212], ThO_2–UO_2 [213, 214], and so on systems. Limited solubility of about 5–10 mol% of ZrO_2 or HfO_2 in CeO_2 or ThO_2 or UO_2 due to ionic radii mismatch while solid solution is in the complete range in ThO_2–CeO_2 system is reported in these studies.

Freshley and Mattys [215] have reported a complete solid solution in between ThO_2–PuO_2. Detailed thermophysical properties of ThO_2, $Th_{1-x}U_xO_2$, and $Th_{1-x}Pu_xO_2$ have also been reported in the literature [216, 221]. The formation of an ideal solid solution or slight deviation from the ideality often depends on the thermal history of the samples in UO_2–ThO_2 system [217–220]. The X-ray diffraction studies on ThO_2–CeO_2 revealed the formation of ideal solid solutions between the two end members [198]. The powder XRD patterns of some of the compositions in ThO_2–CeO_2 system are depicted in Figure 6.6.

The observed unit cell parameters of $Th_{1-x}Ce_xO_2$ compositions show a gradual decreasing trend with the composition due to the ionic radii differences (ionic radii of Th^{4+} and Ce^{4+} are 1.05 and 0.97 Å, respectively, in cubic coordination [221]). The typical variation of the unit cell parameters with CeO_2 content is shown in Figure 6.7. The thermal expansion behavior of this series studied by dilatometer as well as the high-temperature XRD shows almost similar trend with the composition. The variation of the unit cell parameters of $Th_{1-x}Ce_xO_2$ at 1473 K with composition (x) (Figure 6.7) also show a similar trend as that at ambient temperature but with a different slope (−0.017(2) Å per mol of CeO_2) compared to that at ambient temperature (−0.019(2) Å/mol of CeO_2). The deviation in slope can be attributed to the difference in the thermal expansion with compositions. The coefficients of axial thermal expansion for the $Th_{1-x}Ce_xO_2$ compositions are also shown in Figure 6.7.

The variations of unit cell parameters of ThO_2 and CeO_2 with temperatures are shown in Figure 6.8. A comparison of the lattice thermal expansion behavior of ThO_2 and CeO_2 shows that the CeO_2 has higher thermal expansion coefficient than ThO_2.

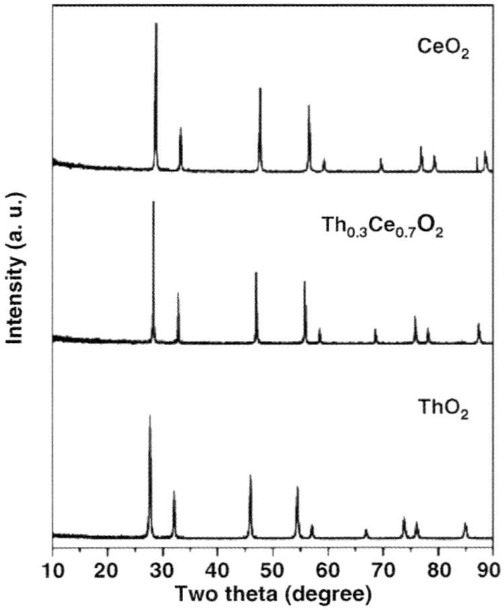

Figure 6.6 Powder XRD patterns (CuKα) of some $Th_{1-x}Ce_xO_2$ compositions.

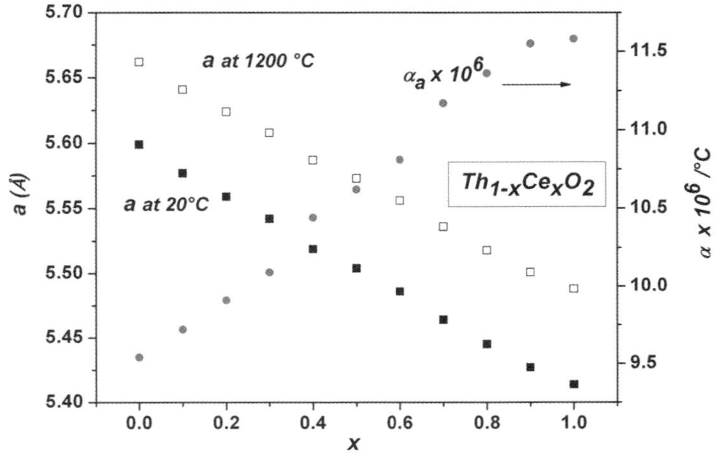

Figure 6.7 Variation of unit cell parameters of $Th_{1-x}Ce_xO_2$ at 293 and 1473 K and α_a with composition.

The difference in the thermal expansion behavior has been attributed to the difference in the melting point of CeO_2 than ThO_2. The thermal expansion studies in the series of compositions between ThO_2 and CeO_2 revealed that the lattice thermal expansion coefficients follow the averaging effect of the two end members.

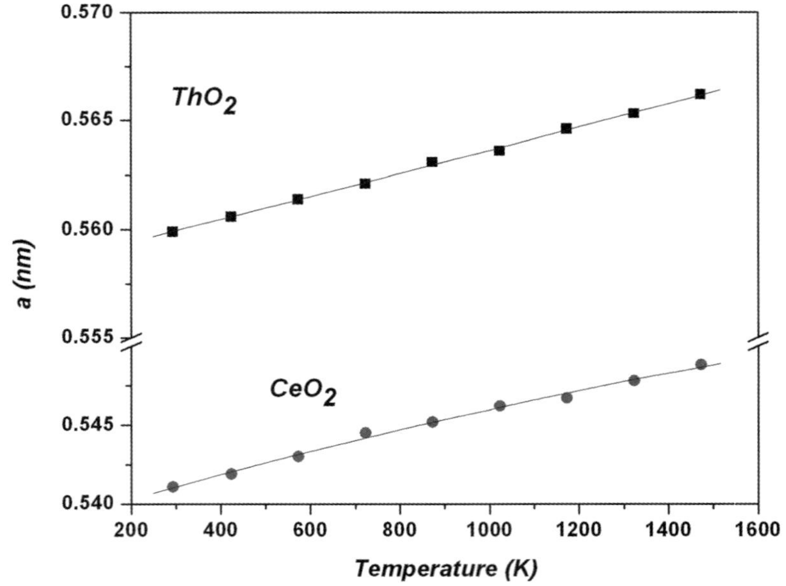

Figure 6.8 Variation of unit cell parameters of ThO_2 and CeO_2 with temperature.

Table 6.3 Comparisons of thermal expansion coefficients of some $Th_{1-x}Ce_xO_2$ obtained by dilatometry and HT-XRD [198].

x	a (nm)	% TD[a)]	$\alpha_l \times 10^6$ (°C^{-1}) RT–850 °C	$\alpha_a \times 10^6$ (°C^{-1}) RT–850 °C	$\alpha_a \times 10^6$ (°C^{-1}) (RT–1200 °C)	mp (°C)
0	0.5599(1)	96	9.04	9.54	9.54	3377
0.04	0.5591(1)	83	9.35	9.76	9.85	
0.08	0.5588(1)	83	9.49	9.96	10.01	
1	0.5411(2)	85	11.58	11.76	12.06	2600

TD, Theoretical density; mp, melting points; α_l, linear thermal expansion coefficients from dilatometry; α_a, lattice thermal expansion coefficients from HT-XRD.
a) sample sintered at 1300 °C for 48 h.

The thermal expansion coefficients of $Th_{1-x}Ce_xO_2$ compositions obtained from dilatometry as well as HT-XRD studies [198] are summarized in Table 6.3. The lower values of thermal expansion coefficients as observed in dilatometry (Table 6.3) can be attributed to lower packing density of the bulk samples.

The variation of unit cell parameters of ThO_2 and CeO_2 with temperature can be fitted with polynomial function, and the typical fitting equations are as follows:

$$\begin{aligned} ThO_2:\quad & a_T = 0.5584 + 5.0906 \times 10^{-6}[T] + 1.1420 \times 10^{-10}[T]^2 \\ CeO_2:\quad & a_T = 0.5387 + 8.3946 \times 10^{-6}[T] - 1.1023 \times 10^{-9}[T]^2, \end{aligned}$$

where T is the temperature in K.

The axial thermal expansion coefficients at various temperatures have been obtained by considering Equations 6.3 and 6.4. The coefficients of volume thermal expansion of these systems can be given by $3 \times \alpha_a$ (as in Equation 6.10). The typical variation of volume thermal expansion coefficients with temperature for ThO_2 is shown in Figure 6.9. With the available literature data of specific heat and compressibility of ThO_2, these data have been further extended to the determination of the Grüneisen parameters at various temperatures by Equation 6.21. For comparison, compressibilities and specific data of ThO_2 at various temperatures are included in Figure 6.9. The typical variation of the Grüneisen parameters of ThO_2 with temperature is shown in Figure 6.10.

Similar phase relations and thermal expansion studies on ZrO_2-, HfO_2-, or UO_2 substituted-ThO_2 or CeO_2 have also been carried out by X-ray diffraction method. As an example, the powder XRD patterns of some representative compositions of $Ce_{1-x}Hf_xO_2$ prepared by solid state reaction of CeO_2 and HfO_2 at 1400 °C for 48 h are shown in Figure 6.11.

The unit cell parameters of $Ce_{0.90}Hf_{0.10}O_2$, $Ce_{0.80}Hf_{0.20}O_2$, and $Ce_{0.70}Hf_{0.30}O_2$ are 5.388(1), 5.382(1), and 5.379(1) Å, respectively. The presence of monoclinic impurity phase of HfO_2 beyond $Ce_{0.90}Hf_{0.10}O_2$ could be confirmed from the variation of unit cell parameters of fluorite unit cell of the successive compositions.

The typical variation of diffraction pattern with temperature for a representative composition of ThO_2-4 wt% UO_2 in ThO_2–UO_2 system is shown in Figure 6.12.

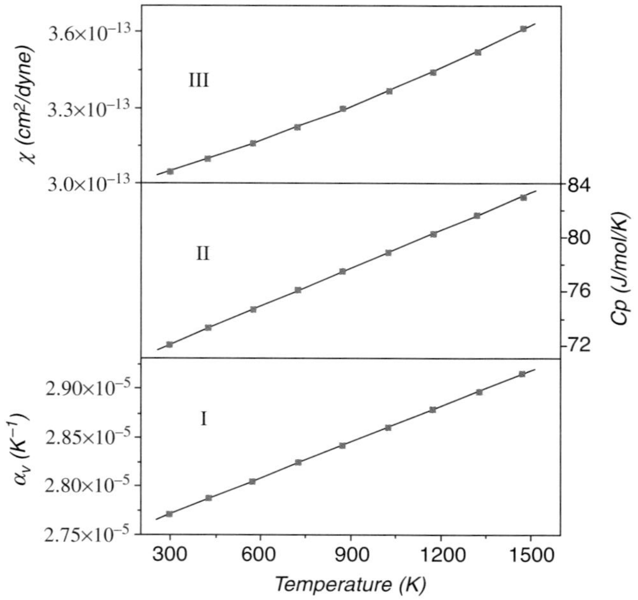

Figure 6.9 Variation of coefficients of volume thermal expansion (I), specific heat (II) (O. Kubaschewski (ed.), 1975, *Atom. Energy Rev.*, Special Issue No. 5, IAEA, Vienna), and compressibilties (III) (J.B. Wachtman, M.L. Wheat, H.J. Anderson, and S.J. Bates, 1965, *J. Nucl. Mater.*, 16(39)) of ThO_2 with temperatures.

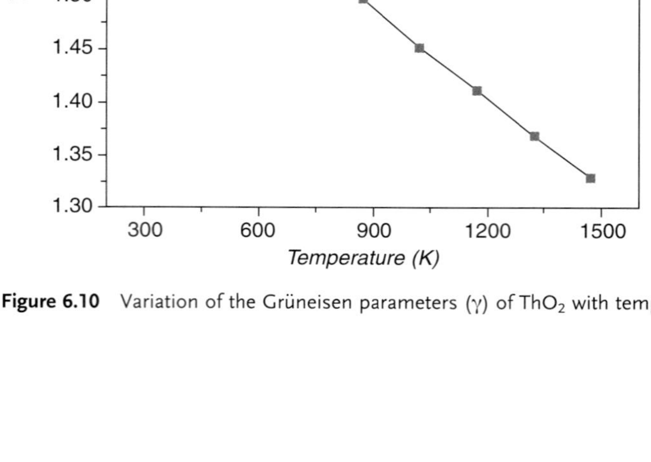

Figure 6.10 Variation of the Grüneisen parameters (γ) of ThO_2 with temperatures.

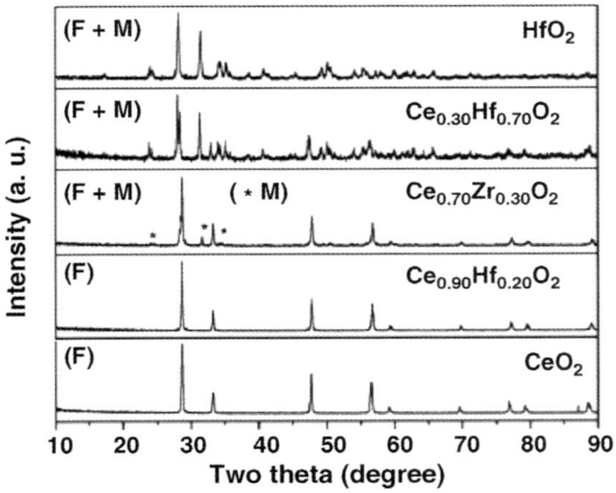

Figure 6.11 Powder XRD patterns (CuKα) of some $Ce_{1-x}Hf_xO_2$ compositions. (M is the monoclinic HfO_2 phase; F is the fluorite-type $Ce_{1-x}Hf_xO_2$ solid solutions.)

The position of the reflections shows a gradual shift toward lower angle with increase in temperature. The thermal expansion coefficients of $Th_{1-x}U_xO_2$ vary linearly with compositions. Besides, the oxidation state of uranium in such solid solutions has also significant influence on them as observed by dilatometric studies. A comparative bulk and lattice thermal expansion studies on the near stoichiometric UO_2 also reveal appreciable contribution of Schottky defects to lattice expansion at higher

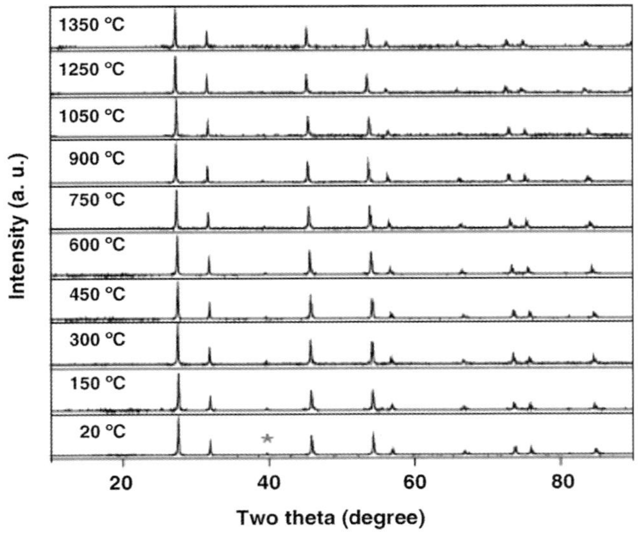

Figure 6.12 Powder XRD patterns (CuKα) of ThO_2-4 wt% UO_2 composition at different temperatures. Asterisk (*) indicates the reflection due to platinum sample holder.

Table 6.4 Axial thermal expansion coefficients (α_a) of some substituted AO_2 compounds.

Composition	α_a ($\times 10^6$) °C^{-1}	Temperature range (°C)	References
ThO_2	9.13, 9.58, 9.67	20–900, −1200, −1350	[213, 214]
ThO_2-2 wt % UO_2	9.74	20–900, −1200, −1350	[213]
ThO_2-2 wt % UO_2	9.35, 9.82	20–900, 20–1350	[214]
ThO_2-4 wt % UO_2	9.35, 10.09	20–900, 20–1350	[214]
ThO_2-6 wt % UO_2	9.97, 10.37	20–900, 20–1350	[214]
$Th_{0.05}Ce_{0.90}Zr_{0.05}O_2$	11.91	20–1200	[211]
$Th_{0.10}Ce_{0.80}Zr_{0.10}O_2$	11.72	20–1200	[211]
$Th_{0.15}Ce_{0.70}Zr_{0.15}O_2$	11.58	20–1200	[211]
$Th_{0.45}Ce_{0.45}Zr_{0.10}O_2$	11.19	20–1200	[211]
$Th_{0.75}Ce_{0.125}Zr_{0.125}O_2$	9.75	20–1200	[211]
$Th_{0.80}Ce_{0.10}Zr_{0.10}O_2$	9.88	20–1200	[211]
$Th_{0.95}Zr_{0.05}O_2$	9.24	20–1200	[211]
$Ce_{0.9}Zr_{0.1}O_2$	13.7	20–1200	[211]
$Ce_{0.8}Zr_{0.2}O_2$	13.9	20–1200	[211]
CeO_2	12.68	20–900	[210]
ZrO_2	8.0	293–1173	[222]

temperatures [217, 221]. An explicit survey of thermal expansion behavior obtained from diffraction measurements shows a strong influence of composition on the temperature dependency of unit cell parameters [217].

The typical axial thermal expansion coefficients of some isovalent substituted AO_2-type compounds are summarized in Table 6.4. The similarity and differences can be attributed either to the measurement methods or to the temperature ranges.

6.10.1.2 **Aliovalent Substituted AO_2 Lattice**

In addition to the above-mentioned isovalent substituted fluorite-type compounds, the aliovalent substituted products also have relevance to nuclear technology for simulating the fission product carrying behavior of nuclear fuels. The major fission products formed in the fission process of U and Pu are rare-earth elements and alkaline-earth elements, such as Sr and Ba. Under different heating conditions, the solid solution limits of various rare-earth ions and other fission product elements have been determined by analyzing the ambient-temperature X-ray diffraction data [223–236]. The phase analysis in the rare-earth substituted products of ThO_2 reveals a fluorite (F-)-type solid solution formation in a limited range and beyond which rare-earth oxide remains as separated phase. While the similar studies in the CeO_2 indicate the fluorite-type solid solution in CeO_2-rich compositions while the C-type (Y_2O_3 group rare-earth oxides) solid solution in other ends. The phase widths of F- and C-type solid solutions depend on ionic radii differences between the rare-earth metals and host lattice metal ions. As an example, the phase analysis of $Ce_{1-x}Gd_xO_{2-x/2}$ compositions in CeO_2–Gd_2O_3 systems revealed the formation of F-type solid solutions up to nominal composition at $Ce_{0.45}Gd_{0.55}O_{1.45}$ and C-type solid solutions at and beyond $Ce_{0.40}Gd_{0.60}O_{1.40}$ [232]. These two solid solution phases

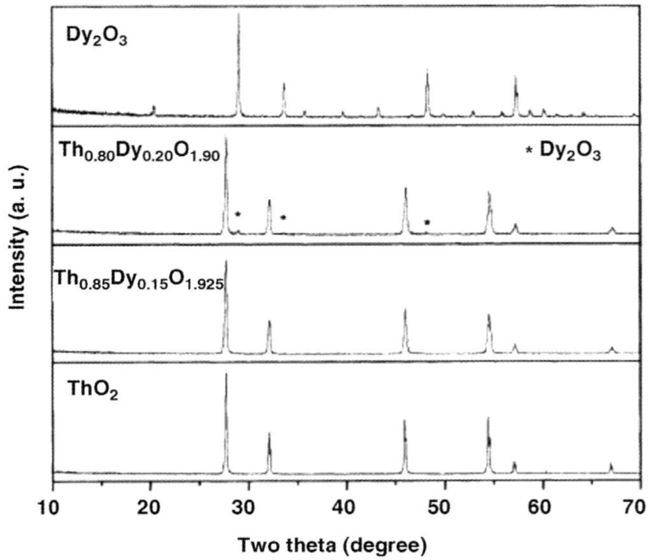

Figure 6.13 Ambient-temperature powder XRD patterns (CuKα) of $Th_{1-x}Dy_xO_{2-x/2}$ compositions.

can have a miscibility gap and its width depends on the ionic radius of the trivalent rare-earth ion [225–231]. Similar results are also observed in ThO_2–Nd_2O_3, ThO_2–Dy_2O_3, CeO_2–Ho_2O_3, and so on systems. The typical XRD patterns depicting the evolution of phases with composition in $Th_{1-x}Dy_xO_{2-x/2}$ system are shown in Figure 6.13.

It can be mentioned here that these aliovalent substituted compositions can have anion excess C-type lattice or anion deficient fluorite lattice. However, the ordering of the anions or distortion of the lattice is also possible in these compositions [243–247]. The presence of such defective fluorite lattice directly affects the lattice thermal expansion and in turn bulk thermal expansion behavior. In addition, the lattice expansion or contraction due to the ionic size and vacancy influence can also lead to the micro- or macrocracks affecting the bulk thermal expansion behavior. The typical high-temperature XRD patterns of $Ce_{0.90}Ho_{0.10}O_{1.95}$ representing a fluorite-type solid solution are shown in Figure 6.14. The unit cell parameters obtained by indexing the peaks of diffraction patterns recorded at different temperatures are shown in Figure 6.15. The typical axial thermal expansion coefficient (α_a) of $Ce_{0.90}Ho_{0.10}O_{1.95}$ is $13.0 \times 10^{-6}\,°C^{-1}$ compared to $12.6 \times 10^{-6}\,°C^{-1}$ of CeO_2.

Similar studies on the thermal expansion behavior of $Th_{1-x}Nd_xO_{2-x/2}$ ($0.0 \leq x \leq 1.0$) shows a systematic increase in the values of thermal expansion coefficients with Nd_2O_3 contents [238]. Salient results of the axial thermal expansion of $Th_{1-x}Nd_xO_{2-x/2}$ ($0.0 \leq x \leq 1.0$) series are given in Table 6.5. However, the opposite trends are observed in the case of Gd^{3+}-, Eu^{3+}-, and Dy^{3+}-substituted ThO_2 lattice (Table 6.6) [240]. High-temperature X-ray diffraction studies on a series of solid solutions with composition $Ce_{0.5}M_{0.5}O_{1.75}$ (M = rare-earth ion) indicate the

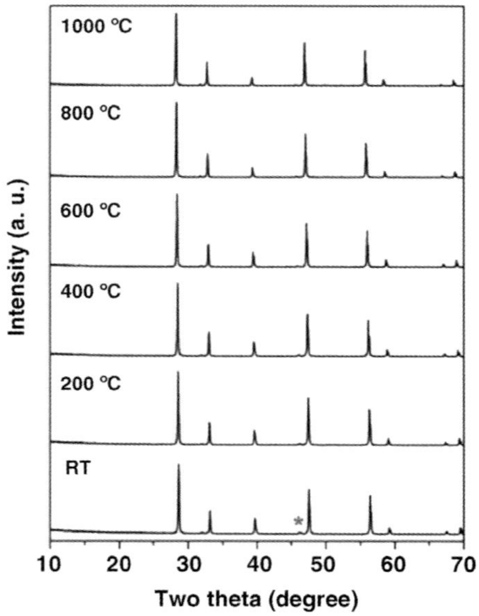

Figure 6.14 Powder XRD patterns (CuKα) of $Ce_{0.90}Ho_{0.10}O_{1.95}$ at different temperatures. Asterisk (*) indicates the reflection due to platinum sample holder.

significant role of the rare-earth oxide structure (Table 6.7) on the lattice thermal expansion behavior.

A comparison of thermal expansion coefficients of $Th_{1-x}M_xO_{2-x/2}$ systems indicates an increasing trend with the increasing contents of trivalent rare-earth ion of the hexagonal rare-earth oxides (Tables 6.5–6.7). The analysis of the structure

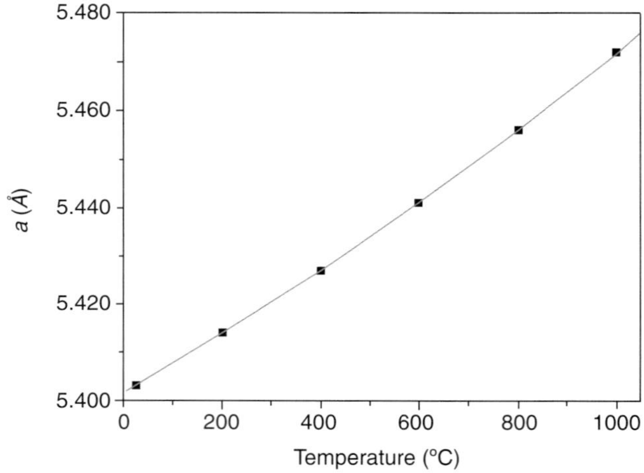

Figure 6.15 Variation of unit cell parameters of fluorite-type $Ce_{0.90}Ho_{0.10}O_{1.95}$.

Table 6.5 Lattice parameters of $Th_{1-x}Nd_xO_{2-x/2}$ ($0 \leq x \leq 1$) solid solutions [232].

S. No.	Composition	Mol % of $NdO_{1.5}$	Phase	Unit cell parameter(s) (Å)	$\alpha \times 10^6$ ($^\circ C^{-1}$)%
1	$Th_{1.00}Nd_{0.00}O_{2.000}$	0.0	F	$a = 5.600(1)$	9.04
2	$Th_{0.95}Nd_{0.05}O_{1.975}$	5.0	F	$a = 5.594(1)$	b)
3	$Th_{0.90}Nd_{0.10}O_{1.950}$	10.0	F	$a = 5.594(1)$	9.40
4	$Th_{0.85}Nd_{0.15}O_{1.925}$	15.0	F	$a = 5.595(1)$	b)
5	$Th_{0.80}Nd_{0.20}O_{1.900}$	20.0	F	$a = 5.594(2)$	9.64
6	$Th_{0.75}Nd_{0.25}O_{1.875}$	25.0	F	$a = 5.598(2)$	b)
7	$Th_{0.70}Nd_{0.30}O_{1.850}$	30.0	F	$a = 5.599(1)$	9.82
8	$Th_{0.65}Nd_{0.35}O_{1.825}$	35.0	F	$a = 5.598(1)$	b)
9	$Th_{0.60}Nd_{0.40}O_{1.800}$	40.0	F	$a = 5.597(1)$	10.12
10	$Th_{0.55}Nd_{0.45}O_{1.775}$	45.0	F	$a = 5.592(1)$	b)
11	$Th_{0.50}Nd_{0.50}O_{1.750}$	50.0	F	$a = 5.592(1)$	10.51
12	$Th_{0.00}Nd_{1.00}O_{1.500}$	100.0	H	$a = 3.831(1)$, $c = 5.998(1)$	11.26

% Temperature range: 20–1200 °C; F, fluorite type; H, hexagonal rare-earth oxide type.
b) No experimental data available.

indicates that the incorporation of the vacancy is likely to hinder the thermal expansion. However, the decreasing bond strength of the average M−O bonds due to smaller charged cations in the eight coordinated polyhedra enhances the thermal expansion due to the bond expansion. These two competing effects govern the overall thermal expansion of the lattice irrespective of the increase or decrease in unit cell parameters.

Besides, the trivalent rare-earth ions, the alkaline-earth metal ions, such as Sr^{2+}- and Ba^{2+}-substituted solid solution compositions, have also been extensively studied for thermal expansion as well as phase relations using X-ray diffraction [240–242]. A typical case of thermal expansion studies in the solid solution region of $Ce_{1-x}Sr_xO_{2-x}$ [235, 236] (Table 6.8) shows that the coefficients of thermal expansion increase with increasing the Sr^{2+} contents. This can also be assigned to the more ionic nature of Sr−O bonds in the Sr^{2+}-substituted fluorite lattice.

Table 6.6 Thermal expansion coefficient (α_a) × 10^6 ($^\circ C^{-1}$) of $Th_{1-x}M'_xO_{2-y}$.

x	Nd	Eu	Gd	Dy	Y	Sr	Ba
0	9.55	9.55	9.55	9.55	—	—	—
0.10	10.60	9.41	9.41	9.27	8.35	9.43	9.65
0.20	10.76	9.26	9.11	—	—	—	—
0.30	10.91	8.81	8.83	—	—	—	—
0.40	11.06	8.68	8.68	—	—	—	—
	20–1200 °C (HT-XRD data)				20–900 °C (dilatometry data)		
References	[224]	[224]	[224]		[234]	[234]	[234]

Table 6.7 Thermal expansion coefficients (α_a) of fluorite-type $Ce_{0.5}M_{0.5}O_{1.75}$ (in the range 293–1473 K) [223].

M	Unit cell of MO_x lattice	α_a ($\times 10^6$) °C^{-1}
Ce	F	12.68
La	H	10.48
Nd	H	11.87
Sm	M	12.10
Eu	M	12.61
Gd	M	12.47
Dy	C	11.77
Ho	C	11.95
Er	C	12.29
Yb	C	12.32
Lu	C	12.45

F, CeO_2 (fluorite); H, hexagonal; M, monoclinic; C, cubic rare-earth oxides lattice.

6.10.2
Framework Materials

It has already been explained that certain crystalline materials show negative or low thermal expansion compared to normal positive thermal expansion [6b, 111–142]. From the precise crystal structure analysis at various temperatures, Sleight and his coworkers explained the role of structure in the negative thermal expansion of ZrW_2O_8 [111, 113]. Following this, a large number of newer materials in phosphates, tungstates, molybdates, and so on with low or negative thermal expansion coefficients have been discovered [124–142]. The low thermal expansion behavior of NZP ($NaZr_2P_3O_{12}$)-type compounds due to anisotropic unit cell expansion has also been reported from the variable-temperature structural studies [144–152]. The most common structural features in all these materials focused on the rigid nature of the polyhedra and their connections. These aspects have been explained in Section 6.1. From the variable-temperature structural studies of such materials, it is revealed that the transverse vibration of bridging atoms of framework crystals shows negative or low thermal expansion. The common flexible feature of the framework of the crystal lies at the atoms that are connected only to two atoms,

Table 6.8 Thermal expansion coefficient (α_a) of fluorite-type $Ce_{1-x}Sr_xO_{1-x}$ solid solutions [235].

Nominal composition	Phases	Unit cell parameter *a* (Å)	α ($\times 10^6$) °C^{-1} (RT–900 °C)	α ($\times 10^6$) °C^{-1} (RT–1200 °C)
$Ce_{1.00}$ $Sr_{0.00}O_{2.00}$	F	5.402(3)	11.58	12.68
$Ce_{0.95}$ $Sr_{0.05}O_{1.95}$	F	5.416(1)	11.62	12.82
$Ce_{0.925}$ $Sr_{0.0755}O_{1.925}$	F	5.422(1)	11.82	13.02

F, cubic (fluorite type).

that is, the bridging atoms [6b,132]. The transverse vibration of the common polyhedral corner atoms leads to the tilting of polyhedron and that in turn lowers the average distance between the polyhedral centers. Several theoretical and dynamics studies support this tilt motions [116–123, 127]. Thus, the search for low thermal expansion materials is mainly focused on such low-density framework-structured materials. In addition, perovskite- and elpasolite-related structured materials also show such tilt motion leading to anisotropic expansion behavior [163–168]. The thermal expansion behaviors of a large number of tungstates, molybdates, phosphates, and so on have been studied by HT-XRD. The detailed crystal structure analyses at various temperatures and their comparison revealed the factors controlling the thermal expansion of such materials. Some of the examples are explained in the following sections.

6.10.2.1 Cristobalite-Type APO_4 (A = Al^{3+}, Ga^{3+}, and B^{3+})

Extensive studies on different crystallographic modifications and phase transitions of silica (SiO_2) revealed quartz, tyrdymite, and cristobalite as the three stable polymorphs at ambient pressure. Further, various displacive transitions in these polymorphs are also known. Considerable amounts of studies on silica have been devoted for understanding the structure of earth's interior. However, technologically important properties such as piezoelectric and dielectric properties of silica with low thermal expansion behavior also attract several studies on SiO_2. Cristobalite, the high-temperature modification of SiO_2, has been known for low thermal expansion coefficients. Several APO_4 (A = trivalent cations)-type phosphates are known to crystallize in almost all modifications of SiO_2 structures and have half of the Si replaced by A and the other half by P atoms [237–242]. The structures of these compounds contain both AO_4 and PO_4 tetrahedral units and the four oxygen atoms of ${PO_4}^{3-}$ are connected to four metal ions [242]. The cristobalite (orthorhombic) lattice for $AlPO_4$ and $GaPO_4$ is observed at elevated temperatures but can be retained at room temperature by annealing for long time above the cristobalite transition temperature. Since $AlPO_4$ and $GaPO_4$ are isostructural and there is small difference in ionic radii of Al^{3+} and Ga^{3+}, a continuous solid solution is formed between these two end members [196, 243, 244]. A series of compositions $Al_{1-x}Ga_xPO_4$ has been studied by HT-XRD in the temperature range of 25–1000 °C. The unit cell details of the series of compositions have been determined by the Rietveld refinement of the powder XRD data. The typical unit cell and structural parameters for $AlPO_4$ and $GaPO_4$ at room temperature are given in Table 6.9.

Similar details of the other compositions have been reported [200]. It can be mentioned here that though the ionic radius of Ga^{3+} is larger than that of Al^{3+}, $GaPO_4$ crystallizes with lattice of a smaller unit cell volume than $AlPO_4$ (unit cell volume of $AlPO_4$ and $GaPO_4$ are 351.2 (1) and 334.6(0) Å^3, respectively, $Z = 4$). A typical variation of unit cell parameters along the composition is shown in Figure 6.16. Besides, the average bond length of Ga–O (1.85 Å) is also higher than the Al–O bond length (1.77Å). The larger unit cell volume of the $AlPO_4$ can be attributed to the Al–O–P angles (average = 145°), which is higher than the Ga–O–P angle in $GaPO_4$ (average = 132°). The higher value of Al–O–P angle causes the

Table 6.9 Crystallographic parameters for $AlPO_4$ and $GaPO_4$ at room temperature.

	$AlPO_4$	$GaPO_4$
Temperature	25 °C	25 °C
Space group	$C222_1$ (No. 20)	$C222_1$ (No. 20)
a, b, c (Å)	7.084(1), 7.082(1), 6.9989(4)	6.9876(5), 6.9624(5), 6.8774(4)
V (Å^3), Z	351.15(9), 4	334.59(4), 4
M: 4b(0, y, 1/4)	0, 0.194(1), 1/4	0, 0.186(1), 1/4
P: 4a(x, 0, 0)	0.295(1), 0, 0	0.316(2), 0, 0
O1: 8c(x, y, z)	0.187(1), 0.058(2), 0.163(1)	0.202(2),0.026(2),0.187(2)
O2: 8c(x, y, z)	0.428(2),0.160(1), 0.946(1)	0.441(2), 0.171(2), 0.969(2)
R_p, R_{wp}, χ^2, R_B	13.4, 18.3, 2.3, 1.7	11.6, 15.9, 1.9. 5.6

larger separation between Al and P atoms as compared to Ga and P, leading to a larger unit cell volume for the former.

The structural details at various temperatures are obtained by comparing the HT-XRD data with the ambient-temperature XRD data. All $Al_{1-x}Ga_xPO_4$ compositions undergo structural transition at elevated temperatures. For example, the XRD pattern (Figure 6.17) of $AlPO_4$ recorded at 300 °C shows phase transformation. The observed reflections for this phase indicate the formation of high-cristobalite phase of

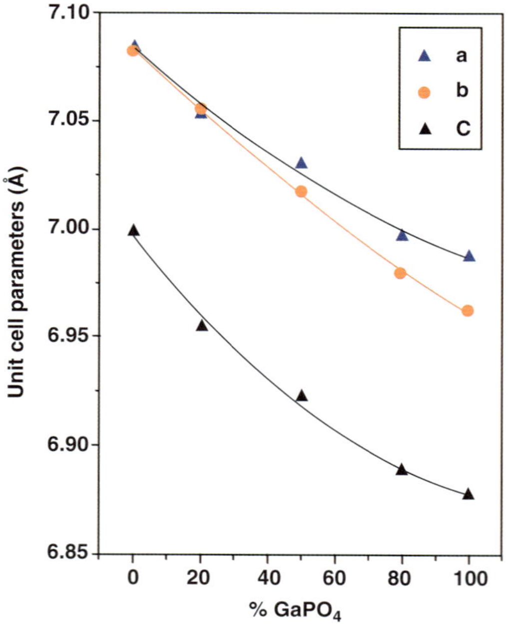

Figure 6.16 Variation of unit cell parameters with composition for $Al_{1-x}Ga_xPO_4$.

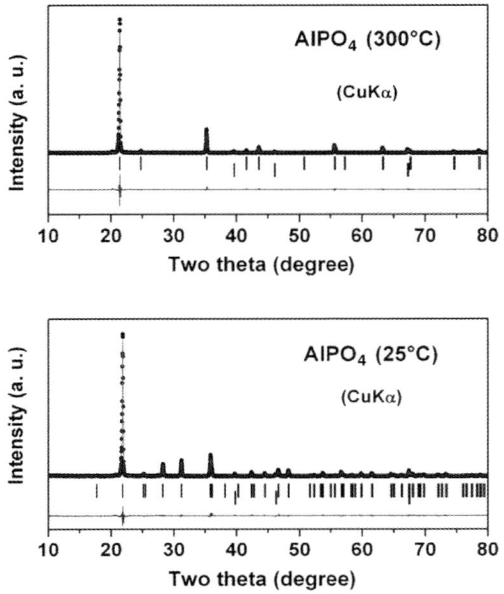

Figure 6.17 The Rietveld refinement plots for low-cristobalite (orthorhombic $C222_1$) and high-cristobalite (Cubic F-43m) phases of $AlPO_4$. Vertical lines indicate the Bragg positions (upper $AlPO_4$ and lower Pt sample holder).

$AlPO_4$, where the oxygen atoms are statistically occupied in 1/3 of the 48*h* sites [196, 243, 245, 246]. The crystallographic data at various temperatures for the high-temperature modifications have been similarly obtained by the Rietveld refinement of the high-temperature powder XRD data. The structural data for two representative compositions are given in Table 6.10. The structural transition has been characterized as reversible and nonquenchable displacive type and the typical transition temperature increases with Ga^{3+} ion content in the composition. A typical crystal structure of low- and high-cristobalite-type $AlPO_4$ is shown in Figure 6.18.

Table 6.10 Crystallographic parameters for the high-cristobalite-type $AlPO_4$ and $GaPO_4$ [196].

	$AlPO_4$	$GaPO_4$
Temperature	300 °C	700 °C
Space group	F-43m (216)	F-43m (216)
a (Å)	7.1969(2)	7.1850(2)
V (Å^3), Z	372.77(1), 4	366.76(2), 4
M: 4*a*	0, 0, 0	0, 0, 0
P: 4*c*	1/4, 1/4, 1/4	1/4, 1/4, 1/4
O1: 48*h*(*x*, *x*, *z*) (occ. 1/3)	0.114(1), 0.114(1), 0.188(1)	0.106(2), 0.106(2), 0.207(5)
R_p, R_{wp}, χ^2, R_B	13.4, 18.3, 2.3, 1.7	11.6, 15.9, 1.9, 5.6

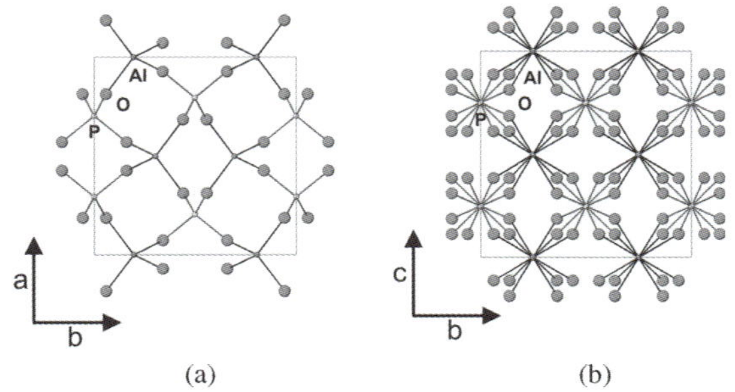

Figure 6.18 Crystal structure of (a) low-cristobalite (orthorhombic $C222_1$) and (b) high-cristobalite (cubic F-43m) phases of $AlPO_4$.

From the high-temperature X-ray diffraction studies, it could also be concluded that the high-temperature behavior of the orthorhombic phase of $GaPO_4$ is different from the rest of the compositions in this series. In the case of $GaPO_4$, the low cristobalite transforms to the β-cristobalite lattice at 700 °C, but relaxes slowly to quartz-type phase [196, 243]. In such cases, the structural details of the high-cristobalite modification in the presence of quartz-type phase could be accurately determined from the high-temperature diffraction studies.

The typical temperature variations of unit cell parameters for $AlPO_4$ and $GaPO_4$ are shown in Figure 6.19. The variation of unit cell volume with temperature for each composition shows that the orthorhombic phase has a significantly larger thermal expansion than the cubic (high-temperature) phase. The phase transition

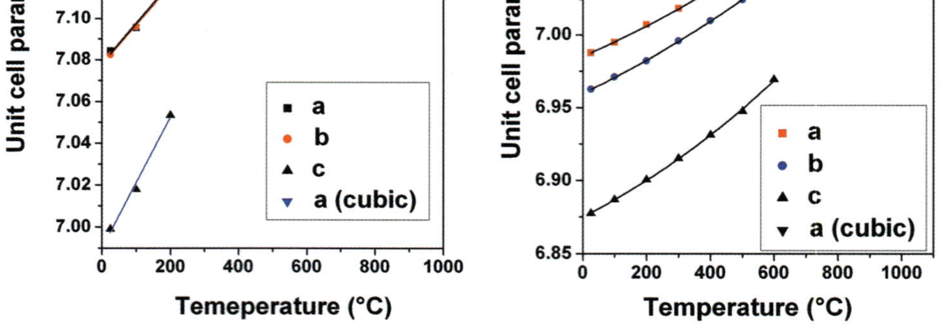

Figure 6.19 Variations of unit cell parameters of $AlPO_4$ and $GaPO_4$ with temperature.

Table 6.11 Thermal expansion coefficient of low (C222$_1$)- and high (F-43m)-cristobalite forms of $Al_{1-x}Ga_xPO_4$.

	C222$_1$ phase α_a, α_b, α_c, $\alpha_V$$^{a)}$	F-43m phase α_a, $\alpha_V$$^{a)}$	D$^{b)}$ × 10^4	T_c (°C)	$\Delta V^{c)}$ (Å^3)
$AlPO_4$	27.4, 27.2, 44.5, 99.8	1.93, 5.75	1.4	202	15.2
$Al_{0.8}Ga_{0.2}PO_4$	23.0, 20.3, 35.4, 79.1	4.11, 12.3	2.1	273	15.4
$Al_{0.5}Ga_{0.5}PO_4$	18.7, 22.4, 30.4, 71.9	5.52, 16.6	9.2	340	19.7
$Al_{0.2}Ga_{0.8}PO_4$	18.8, 20.3, 25.9, 65.6	6.40, 19.3	12.8	506	21.6
$GaPO_4$	18.5, 19.8, 23.2, 62.2	7.17, 21.4	17.9	605	24.8

T_c: transition temperature.
a) ($\times$ 10^6 °C^{-1}).
b) D: orthorhombicity ($|b - a/b + a|$).
c) Excess volume (ΔV) change in unit cell volume at the transition temperature.

accompanied by a significant increase in unit volume leads to the formation of a less dense structure at higher temperatures. The thermal expansion coefficients of both low- and high-cristobalite phases of various compositions along with the excess volume are shown in Table 6.11. The coefficient of volume thermal expansion of the high-cristobalite phase increases with the increase in Ga^{3+} content in the unit cell, while that for the low-cristobalite phase decreases with the Ga^{3+} content. The phase-transition temperatures and associated enthalpy are related to the change in unit cell volume and the orthorhombicity and strain in the lattice of the respective phosphates.

Similar studies on BPO_4 that crystallize in tetragonal lattice within the high-cristobalite frame have also been carried out in the temperature range of 25–900 °C [195]. The refined crystallographic data for two representative temperatures are given in Table 6.12 and the corresponding powder XRD patterns are shown in Figure 6.20. The typical variation of unit cell parameters along with temperature is shown in Figure 6.21. The variation of unit cell parameters with temperature shows a significant anisotropic expansion along the a-axis compared to c-axis.

Table 6.12 Structural details of BPO_4 at 25 and 900 °C (space group: I-4, No. 82) [195].

Temperature	a (Å)	c (Å)	V (Å)^3	Oxygen (x, y z)	(B) B (Å^2)	(P) B (Å^2)	(O) B (Å^2)
25 °C	4.3447(2)	6.6415(5)	125.37(2)	0.139(1) 0.259(1) 0.1275(5)	2.9(6)	2.4(1)	3.7(2)
900 °C	4.3939(2)	6.6539(6)	128.46(1)	0.127(1) 0.257(1) 0.1250(7)	3.2(7)	3.6(2)	5.2(3)
α (°C^{-1})	12.9 × 10^{-6}	2.1 × 10^{-6}	28.2 × 10^{-6}				

O: 8g (x, y, z); B: 2c (0, 1/2, 1/4); P: 2a (0, 0, 0).

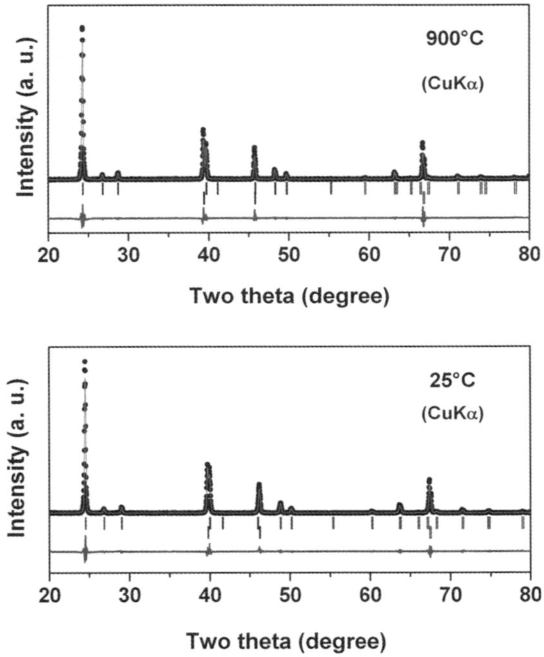

Figure 6.20 The Rietveld refinement plots for low-cristobalite-type BPO_4 at RT and 900 °C. Vertical lines indicate the Bragg positions (upper BPO_4 and lower Pt sample holder).

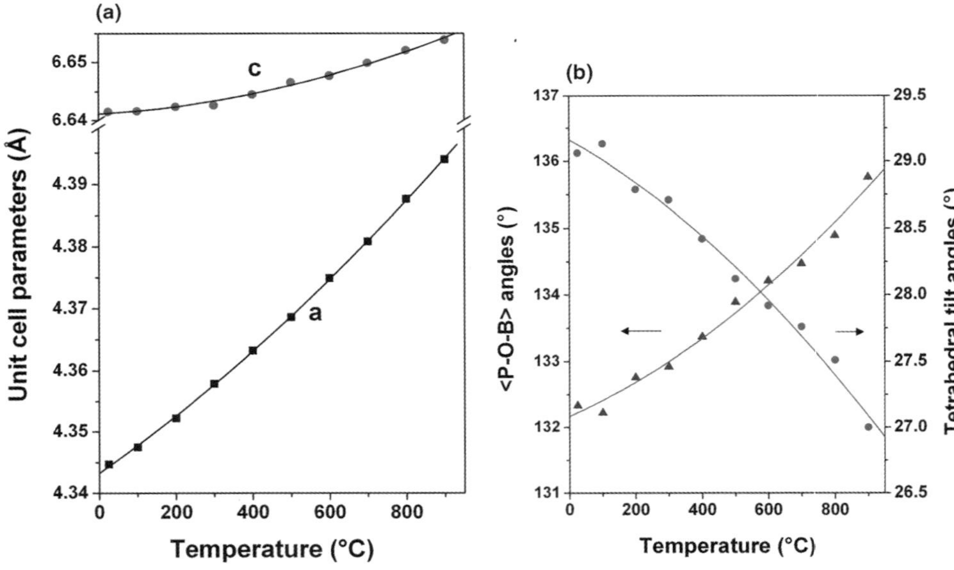

Figure 6.21 Variation of unit cell parameters (a), and B-O-P angles and polyhedral tilt angles (b) of BPO_4 with temperature.

The variations of the structural parameters, in particular, the thermal parameters of various atoms and derived structural parameters like bond angle and bond lengths, have been used to explain the thermal expansion behavior of the unit cell [195]. It has been observed that with increasing temperature, thermal parameters of the O atoms increases significantly compared to that of B and P atoms. A systematic increase of B—O—P bond angle with increasing temperature leads to gradual increase in the interpolyhedral separation and hence dilation in the unit cell axes. This is further reflected in the anisotropic expansion of unit cell axes ($c/a = 1.529$ at 25 °C and 1.514 at 900 °C). The typical B—O—P bond angle at 25 and 900 °C are 132.3(3)° and 135.8(4)°, respectively. It can also be emphasized here that the typical BO_4 and PO_4 units act as the typical rigid polyhedra, as reported by Hazen and Prewitt [180] for highly charged central cation polyhedra. Thus, with the increasing temperature, the tilts (ϕ) [247] between the BO_4 and PO_4 tetrahedra are lowered. The polyhedral tilt is common in the temperature- or pressure-induced structural variation of framework compounds. Such tilting variation has been used to explain the NTE or displacive phase transitions in perovskites [248–253]. The variations of the interpolyhedral angle (B—O—P) and tilt angles with temperature are shown in Figure 6.21.

A comparison of the thermal expansion coefficients of various high-cristobalite-type phosphates revealed that the higher the value of M—O—P bond angle, the lower the thermal expansion coefficient. The coefficient of volume thermal expansion (α_V) can be given by α_V (°C^{-1}) $= -191.32 + 4.33 \times [\theta] - 0.02 \times [\theta]^2$. The temperature variation of crystal structures of cristobalite-type compounds suggests that the M—O—P bond angle has significant role in thermal expansion as in other framework-type materials [126–132].

6.10.2.2 Molybdates and Tungstates

The MX_2O_8 (M = Zr, Hf and X = W Mo) and $M_2(XO_4)_3$, M = trivalent ions such as Y, Sc, Lu, Al, Cr, Fe, and so on, type molybdates and tungstates have drawn significant attention to thermal expansion measurements after the discovery of NTE behavior in several of them [111–115, 126–142]. All these compounds crystallize in framework-structured architect of corner-shared octahedral MO_6 or tetrahedral XO_4 units. Structural studies on MX_2O_8 compounds indicate that tungstates form cubic framework structure with MO_6 octahedral and XO_4 units, while the corresponding molybdates form layered structure with the sheets of MO_6 octahedral and XO_4 tetrahedral units. The studies on $A_2(MO_4)_3$ (A = trivalent cation and M = Mo^{6+} or W^{6+}) [254, 255] indicate the existence of a series of lattice types, namely, tetragonal, orthorhombic, monoclinic, and so on depending on the nature and ionic radii of trivalent countercations. Abrahams *et al.* [255] reported that the countermetal ions in these compounds exist in eight- or sixfold coordinations with orthorhombic $Sc_2(WO_4)_3$ type or monoclinic $Eu_2(WO_4)_3$ type depending upon the radius ratio of A^{3+} and O^{2-} ions. The thermal expansions of such materials have been determined extensively by the X-ray and dilatometric techniques and some of them are explained in the following sections.

6.10.2.2.1 **$A_2(MoO_4)_3$** The thermal expansion behavior of $A_2(MoO_4)_3$, A being the heavier rare-earth ions, and other trivalent ions such as Cr^{3+}, Fe^{3+}, and Al^{3+} has been studied in detail by dilatometry and HT-XRD [128–132, 255–261]. Dilatometric studies on dense pellets of $Al_2(MoO_4)_3$, $Cr_2(MoO_4)_3$, and $Fe_2(CrO_4)_3$ carried out over the temperature range from RT to 850 °C shows a common feature of expansion in all these three samples [130]. Detailed structural studies of these samples at ambient temperature indicate that all samples crystallize in monoclinic ($P2_1/c$) lattices [130, 262–266], with unit cell parameters: $a = 15.687(4)$, $b = 9.233(2)$, $c = 18.212(5)$ Å, and $\beta = 125.27(1)°$ for $Fe_2(MoO_4)_3$; $a = 15.554(6)$, $b = 9.151(3)$, $c = 18.102(9)$ Å, and $\beta = 125.32(2)°$ for $Cr_2(MoO_4)_3$; and $a = 15.368(1)$, $b = 9.034(1)$, $c = 17.864(1)$ Å, and $\beta = 125.35(1)°$ for $Al_2(MoO_4)_3$. In the thermal expansion studies, a significant positive expansion followed by a negative expansion is observed in each case (Figure 6.22) [130]. The typical observed thermal expansion coefficients in these three samples are given in Table 6.13. The discontinuity in thermal expansion behavior indicates a structural transition that is further confirmed by DSC [130]. Further thermal expansion behaviors of these are studied by high-temperature diffraction studies. The detailed high-temperature X-ray diffraction studies show that the ambient monoclinic lattice transforms to high-temperature orthorhombic lattice (Pnca). The typical powder XRD pattern of the $Cr_2(MoO_4)_3$ at temperatures below and above the transition temperature is shown in Figure 6.23.

The typical unit cell parameters for $Cr_2(MoO_4)_3$ at 400 °C are $a = 9.258(1)$, $b = 12.732(1)$, and $c = 9.177(1)$ Å and $V = 1081.8(1)$ Å^3 (space group: Pnca). The temperature variation of unit cell parameters of both ambient and high-temperature phases of $Cr_2(MoO_4)_3$ is shown in Figure 6.24. In contrast to the observed negative thermal expansion behavior in dilatometry, a subtle but positive thermal expansion

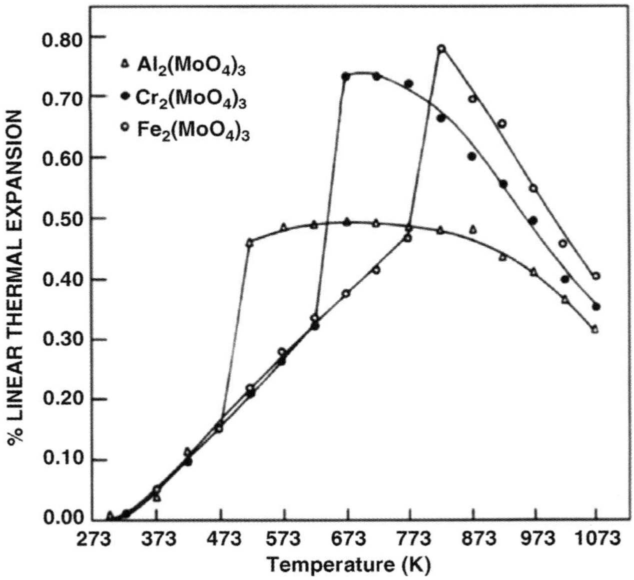

Figure 6.22 Variation in the percentage of linear thermal expansion with temperature [130].

Table 6.13 Coefficients of thermal expansion ($\alpha_l \times 10^6$ in $°C^{-1}$) data of $A_2(MoO_4)_3$ [130].

A^{3+}	Temperature range (°C)	$\alpha_l \times 10^6$ (in $°C^{-1}$)	T_c (°C)	
Fe^{3+}	27–500	9.7		500
	550–800		−14.8	
Cr^{3+}	27–350	9.8		380
	400–800		−9.4	
Al^{3+}	27–200	8.7		200
	250–800		−2.8	

T_c: Monoclinic to orthorhombic phase transition temperature.

is observed in high-temperature orthorhombic phase. The typical axial thermal expansion coefficients of $Cr_2(MoO_4)_3$ in orthorhombic and monoclinic phases are α_a ($°C^{-1}$) $= -0.7 \times 10^{-6}$, α_b ($°C^{-1}$) $= 6.8 \times 10^{-6}$, and α_c ($°C^{-1}$) $= -1.5 \times 10^{-6}$ and α_a ($°C^{-1}$) $= 8.87 \times 10^{-6}$, α_b ($°C^{-1}$) $= 9.12 \times 10^{-6}$, and α_c ($°C^{-1}$) $= 13.05 \times 10^{-6}$, respectively. Further, the analysis of the unit cell parameters indicates that about 1.5% of unit cell volume increases at the transition temperature. The drastic expansion of unit cell volume at the transition temperature develops micro- and macrocracks in the pellet and hence masks the thermal expansion of the actual lattice.

Similar variable-temperature crystal structure analyses on $Sc_2(MoO_4)_3$ (in the temperature range of 4–300 K) indicate that a monoclinic ($\alpha_v = 21.9 \times 10^{-6}\,K^{-1}$) to orthorhombic ($\alpha_v = -6.3 \times 10^{-6}\,K^{-1}$) phase transition occurs at 180 K [129]. The monoclinic phases of $Sc_2(MoO_4)_3$, $Fe_2(MoO_4)_3$, and $Al_2(MoO_4)_3$ are isostructural to ambient-temperature $Cr_2(MoO_4)_3$ type, while the orthorhombic phase is isostructural to $Sc_2(WO_4)_3$ [129, 255, 263, 267, 268]. In all these cases, the crystal structure is made up of AO_6 octahedra and MoO_4 tetrahedra. The detailed studies on the monoclinic and orthorhombic phases indicate that both the structures are related and the structural change is brought by small displacement of the atoms, in particular

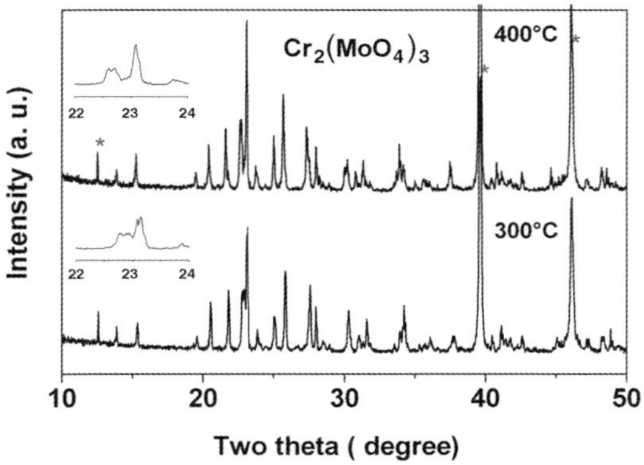

Figure 6.23 Typical XRD pattern of $Cr_2(MoO_4)_3$ at 573 and 673 K, representing the $P2_1/a$ and Pnca phases. (Reflections due to sample holder is marked as *.)

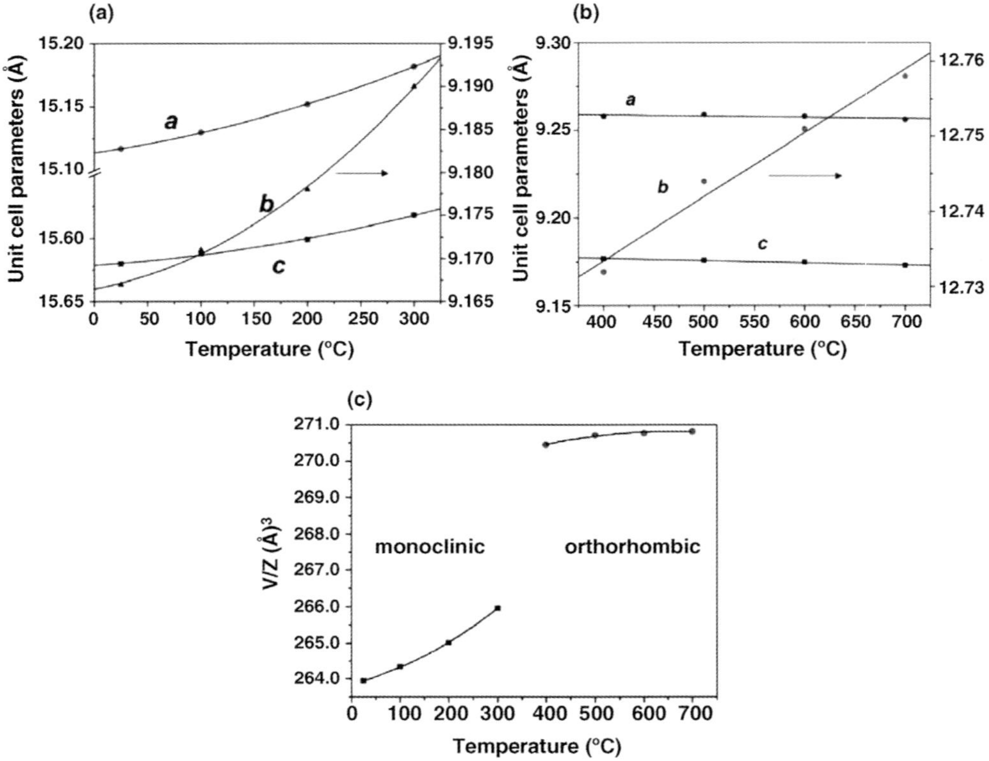

Figure 6.24 Variation of unit cell parameters, (a) monoclinic, (b) orthorhombic, and (C) molar volume (V/Z) of $Cr_2(MoO_4)$ with temperature.

in the oxygen sublattices. This transition is very commonly observed in several molybdates and is a ferroelastic to paraelastic displacive transition [129]. The low-temperature monoclinic modification has six distinguishable Mo sites, whereas the high-temperature orthorhombic phase has two distinguishable Mo sites. In both lattices, the Mo atoms have tetrahedral coordination and trivalent cations have an octahedral coordination and have flexible framework-type arrangements. Though both the unit cells are made from similar structural arrangement, the drastic expansion of the unit cell volume leads to poorly packed lattice and hence lower thermal expansion coefficient is observed for orthorhombic phase compared to the monoclinic phase. Similar monoclinic to orthorhombic transition is reported for other $M_{2-x}M'_x(MoO_4)_3$ analogues [260–263] with drastic difference in thermal expansion behaviors.

A comparison of thermal expansion coefficients of isostructural molybdates of various trivalent ions as observed by dilatometry or high-temperature XRD shows that the ionic radii of trivalent ions have controlling role on them. The typical thermal expansion coefficients of various orthorhombic molybdates are summarized in Table 6.14. It can also be mentioned here that the heavier lanthanide analogues

Table 6.14 Thermal expansion coefficients (°C^{-1}) of few orthorhombic $A_2(MoO_4)_3$-type molybdates.

A	Temperature range (°C)	$\alpha_a \times 10^6$ (°C^{-1})	$\alpha_b \times 10^6$ (°C^{-1})	$\alpha_c \times 10^6$ (°C^{-1})	$\alpha_l \times 10^6$ (°C^{-1})	References
Sc^{3+}	−90 to +25	−8.41	10.82	−8.83	−6.30	[129]
La^{3+}	25–800	3.02	3.02	12.83	6.36	[260]
Y^{3+}	25–800	−11.69	−6.57	−10.04	−9.36	[260]
Er^{3+}	25–800	−10.84	−3.34	−8.57	−7.56	[260]
Yb^{3+}	25–800	−10.02	−2.99	−5.21	−6.04	[260]
Lu^{3+}	25–800	−8.69	−1.64	−7.75	−6.02	[260]
Fe^{3+}	540–800	−1.94	4.57	0.71	1.14	[260]
Fe^{3+}	550–750	−2.28	7.35	−0.70	1.34	[258]
Al^{3+}	250–650	−0.94	7.34	0.57	1.8	[258]
Cr^{3+}	420–750	−2.40	6.26	−1.75	0.7	[258]
Cr^{3+}	400–700	−0.72	6.80	−1.54	1.52	a)
$(Yb_{1-x}Cr_x)^{3+}$						
0.1	200–800	−8.530	0.203	−5.886	−4.729	[257]
0.2	200–800	−5.378	3.512	−3.201	−1.691	[257]
0.9	250–800	−1.112	5.069	−0.551	1.134	[257]
1.0	450–800	−2.424	6.293	−1.738	0.708	[257]

All the reported data are given in Pnca setting.
a) Present data.

exhibit similar phase transition as well as thermal expansion behavior. However, at ambient temperature, $Gd_2(MoO_4)_3$-type orthorhombic lattices are formed in heavier lanthanide molybdates. The thermal expansion behaviors of these lattices also show a significant expansion in the lattice with different thermal expansion coefficients.

6.10.2.2.2 **$A_2(WO_4)_3$** Similar to $A_2(XO_4)_3$-type molybdates, the analogous tungstates have been studied extensively by high-temperature X-ray diffraction. The low and/or negative thermal expansion behaviors are observed in most of them. For this reason, most of the studies are mainly focused on the smaller trivalent ions, such as heavier rare-earth or transition metal ions. Often, the heavier rare-earth tungstates are hygroscopic at ambient temperature and on dehydration they form orthorhombic $Sc_2(WO_4)_3$-type structures [261]. However, with other cations, namely, Al^{3+}, Ga^{3+}, Sc^{3+}, In^{3+}, and so on, these tungstates do not show hygroscopic behavior. The crystal structure of monoclinic $Eu_2(WO_4)_3$ is closely related to scheelite and in turn to fluorite lattice. In these structures, the larger metal ion (Ln^{3+}) form eight coordinated polyhedra that share their edges similar to the normal fluorite-type lattice. The tungsten atoms form tetrahedra and penta-coordinated polyhedra and they link these LnO_8 units by sharing edges and corners. The stabilities of these structures are related to the stability of the eight and six coordinated polyhedra of the rare-earth cations. The crystal structures of these two types are shown in Figure 6.25. Besides, the orthorhombic $Sc_2W_3O_{12}$-type structures have relatively lower packing density

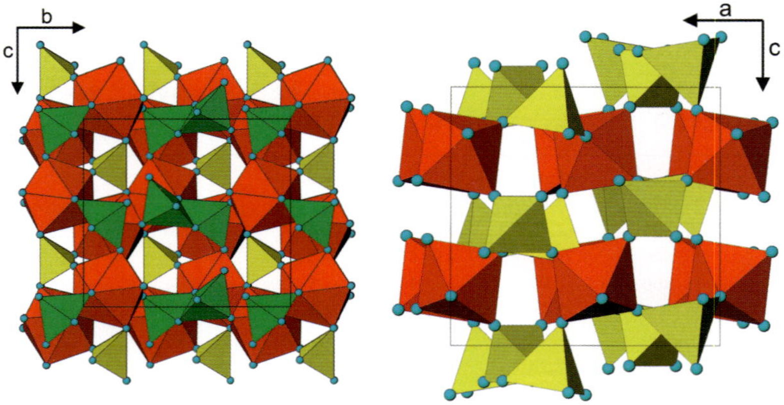

Figure 6.25 (a) Crystal structure of $Eu_2(WO_4)_3$ (EuO_8: red, WO_5: green, WO_4: yellow). (b) Crystal structure of $Sc_2(WO_4)_3$ (ScO_6: red, WO_4: yellow).

compared to the monoclinic analogues. Thus, the lower values of thermal expansion coefficients are expected only in the orthorhombic class tungstates.

As a typical example, the thermal expansion behavior of orthorhombic $Al_2(WO_4)_3$ studied by dilatometry and high-temperature XRD studies [134–136] is discussed in this section. At ambient temperature, unit cell parameters of $Al_2(WO_4)_3$ are $a = 9.134(2)$, $b = 12.575(2)$, and $c = 9.050(1)$ Å, $V = 1039.6(2)$ Å^3, and $Z = 4$ (space group: Pnca). The typical ambient-temperature XRD pattern of $Al_2(WO_4)_3$ is shown in Figure 6.26. The crystal structure reveals the presence of AlO_6 octahedra (Al—O bonds ranging from 1.82(2) to 2.01(2) Å) and two types of WO_4 tetrahedra, namely, $W(1)O_4$ (W—O bonds $1.74(2) \times 2$ and $1.77(2) \times 2$) and $W(2)O_4$ (W—O bonds 1.77(2) to 1.79(2) Å). The structure is isostructural to the high-temperature orthorhombic modification of the above-mentioned molybdates. A pellet (~12 mm diameter and

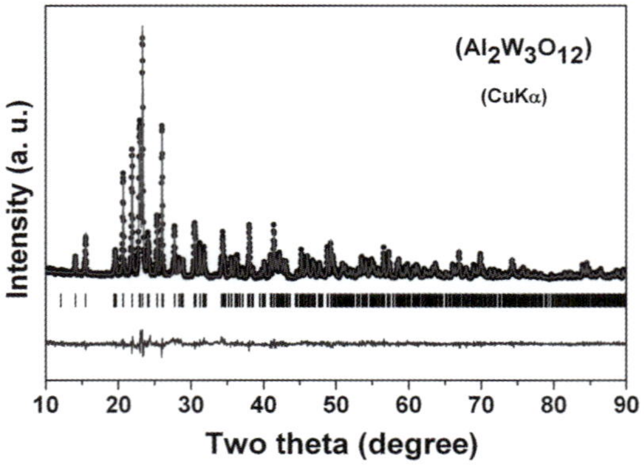

Figure 6.26 Ambient-temperature powder XRD pattern of $Al_2(WO_4)_3$. R_p: 9.52%; R_{wp}: 13.1% and χ^2: 1.84; R_B: 5.92%; and R_F: 3.41%).

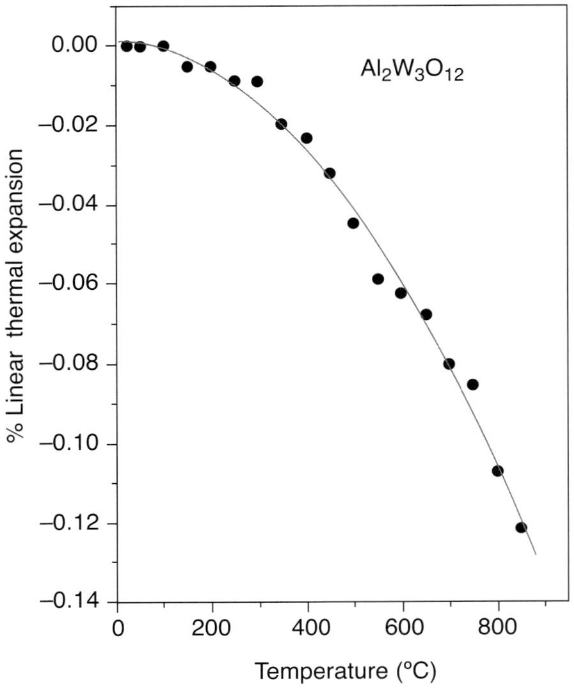

Figure 6.27 Variation in the percentage of linear thermal expansion with temperature of $Al_2(WO_4)_3$.

10 mm height) with approximately 88% of theoretical density obtained by heating a pellet of preformed powder at 1100 °C for 18 h was used for the dilatometric studies. The bulk thermal expansion of $Al_2(WO_4)_3$ as observed from dilatometry is depicted in Figure 6.27.

The variation in the percentage of the linear thermal expansion with temperature remains constant initially and then decreases. The observed coefficient of average linear thermal expansion (α_l) is $-1.5 \times 10^{-6}\,°C^{-1}$ in the region of 25–850 °C [136]. Further, it may be added here that a negative value of about $-3 \times 10^{-6}\,°C^{-1}$ has also been reported for average linear thermal expansion coefficients [126]. However, on the basis of the diffraction studies, this compound was shown to have an anisotropic expansion with a net positive volume expansion, $\alpha_v = +2.2 \times 10^{-6}\,°C^{-1}$ [126] and $+4.5 \times 10^{-6}\,°C^{-1}$ [134]. In the temperature range of 25–900 °C, the thermal expansion behavior of this is further determined by *in situ* high-temperature X-ray diffraction studies. The typical powder XRD patterns recorded at various temperatures are shown in Figure 6.28a. The observed reflections at each temperature are indexed to get the high-temperature unit cell parameters. The typical unit cell parameters of $Al_2(WO_4)_3$ at 900 °C are $a = 9.113(3)$, $b = 12.645(4)$, and $c = 9.051(2)$ Å and $V = 1043.0(3)$ Å^3. The variations of unit cell parameters with temperature are shown in Figure 6.28b. The typical coefficients of thermal expansion

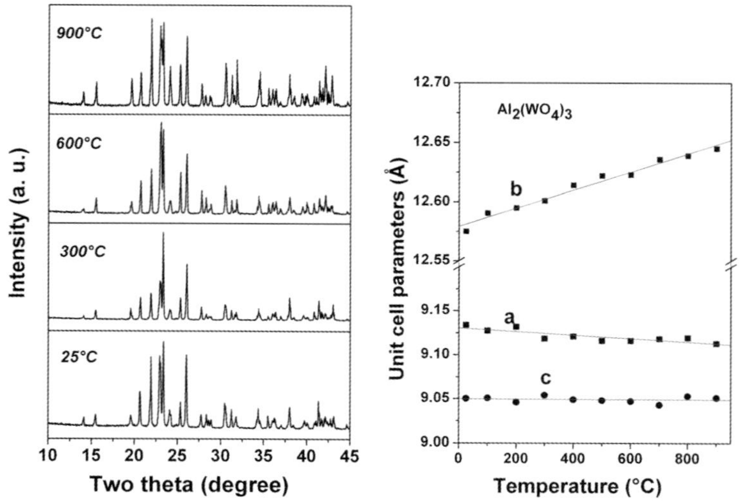

Figure 6.28 (a) Typical powder XRD pattern (CuKα) of $Al_2(WO_4)_3$ at different temperatures. (b) Variation of unit cell parameters of $Al_2(WO_4)_3$ at different temperatures.

along the different axes are $\alpha_a = -2.63 \times 10^{-6}\,°C^{-1}$, $\alpha_b = 6.36 \times 10^{-6}\,°C^{-1}$, $\alpha_c = 0.088 \times 10^{-6}\,°C^{-1}$, and $\alpha_V = 3.74 \times 10^{-6}\,°C^{-1}$.

The closely related crystal structures of monoclinic lighter rare-earth tungstate as well as orthorhombic heavier rare-earth tungstates, but with opposite thermal expansion behaviors, are promising for the creation of low or tunable thermal expansion ceramic materials. For example, at ambient temperature, $Nd_2(WO_4)_3$ has monoclinic (space group: C2/c) lattice with unit cell parameters: $a = 7.753$, $b = 11.602$, and $c = 11.538$ Å, $\beta = 109.77°$, $V = 977$ Å^3, and $Z = 4$. The dilatometric study on sintered bulk ceramic sample of $Nd_2(WO_4)_3$ shows a positive thermal expansion coefficient ($\alpha_l = 7.9 \times 10^{-6}\,K^{-1}$). Similar studies on $Y_2(WO_4)_3$ show an appreciable negative thermal expansion coefficient ($\alpha_l = -5.8 \times 10^{-6}\,K^{-1}$). The ambient-temperature unit cell parameters of $Y_2(WO_4)_3$ are $a = 9.166$, $b = 12.610$, and $c = 8.968$ Å, $V = 1036$ Å^3, and $Z = 4$ (space group: Pnca) and is isostructural to the above-explained $Al_2(WO_4)_3$ structure. In general, a continuous solid solution between these two structures do not exists [271, 272]. Tunable thermal expansion behaviors have been reported in such systems, even though the two structures do not form a continuous solid solution.

The thermal expansion behavior of a large number of tungstates having orthorhombic crystal structure have been studied by dilatometry as well as high-temperature diffraction methods [6, 112–115, 126–136]. The typical thermal expansion data of some of the tungstates are given in Table 6.15. In all of them, anisotropic thermal expansions similar to $Al_2(WO_4)_3$ have been observed. In addition, the magnitudes of thermal expansion coefficient of both monoclinic and orthorhombic phases are strongly influenced by the ionic radii of the rare-earth ions. In addition to the solid solutions of the orthorhombic tungstates, several solid solution

Table 6.15 Thermal expansion coefficients ($^\circ C^{-1}$) of few orthorhombic $A_2(WO_4)$-type tungstates. (All the reference data are mentioned here in Pnca setting.).

A^{3+}	Temperature range (°C)	$\alpha_a \times 10^6$	$\alpha_b \times 10^6$	$\alpha_c \times 10^6$	$\alpha_V \times 10^6$	References
Y	20–800	−10.4	−3.1	−7.6	−20.9	[134]
Y	−258–800	−10.6	−2.53	−7.97	−20.9	[127]
Y	200–800	−9.78	−5.13	−6.68	−22.0	[269]
Lu	20–800	−9.9	−2.2	−8.3	−20.4	[140]
Yb	200–800	−10.20	−2.65	−6.41	−19.1	[269]
Sc	20–800	−6.3	7.5	−5.5	−6.5	[128]
Al	20–800	−1.31	5.94	−0.099	4.51	[134]
Al	25–900	−2.63	6.36	0.088	3.74	a)
Er	200–800	−10.14	−3.35	−6.70	−20.1	[269]
Lu	200–800	−9.70	−2.89	−5.74	−18.5	[269]
Lu	120–620	−9.9	−2.2	−8.3	−20.4	[135]

a) Present data.

compositions between the orthorhombic and monoclinic lattices have also been investigated [269–283].

6.10.2.2.3 **AX_2O_8** Thermal expansion measurements on ZrW_2O_8 and HfW_2O_8 by diffraction techniques have been well established in literature. The negative thermal expansion behavior of ZrW_2O_8 has been known since long ago [273], but it has attracted much attention only recently [113]. Both ZrW_2O_8 and HfW_2O_8 crystallize in cubic ($P2_13$) lattice at ambient temperature. So the observed isotropic axial thermal expansion data correlate with dilatometric studies [111–113]. Besides, the diffractometric studies could also characterize the discontinuity (at ~157 °C) in the variation of linear or axial expansion with temperature. The symmetry change has been attributed to an order–disorder ($P2_13$–Pa3) transition involving the migration of oxygen atom leading to a dynamic equilibrium state of the WO_4 tetrahedra along the threefold axis [111]. The typical thermal expansion coefficients of the two lattices of ZrW_2O_8 are $-9.4 \times 10^{-6}\,^\circ C^{-1}$ (in −272.7 to 127 °C) and $-5.5 \times 10^{-6}\,^\circ C^{-1}$ (in 157–777 °C) [111, 113]. These interesting intricacies prompt preparation and thermal expansion measurement on similar analogues. Thermal expansion data have been reported for similar cubic $ZrMo_2O_8$ and $HfMo_2O_8$ and other closely related compounds, such as ZrP_2O_7 and ZrV_2O_7. In the temperature range of −262 to 300 °C, no such order–disorder transition in the metastable cubic $ZrMo_2O_8$ or $HfMo_2O_8$ is observed in high-temperature diffraction studies [133]. The typical axial thermal expansion coefficients for cubic $ZrMo_2O_8$ and $HfMo_2O_8$ are $-4.9 \times 10^{-6}\,^\circ C^{-1}$ (in the range −262 to 300 °C) [133] and $-4.0 \times 10^{-6}\,^\circ C^{-1}$ (in the range −196 to 300 °C) [274].

In addition to the above-mentioned cubic polymorphs, several other crystallographic modifications are known for AX_2O_8 compounds. At ambient temperature and pressure, trigonal $ZrMo_2O_8$ and $HfMo_2O_8$ are the stable polymorphs for molybdates [274–278], while no other stable phases exist for tungstates [279]. A denser monoclinic polymorph of $ZrMo_2O_8$ is also known [280]. Under high pressure and high pressure and high temperature, orthorhombic and trigonal phases of ZrW_2O_8 and HfW_2O_8 are reported [281–283]. Similarly, a large number of monoclinic and triclinic polymorphs of $ZrMo_2O_8$ and $HfMo_2O_8$ under high pressure or high pressure and high temperature are also reported [125, 283–288]. Thermal expansion behaviors of only few noncubic polymorphs of AX_2O_8 have been studied by high-temperature X-ray diffraction method [132, 289] (Anitha, M., Achary, S.N., and Tyagi, A.K., personal communication). Thermal expansions of the trigonal and monoclinic $HfMo_2O_8$ [288] have been studied by high-temperature X-ray diffraction and the typical structural implications on the thermal expansion are explained here as an example.

The structure of trigonal (α) $HfMo_2O_8$ and $ZrMo_2O_8$ consists of three-dimensional framework of corner-sharing AO_6 and MoO_4 polyhedra, where each tetrahedral molybdate group shares three of the corner oxygen atoms with the three different AO_6 octahedra. The fourth oxygen atom in the MO_4 tetrahedra is nonbridged and points to the interlayer region. Thus, the trigonal MX_2O_8 produces a layered structure along *c*-axis and enables large amplitude rocking motion of MO_4 tetrahedra and nonbridged oxygen atoms. The monoclinic (β) form of $HfMo_2O_8$ is a high pressure and high temperature polymorph and can be obtained by quenching the trigonal $HfMo_2O_8$ from 2.15 GPa and 560 °C [288]. The detailed crystal structure analysis of the monoclinic form revealed that the structural transformation occur with an increase in the coordination number of Hf and Mo atoms and dimerization of isolated MoO_4 units. In addition, about 20% increase in density occurs during this transformation. The typical crystal structures of trigonal and monoclinic $HfMo_2O_8$ are shown in Figure 6.29.

The powder XRD patterns for trigonal and monoclinic $HfMo_2O_8$ recorded at different temperatures are shown in Figure 6.30a and b. It is observed that the powder XRD patterns recorded up to 700 °C do not indicate any difference from that at the

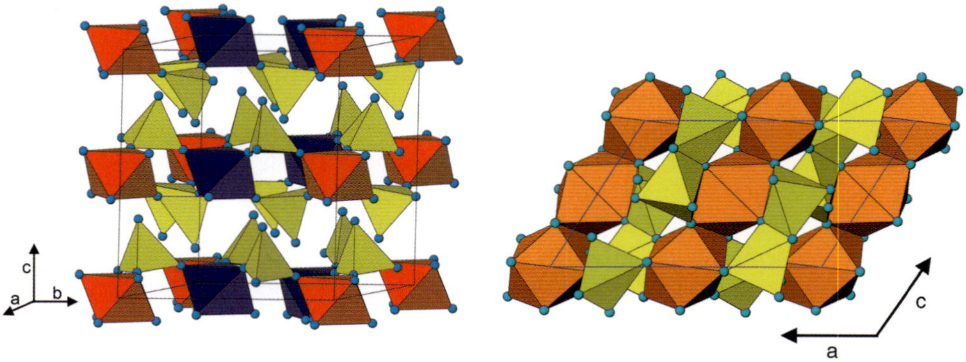

Figure 6.29 (a) Crystal structure of trigonal $HfMo_2O_8$ (HfO_6 and MoO_4 are shown): $a = 10.1086(3)$, $c = 11.7509(4)$ Å, $V = 1039.89(5)$ Å^3, space group: P-31c. (b) Crystal structure of monoclinic $HfMo_2O_8$ (HfO_8 and Mo_2O_{10} are shown): $a = 11.4264(8)$, $b = 7.9095(6)$, $c = 7.4461(5)$ Å, $\beta = 122.368(5)°$, $V = 568.4(1)$ Å^3, space group: C2/c.

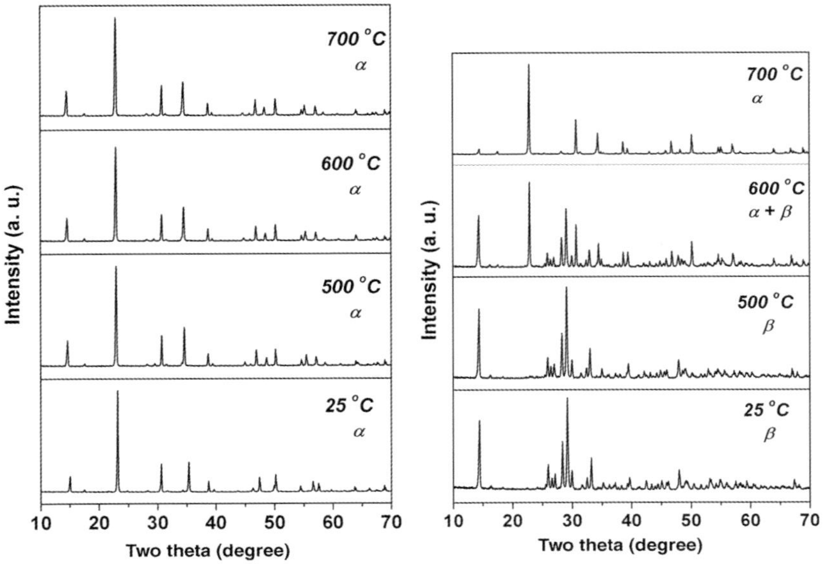

Figure 6.30 (a) XRD patterns of trigonal $HfMo_2O_8$ at different temperatures. (b) XRD patterns of monoclinic $HfMo_2O_8$ at different temperatures (Recorded with CuKα radiation).

ambient temperature. The unit cell parameters at each temperature have been determined and they are shown in Figure 6.31a and b. The small discontinuity in the variation of unit cell parameters around 200 °C is attributed to order–disorder transition of the trigonal phase. No such discontinuity in the case of monoclinic $HfMo_2O_8$ indicating any phase transition is observed.

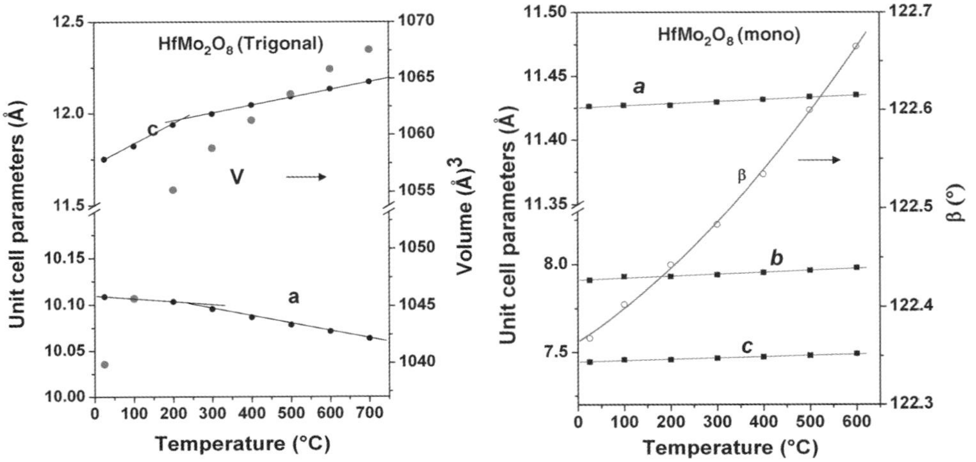

Figure 6.31 (a) Variation of unit cell parameters of trigonal $HfMo_2O_8$ with temperature. (b) Variation of unit cell parameters of monoclinic $HfMo_2O_8$ with temperature.

The typical variation of the unit cell parameters of trigonal $HfMo_2O_8$ shows negative expansion along *a*- and *b*-axes, whereas a significant positive expansion along *c*-axis similar to $ZrMo_2O_8$ [290]. The coefficients of axial thermal expansion for $HfMo_2O_8$ are $\alpha_a = \alpha_b = -6.42 \times 10^{-6}\,°C^{-1}$, $\alpha_c = 56.6 \times 10^{-6}\,°C^{-1}$, $\alpha_v = 43.4 \times 10^{-6}\,°C^{-1}$ (in 25–700 °C). It can be recalled here that the crystal structure of the trigonal form is made up of sheets of HfO_6 octahedra linked with MoO_4 tetrahedra at the corners. The apex of the MoO_4 units points toward the space between the layers and thus there is a strong O–O repulsion, leading to a weak interaction between the layers. Thus, there is a strong expansion along this direction. The negative expansion along other directions can be realized by the cooperative rotation of the MoO_4-HfO_6 framework. The thermal expansion coefficients of the monoclinic phase are $\alpha_a = 1.3 \times 10^{-6}\,°C^{-1}$, $\alpha_b = 15.6 \times 10^{-6}\,°C^{-1}$, $\alpha_c = 10.7 \times 10^{-6}\,°C^{-1}$, and $\alpha_v = 21.9 \times 10^{-6}\,°C^{-1}$. The monoclinic phase has a relatively higher packed structure compared to the trigonal phase [288]. Thus, in this case, the expansion is mainly controlled by the expansion of the chemical bonds, which results in a positive expansion but the average thermal expansion is lower than that of the trigonal phase.

6.10.3
Scheelite- and Zircon-Type ABO_4 Compounds

The ionic radii and charge combination of A and B cations govern the crystal structure of the ABO_4-type compounds [291, 292]. A large variety of structure types, namely, scheelite, monazite, zircon, wolframite, baryte, anhydrite, $MnSO_4$, $CrPO_4$, and so on, and silica polymorphs types are known for ABO_4 compositions. The thermal expansion of silica analogue ABO_4 compounds have already been explained. Among the ABO_4-type compounds, the zircon-type compounds are well known for the low thermal expansion and incompressible behavior, while scheelite lattice show appreciably higher thermal expansion [45, 46, 105–110, 293–306] (Anitha, M., Achary, S.N., and Tyagi, A.K., personal communication). The thermal expansion behavior of $CaMoO_4$ and $CaWO_4$ (scheelite types) and $LuVO_4$ and $GdVO_4$ (zircon types) as observed from high-temperature X-ray diffraction studies are explained here.

Both scheelite and zircon lattice of ABO_4 have tetragonal unit cell built from the AO_8 polyhedra (bisdisphenoid) and BO_4 (tetrahedra). However, the typical connections of the polyhedra are different in these two structures. In a general view, the scheelite structure is related to the fluorite lattice, while the zircon is related to the rutile lattice [297].

6.10.3.1 $CaMoO_4$ and $CaWO_4$

Both $CaMoO_4$ and $CaWO_4$ have tetragonal unit cell (space group: $I4_1/a$), Ca, Mo or W, and O atoms are located at (*4b: 0, 1/4, 5/8*), (*4a:0, 1/4, 1/8*), and (*16f: x, y, z*) sites, respectively [297]. The thermal expansion of $CaMoO_4$ has been studied by Deshpande and Suryanarayana [298] in the temperature range of 25–350 °C. The typical values of average thermal expansion coefficients along *a*- and *c*-axes (α_a and α_c) are 10.71×10^{-6} and $16.17 \times 10^{-6}\,°C^{-1}$. Bayer [105] has reported the average axial

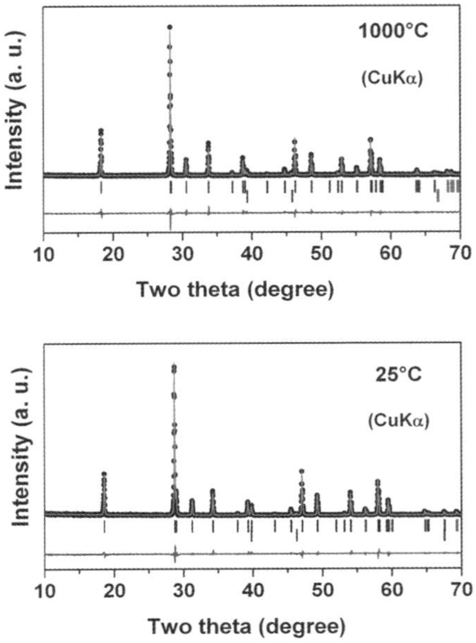

Figure 6.32 The Rietveld refinement plots of $CaMoO_4$ at 25 and 1000 °C (vertical marks indicate the Bragg positions; lower vertical marks indicate platinum-base metal reflections).

thermal expansion for $CaWO_4$ as 13.7×10^{-6} and $21.5 \times 10^{-6}\,°C^{-1}$ (in the temperature range of 20–1020 °C) and 11.5×10^{-6} and $19.2 \times 10^{-6}\,°C^{-1}$ (in the temperature range of 20–520 °C). The thermal expansion behaviors of $CaWO_4$ and $CaMoO_4$ have been studied by HT-XRD in the temperature range of 25–1000 °C [299]. The unit cell and other structural parameters are obtained by the Rietveld refinement of the powder X-ray data recorded at various temperatures. The typical powder XRD data of $CaMoO_4$ recorded at ambient temperature and 1000 °C are shown in Figure 6.32. The refined unit cell of $CaMoO_4$ and $CaWO_4$ at ambient temperature and 1000 °C are given in Table 6.16.

Smooth variations of unit cell parameters with temperature are observed for $CaMoO_4$ and $CaWO_4$ (Figure 6.33). The variations of unit cell parameters with temperature are fitted with polynomial function and the typical fitting parameters are included in Table 6.15. Further, the variation of c/a of $CaMoO_4$ and $CaWO_4$ reflects the significantly higher expansion along the c-axis compared to that along a-axis. The typical values of c/a of $CaMoO_4$ and $CaWO_4$ at ambient temperature are 2.188 and 2.169, respectively, and those at 1000 °C are 2.207 and 2.188, respectively.

The detailed analyses of the structural data observed for $CaMoO_4$ at various temperatures have been used to explain the thermal expansion behavior of both. A comparison of various interatomic distances indicates a significantly larger expansion of Ca–O bonds compared to the Mo–O bond. The typical Ca–O bond lengths at ambient temperature are 2.450(8) and 2.474(8) Å and those at 1000 °C are

Table 6.16 Unit cell parameters of $CaMoO_4$ and $CaWO_4$ at various temperatures.

Temperature (°C)	$CaMoO_4$			$CaWO_4$		
	a (Å)	*c* (Å)	*V* (Å)^3	*a* (Å)	*c* (Å)	*V* (Å)^3
25	5.2261(1)	11.4329(3)	312.25(1)	5.2422(1)	11.3723(2)	312.52(1)
1000	5.2950(1)	11.6869(3)	327.66(1)	5.3074(1)	11.6101(4)	327.04(1)
a) x_0	5.2261(1)	11.4332(3)		5.2422(1)	11.3722(4)	
x_1	6.0×10^{-5}	21.0×10^{-5}		5.0×10^{-5}	$19.0(4) \times 10^{-5}$	
x_2	1.1949×10^{-8}	3.5321×10^{-8}		1.5196×10^{-8}	5.6041×10^{-8}	
x_3	1.3772×10^{-12}	1.7316×10^{-11}		2.5386×10^{-12}	-1.2401×10^{-12}	
b) α(°C^{-1})	13.5×10^{-6}	22.8×10^{-6}	50.6×10^{-6}	12.7×10^{-6}	21.4×10^{-6}	47.5×10^{-6}
c) α_l(°C^{-1})	15.9×10^{-6}			14.1×10^{-6}		

a) $x_t = x_0 + x_1(T-25) + x_2(T-25)^2 + x_3(T-25)^3$, where x are the unit cell parameters in Å and T are the temperatures in °C.
b) Average thermal expansion coefficients.
c) Average linear thermal expansion coefficients in temperature range of 25–850 °C as obtained by dilatometry [299].

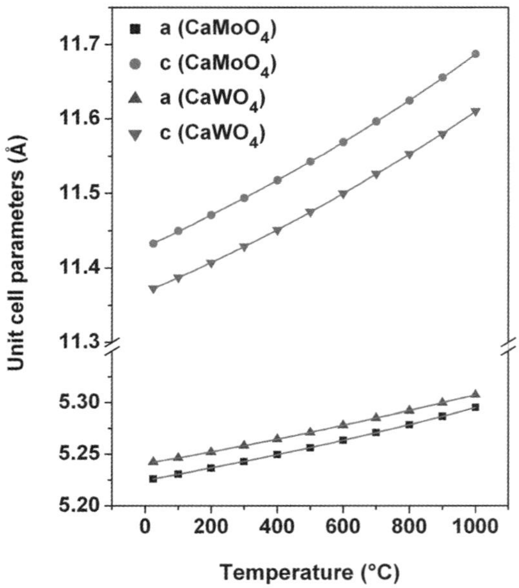

Figure 6.33 Variations of unit cell parameters with temperature [8].

2.52(1) and 2.50(1) Å. The expansion of the average Ca–O bonds with temperature can be fitted as

$$d_{\langle \text{Ca-O} \rangle_{\text{av}}}(\text{Å}) = 2.461(2) + 5.13(3) \times 10^{-5}[T],$$

where T is the temperature in °C.

Similarly, the polyhedral volume of the CaO_8 units shows a significant expansion compared to the MoO_4 units (Figure 6.34). The average thermal expansion of the Ca–O bonds is about $0.019 \times 10^{-6}\,°C^{-1}$, which is comparable to the reported value of $0.016 \times 10^{-6}\,°C^{-1}$ in similar eight coordinated Ca^{2+} polyhedra [180]. A closer insight into the unit cell of $CaMoO_4$ indicates that the unit cell is built by the CaO_8 polyhedral units by sharing four of their edges with four other CaO_8 units. Along the *a*- and *b*-axes, the CaO_8 units are linked by the MoO_4 units. The lower thermal expansion of the MoO_4 tetrahedral units results in the lower thermal expansion of the *a*-axis compared to the *c*-axis.

6.10.3.2 $LuPO_4$, $LuVO_4$, and $GdVO_4$ (Zircon Type)

In general, the zircon group materials show drastically different thermal expansion behaviors depending upon the cation charge and ionic radii combination [108, 109]. The low thermal expansion of $ZrSiO_4$ has been reported from high-temperature diffraction data [296–299]. Subbarao *et al.* [109] and Bayer [105] have compared a large number of compounds with zircon structure, where low thermal expansion is generally observed in the case of silicates compared to others. Phosphates and vanadates of smaller rare-earth ions form stable zircon-type lattices. The thermal

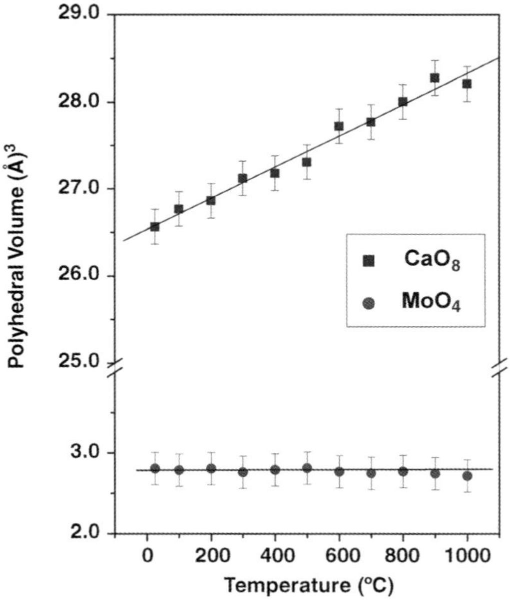

Figure 6.34 Variation of polyhedral volumes of CaO_8 and MoO_4 units of $CaMoO_4$ with temperature [8].

expansion measurements of rare-earth phosphates or vanadates have been carried out extensively from the high-temperature X-ray diffraction and dilatometric studies [105–110, 300–306] (Anitha, M., Achary, S.N., and Tyagi, A.K., personal communication). The typical coefficients of thermal expansion reported for such vanadates and phosphates are summarized in Table 6.17.

A comparison of the thermal expansion data shows higher expansion and higher anisotropy in vanadates than in the phosphates. Detailed analysis of thermal expansion behavior of $LuVO_4$ and $LuPO_4$ has been carried out. Structural details at higher temperature are obtained by the Rietveld refinement of the powder XRD data. The typical variations of unit cell parameters at various temperatures are shown in Figure 6.35a. In order to compare the anisotropy in the thermal expansion, the variation of c/a with temperature is shown in Figure 6.35b. The temperature evolution of unit cell parameters for $LuVO_4$ and $LuPO_4$ can be expressed by the polynomial relation:

$$
\begin{aligned}
\mathrm{LuPO_4}: \quad & a(\text{Å}) = 6.7875(4) + 4.21(7) \times 10^{-5}[T] \\
& c(\text{Å}) = 5.9542(4) + 4.41(8) \times 10^{-5}[T] \\
& V(\text{Å})3 = 274.30(5) + 0.0055(9) \times [T] \\
\mathrm{LuVO_4}: \quad & a(\text{Å}) = 7.0214(3) + 2.60(5) \times 10^{-5}[T] \\
& c(\text{Å}) = 6.2277(3) + 7.42(5) \times 10^{-5}[T] \\
& V(\text{Å})3 = 307.02(4) + 0.0060(7) \times [T],
\end{aligned}
$$

where T is the temperature in °C.

Table 6.17 Coefficients of axial thermal expansion of some rare-earth phosphates and vanadates.

APO_4	Temperature range (°C)	$\alpha_{av} \times (10^6)$ (°C^{-1})	$\alpha_a \times (10^6)$ (°C^{-1})	$\alpha_c \times (10^6)$ (°C^{-1})	Experiment	References
$LuPO_4$	25–1000	6.44	6.04	7.25	HT-XRD	[306]
$LuPO_4$	20–1000	6.2			Dilatometer	[107]
$ErPO_4$	20–1000	6.0			Dilatometer	[107]
$YbPO_4$	20–1000	6.0			Dilatometer	[107]
YPO_4	20–1000	6.2			Dilatometer	[107]
YPO_4	20–520		5.0	5.9	HT-XRD	[105]
YPO_4	20–1020		5.4	6.0	HT-XRD	[105]
YPO_4	25–1000	5.7	—	—	Dilatometer	[110]
$ScPO_4$	20–1200	5.5	4.1	8.4	HT-XRD	[109]
YVO_4	20–520		3.7	10.1	HT-XRD	[105]
YVO_4	20–1020		4.0	10.5	HT-XRD	[105]
$ScVO_4$	20–1200		4.05	12.95	HT-XRD	[307]
$ScVO_4$	20–1200		3.9	14.	HT-XRD	[308]
YVO_4	27–670		3.1	7.21	HT-XRD	[302]
$GdVO_4$	25–540		2.16	8.92	HT-XRD	[303]
$GdVO_4$	25–1000		3.6	10.8	HT-XRD	Anitha, M., Achary, S.N., and Tyagi, A.K., personal communication
$GdVO_4$	25–600		2.97	9.97	HT-XRD	Anitha, M., Achary, S.N., and Tyagi, A.K., personal communication
$NdVO_4$	20–520		3.7	10.1	HT-XRD	[105]
YVO_4	20–1020		4.0	10.5	HT-XRD	[105]
$LuVO_4$	25–1000	6.34	3.62	11.80	HT-XRD	[306]
$LuVO_4$	12–300 K		0.7	5.6	LTND	[45, 46]

The ambient-temperature unit cell parameters of $LuVO_4$ and $LuPO_4$ are $a = 7.0230(1)$ and $c = 6.2305(1)$ Å and $V = 307.31(1)$ Å^3 and $a = 6.7895(3)$ and $c = 5.9560(4)$ Å and $V = 274.56(2)$ Å^3 (space group: $I4_1/amd$, $Z = 4$), respectively. In the temperature range of 25–1000 °C, the typical coefficients of axial thermal expansion of these are $\alpha_a = 6.04 \times 10^{-6}\,°C^{-1}$ and $\alpha_a = 7.25 \times 10^{-6}\,°C^{-1}$ (for $LuPO_4$) and $\alpha_a = 3.62 \times 10^{-6}\,°C^{-1}$ and $\alpha_a = 11.80 \times 10^{-6}\,°C^{-1}$ (for $LuVO_4$). Similar studies on $GdVO_4$ also revealed the anisotropic thermal expansion coefficients, namely, $\alpha_a = 3.6 \times 10^{-6}\,°C^{-1}$ and $\alpha_c = 10.80 \times 10^{-6}\,°C^{-1}$.

The strain coefficients for $LuVO_4$ and $LuPO_4$ have been obtained at various temperatures from the determined unit cell parameters, as explained in Equation 6.12. Since the present systems are tetragonal, only principal strain components (ε_{11}, ε_{22}, and ε_{33}) are nonzero in the strain matrix given in Equation 6.12. Besides, the principal strains ε_{11} and ε_{22} are equal due to the tetragonal symmetry. Between

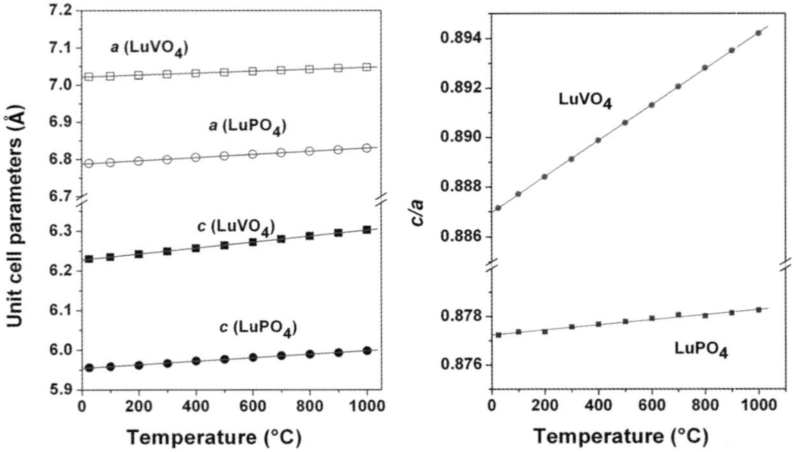

Figure 6.35 Variation of unit cell parameters (a) and c/a (b) of $LuPO_4$ and $LuVO_4$ with temperature.

ambient temperature and 1000 °C, the principal axes strain coefficients are $\varepsilon_{11} = \varepsilon_{22} = 0.35 \times 10^{-2}$ and $\varepsilon_{33} = 1.15 \times 10^{-2}$ (for $LuVO_4$); $\varepsilon_{11} = \varepsilon_{22} = 0.59 \times 10^{-2}$ and $\varepsilon_{33} = 0.71 \times 10^{-2}$ (for $LuPO_4$). The trace of the matrix represents the volume strain. The temperature variation of principal strain components for both $LuVO_4$ and $LuPO_4$ are shown in Figure 6.36. It can be observed that the anisotropy of the strain components is significant in the case of $LuVO_4$ compared to that of $LuPO_4$. Thus, the thermal expansion ellipsoid (shown in Figure 6.37) for $LuPO_4$ is more spherical compared to that of $LuVO_4$.

The crystal structures of $LuPO_4$, $LuVO_4$, or $GdVO_4$ consist of LuO_8 or GdO_8 polyhedra (bisdisphenoid) and PO_4 or VO_4 tetrahedra. Along the c-axis, the AO_8

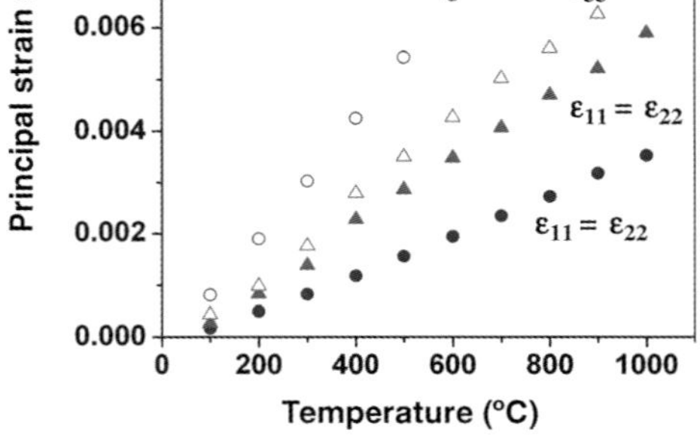

Figure 6.36 Variations of principal strain components of $LuVO_4$ and $LuPO_4$ with temperature.

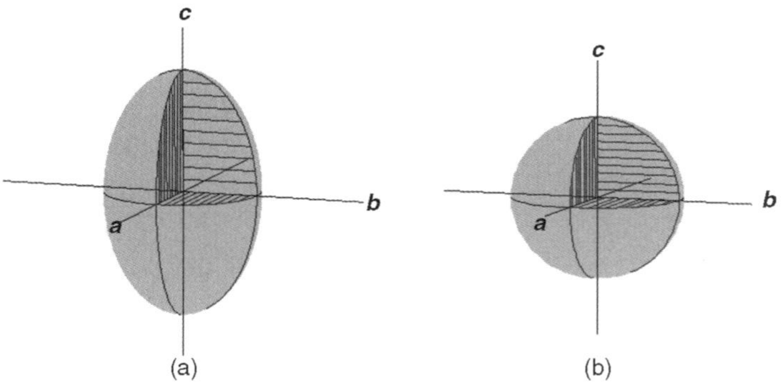

Figure 6.37 Thermal expansion ellipsoids $LuVO_4$ (a) and $LuPO_4$ (b).

and PO_4 or VO_4 units are linked by sharing their edges. Along *a*- and *b*-directions, the chains of these units are joined by sharing other edges of the AO_8 polyhedra. The differences in these phosphate or vanadate structures arise from the difference between PO_4 and VO_4 tetrahedra. This leads to the difference in the axial ratio and the distortion in AO_8 units and hence the differences in thermal expansion behaviors. A typical AO_8 polyhedral unit of zircon-type structure is shown in Figure 6.38.

LuO_8 polyhedra in both the structures are formed by two sets of Lu–O bonds, namely, LuO_a and LuO_c (marked in Figure 6.38). The significant difference in the Lu–O (sets of longer four Lu–O bond lengths) in these two structures due to the change in B cation from V to P is reflected in all the differences in the crystal structure. The longer Lu–O_c bonds show appreciable expansion with temperature compared to the shorter bonds in $LuVO_4$. The typical variation of interatomic

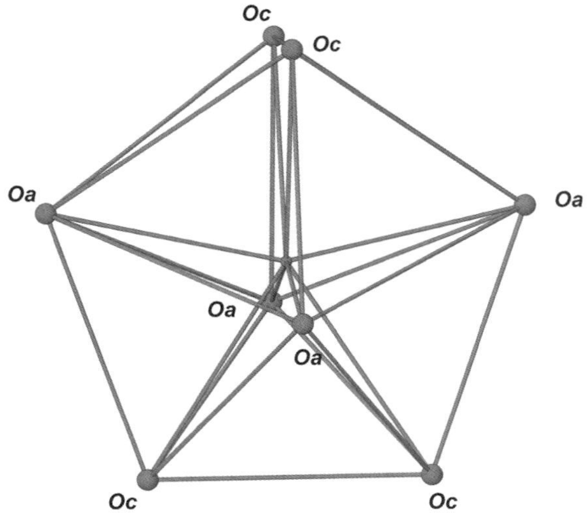

Figure 6.38 Typical AO_8 polyhedral unit of zircon-type structure.

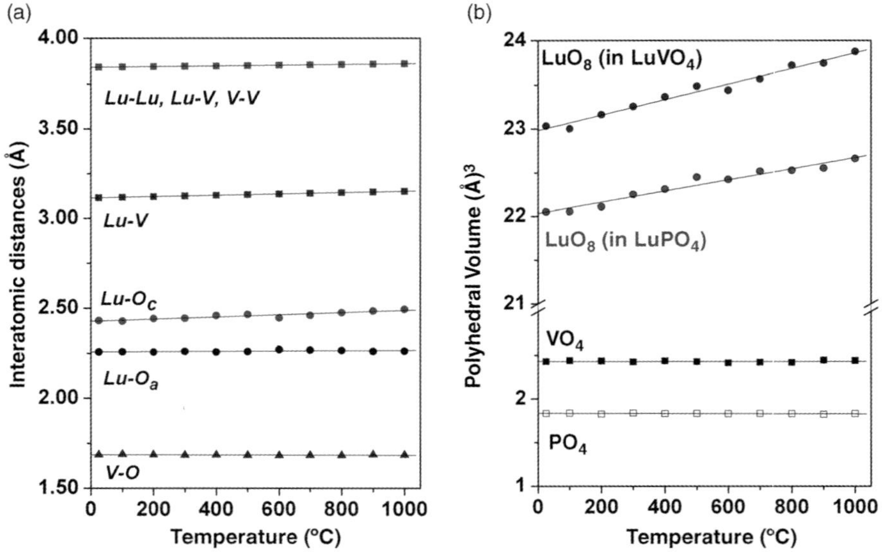

Figure 6.39 (a) Variation of typical interatomic distances in $LuVO_4$ with temperature. (b) Variation of volume of LuO_8 units of $LuPO_4$ and $LuVO_4$ with temperature.

distances with temperature for $LuVO_4$ is shown in Figure 6.39a. The typical polyhedral parameters of these two materials at two extreme temperatures of study are given in Table 6.18. This suggests that the AO_8 units of LuPO4 are more spherical than those of $LuVO_4$.

Table 6.18 Typical interatomic distances and polyhedral parameters of LuO_8 units [306].

	$LuVO_4$			$LuPO_4$		
	T (°C)		$\alpha \times 10^6\,°C^{-1}$	*T* (°C)		$\alpha \times 10^6\,°C^{-1}$
	25	1000		25	1000	
Lu-O_a	2.262(8)	2.262(8)	3.4	2.255(7)	2.277(8)	10.8
S	0.46	0.456		0.469	0.438	
$\langle\lambda\rangle$	1.511	1.437		1.422	1.408	
σ^2	1048	955		935	916	
Lu-O_c	2.431(5)	2.493(8)	24.7	2.364(6)	2.386(7)	8.9
S	0.289	0.244		0.346	0.326	
$\langle\lambda\rangle$	1.355	1.371		1.435	1.448	
σ^2	1321	1372		1562	1601	
V_{LuO8} (Å^3)	23.04(10)	23.88(10)	38	22.05(10)	22.66(10)	29
D	13.6×10^{-4}	23.7×10^{-4}		5.6×10^{-4}	5.5×10^{-4}	
Lu-Lu/B	3.8415(1)	3.8601(1)	5.1	3.7070(1)	3.7295(1)	6.4
Lu-B	3.1152(1)	3.1511(1)	11.9	2.9780(1)	2.9990(1)	7.4

S, bond valence; D, distortion in LuO_8 units; $\langle\lambda\rangle$ and σ^2 are quadratic elongation and bond angle variance (defined as Equation 6.41), respectively.

In addition, the larger polyhedral volume of LuO_8 unit in $LuVO_4$ (23.04 Å^3) than that in $LuPO_4$ (22.05 Å^3) is a consequence of this difference. The variations of polyhedral volumes with temperature for these two compounds are shown in Figure 6.39b. The difference in anisotropy in thermal expansion of zircon-type vanadates and phosphates is related to the sphericity and distortion of AO_8 units as well as spatial distribution of the A−O bonds.

In both ABO_4-type scheelite and zircon structures, the BO_4 polyhedra act as typical rigid units, with virtually no expansion. The overall expansion behavior is governed by the AO_8 units. The typical thermal expansion behavior of scheelite lattice is basically controlled by the temperature-induced expansion of the A−O bonds and their orientation. The lower expansion of zircon lattice compared to scheelite is due to the sharing of edges of AO_8 with a rigid BO_4 unit.

6.11 Conclusion

In this chapter, the thermal expansion and its origin are briefly explained. Subsequently, the effects of various parameters and in particular the structural effect on the thermal expansion are explained. Various techniques used in thermal expansion measurements are also touched upon. Further, the X-ray diffraction in determining the thermal expansion is also explained in detail. The thermal expansion measurements by X-ray diffraction providing the structural origin as plausible reason are explained with selected examples. Besides the mentioned examples, a large number of studies delineating the crystal structure and thermal expansion have been carried out in recent years. However, the perovskite systems, where the diffraction techniques have immense role in understanding the thermal expansion and phase transition, are not explained in this chapter.

References

1 Yates, B. (1972) *Thermal Expansion*, Plenum Press, New York.
2 Krishnan, R.S., Srinivasan, R., and Devanarayann, S. (1979) *Thermal Expansion of Crystal*, Pergamon Press, Oxford.
3 Shiga, M. (1994) Invar alloys, in *Materials Science and Technology*, vol. 3B (eds R.W. Cahn, P. Haasen, and E.J. Kramer), VCH, Weinheim, p. 159.
4 Krause, D. and Bach, H. (2005) *Low Thermal Expansion Glass Ceramics*, 2nd edn, Springer, Berlin.
5 Birch, K.P. and Wilton, P.T. (1988) *Appl. Opt.*, **27** (14), 2813–2815.
6 (a) Roy, R., Agrawal, D.K., and McKinstry, H.A. (1989) *Annu. Rev. Mater. Sci.*, **19**, 59; (b) Sleight, A.W. (1998) *Annu. Rev. Mater. Sci.*, **28**, 29.
7 Goodenough, J.B. (2003) *Annu. Rev. Mater. Res.*, **33**, 9.
8 Kim, D.-S., Schweiger, M.J., Buchmiller, W.C., Vienna, J.D., Day, D.E., Zhu, D., Kim, C.W., Day, T.E., Neidt, T., Peeler, D.K., Edwards, T.B., Reamer, I.A., and Workman, R.J.Pacific Northwest National Laboratory Report No. PNNL14251, for US-DOE contract DE-AC06-76RL01830.

9 (a) Shen, Y.-L., Needleman, A., and Suresh, S. (1994) *Metall. Mater. Trans. A*, **25**, 839; (b) Holzer, H. and Dunand, D.C. (1999) *J. Mater. Res.*, **14**, 780.

10 (a) Wilson, A.J.C. (1941) *Proc. Phys. Soc.*, **53**, 235; (b) Stokes, S. and Wilson, A.J.C. (1941) *Proc. Phys. Soc.*, **53**, 658.

11 (a) Simmons, R.O. and Balluffi, R.W. (1960) *Phys. Rev.*, **117**, 52; **119**, 600, 1960; **125**, 862, 1962; **129**, 1533, 1963; (b) Sher, A., Solomon, R., Lee, K., and Muller, m.V. (1966) *Phys. Rev.*, **144**, 593.

12 (a) Kelly, A., Stearn, R.J., and McCartney, L.N. (2006) *Compos. Sci. Technol.*, **66**, 154; (b) Yang, X., Cheng, X., Yan, X., Yang, J., Fu, T., and Qiu, J. (2007) *Compos. Sci. Technol.*, **67**, 1167.

13 Nye, J.F. (1985) *Physical Properties of Crystals: Their Representation by Tensor and Matrices*, Oxford Press, Oxford.

14 Thurston, R.N. and Burger, K. (1964) *Phys. Rev.*, **157**, 524.

15 Baron, T.H.K. and Munn, R.W. (1970) *Pure and Appl. Chem.*, **22**, 527.

16 Wallace, D.C. (1970) *Solid State Phys.*, **25**, 301.

17 Hatch, D.M., Ghose, S., and Bjorkstam, J.L. (1994) *Phys. Chem. Min.*, **21**, 67.

18 Kittle, C. (1976) *Introduction to Solid State Physics*, 5th edn, John Wiley & Sons, Inc., New York.

19 Barron, T.H.K. and White, G.K. (1999) *Heat Capacity and Thermal Expansion at Low Temperatures*, Kluwer, New York.

20 Wallace, D.C. (1972) *Thermodynamics of Crystal*, John Wiley & Sons, Inc., New York.

21 Pippard, A.B. (1964) *The Elements of Classical Thermodynamics*, Cambridge University Press, Cambridge.

22 Grima, J.N., Zammit, V., and Gatt, R. (2006) *Xjenza*, **11**, 1.

23 Barron, T.H.K. (1998) Chapter 1, in *Thermal Expansion of Solids, CINDAS Data Series on Material Properties*, vol. I-4 (eds C.Y. Ho and R.E. Taylor), ASM International, Materials Park, OH.

24 (a) Grüneisen, E. (1912) *Annu. Phys.*, **39**, 257; (b) Grüneisen, E. (1926) *Hand. Phys.*, **10**, 1.

25 Grüneisen, E. (1908) *Annu. Phys.*, **26**, 393.

26 Allan, N.L., Braithwite, M., Cooper, D.L., Mackrodt, W.C., and Wright, S.C. (1991) *J. Chem. Phys.*, **95**, 6792.

27 Kobayashi, M. (1979) *J. Chem. Phys.*, **70**, 509.

28 Ashcroft, N.W. and Mermin, N.D. (1976) *Solid State Physics*, Holt, Rinehart and Winston, New York.

29 Barrera, G.D., Bruno, J.A.O., Barron, T.H.K., and Allan, N.L. (2005) *J. Phys.: Condens. Matter*, **17**, R217.

30 Mittal, R., Chaplot, S.L., and Choudhury, N. (2006) *Prog. Mater. Sci.*, **51**, 211.

31 Chaplot, S.L., Choudhury, N., Ghose, S., Rao, M.N., Mittal, R., and Goel, P. (2002) *Eur. J. Mineral.*, **14**, 291.

32 Venkataraman, G., Feldkamp, L., and Sahni, V.C. (1975) *Dynamics of Perfect Crystals*, MIT Press, Cambridge.

33 Born, M. and Huang, K. (1954) *Dynamical Theory of Crystal Lattices*, Oxford University Press, London.

34 Barron, T.H.K., Collin, J.G., and White, G.K. (1980) *Adv. Phys.*, **29**, 609.

35 Azaroff, L.V. (1977) *Introduction to Solids*, 3rd edn, McGraw Hill Publishing Company Ltd., New Delhi.

36 Dekkar, A.J. (1957) *Solid State Physics*, Prentice Hall, Englewood Cliffs, NJ.

37 Fei, Y. (1995) Thermal expansion, in *Mineral Physics and Crystallography: A Handbook of Physical Constants* (ed. T.J. Arhens), American Geophysical Union, Washington, DC.

38 Baron, T.H.K. and Munn, R.W. (1967) *Phil. Mag.*, **15**, 85.

39 Philips, N.E. (1971) *Crit. Rev. Solid State Sci.*, **2**, 467.

40 Harding, G.L., Lanchester, P.C., and Street, R. (1971) *J. Phys, C. Solid State Phys.*, **4**, 2923.

41 Guillaume, C.E. (1920) *C. R. Acad. Sci.*, **125**, 235.

42 Khmelevskyi, S. and Mohn, P. (2004) *J. Magn. Magn. Mater.*, **272–276**, 525.

43 Berthold, J.W., Jacobs, S.F., and Norton, M.A. (1977) *Metrologia*, **13**, 9.

44 Lagarec, K., Rancourt, D.G., Bose, S.K., Sanyal, B., and Dunlap, R.A. (2001) *J. Magn. Magn. Mater.*, **236**, 107.

45 Skanthakumar, S., Loong, C.K., Soderholm, L., Nipko, J., Richardson, J.W. Jr., Abraham, M.M., and Boatner, L.A. (1995) *J. Alloys Comp.*, **225**, 595.

46 Skanthakumar, S., Loong, C.K., Soderholm, L., Richardson, J.W. Jr.,

Abraham, M.M., and Boatner, L.A. (1995) *Phys. Rev. B*, **51**, 5644.

47 de Visser, A., Lacerda, A., Haen, P., Flouquet, J., Kaysel, F.E., and France, J.J.M. (1989) *Phys. Rev. B.*, **39**, 11301.

48 Barron, T.H.K. (1992) *J. Phys.: Condens. Matter.*, **39**, L455.

49 Rufino, G. (1998) Chapter 3, in *Thermal Expansion of Solids, CINDAS Data Series on Material Properties*, vol. I-4 (eds C.Y. Ho and R.E. Taylor), ASM International, Materials Park, OH.

50 Touloukian, Y.S., Kirby, R.K., Taylor, R.E., and Lee, T.Y.R. (1977) Methods for measurement of thermal expansion of solids, in *Thermal Expansion Nonmetallic Solids. Part I*, IFL/Plenum Press, New York.

51 Kirby, R.K. (1992) Methods of measuring thermal expansion, in *Compendium of Thermophysical Measurement Methods*, vol. II (eds D.D. Maglic, A. Cezarirliyan, and V.E. Peletsky), Plenum Press, New York.

52 Hidnert, P. and Souder, W. (1950) National Bureau of Standards Circular 486.

53 Austin, J.B. (1952) *J. Am. Ceram. Soc.*, **35**, 243.

54 Srinivasan, R. and Krishnan, R.S. (1958) *Progress in Crystal Physics*, vol. 1 (ed. S. Viswanathan). Central Arts Press, Madras, India.

55 Kirchher, H.P. (1964) *Prog. Solid State Chem.*, **1**, 1.

56 Gall, P.S. (1998) Chapter 5, in *Thermal Expansion of Solids, CINDAS Data Series on Material Properties*, vol. I-4 (eds C.Y. Ho and R.E. Taylor), ASM International, Materials Park, OH.

57 Wagner, P., Gonzale, A.L., Minor, R.C., and Armstrong, P.E. (2009) *Rev. Sci. Instrum.*, **37** (2), 180.

58 Sparks, P.W. and Swenson, C.A. (1967) *Phys. Rev.*, **163**, 779.

59 Jenkin, F.A. and White, H.E. (1976) *Fundamentals of Optics*, McGraw-Hill, New York.

60 James, J.D., Spittle, J.A., Brown, S.G.R., and Evans, R.W. (2001) *Meas. Sci. Technol.*, **12**, R1.

61 Bianchini, G., Barucci, M., Del Rosso, T., Pasca, E., and Ventura, G. (2006) *Meas. Sci. Technol.*, **17**, 689.

62 Merritt, G.E. (1939) *J. Res. NBS*, **23**, 179 RP 1227.

63 Saunders, J.B. (1939) *J. Res. NBS*, **23**, 579 RP 1253.

64 Work, R.N. (1951) *J. Res. NBS*, **47**, 80 RP 2230.

65 Fizeau, M. (1864) *Comptes Rendus*, **58**, 923.

66 Fraser, D.B. and Hollis Hallet, A.C. (1965) *Can. J. Phys.*, **43**, 193.

67 Gray, H.W. (1914) Micrometer microscopes. NBS Science Paper 215.

68 Brixner, B. (1965) *Rev. Sci. Instrum.*, **36**, 1896.

69 Azaroff, L.V. and Buerger, M.J. (1958) *The Powder Method in X-Ray Crystallography*, McGraw-Hill Book Company, New York.

70 Cullity, B.D. (1959) *Elements of X-Ray Diffraction*, Addison-Wesley, Reading, MA.

71 Klug, H.P. and Alexander, L.E. (1974) *X-Ray Diffraction Procedures*, Wiley-Interscience, New York.

72 Warren, B.E. (1969) *X-Ray Diffraction*, Addison Wesley.

73 Buerger, M.J. (1960) *Crystal Structure Analysis*, John Wiley & Sons, Inc., New York.

74 Lipson, H. and Cochran, W. (1966) *Determination of Crystal Structure*, G. Bell & Sons. Ltd., London.

75 Ito, T. (1950) *X-Ray Studies on Polymorphism*, Maruzen Co. Ltd., Tokyo, pp. 187–228.

76 de Wolff, P.M. (1968) *J. Appl. Crystallogr.*, **1**, 108.

77 Smith, G.S. and Snyder, R.L. (1979) *J. Appl. Crystallogr.*, **12**, 60.

78 Rietveld, H.M. (1967) *Acta Crystallogr.*, **22**, 151.

79 Rietveld, H.M. (1969) *J. Appl. Crystallogr.*, **2**, 65.

80 Young, R.A. (2000) *The Rietveld Method*, IUCr Publications.

81 Warren, B.E. (1969) *X-Ray Diffraction*, Addison-Wesley, Reading, MA.

82 Touloukian, Y.S., Kirby, R.K., Taylor, R.E., and Desai, E.D. (1975) Thermal expansion: metallic elements and alloys, in *Thermophysical Properties of Matter*, vol. 12 (eds Y.S. Touloukian and C.Y. Ho), Plenum, New York.

83 Touloukian, Y.S., Kirby, R.K., Taylor, R.E., and Lee, T.Y.R. (1977) Thermal expansion: nonmetallic solids, in *Thermophysical Properties of Matter*, vol. 13 (eds Y.S. Touloukian and C.Y. Ho), Plenum, New York.

84 Touloukian, Y.S. (1976) *Thermophysical Property Research Center (TPRC) Data Series*, vol. 13, Purdue University, West Lafayette.

85 Skinner, B.J. (1966) Thermal expansion, in *Handbook of Physical Constants* (ed. S.P. Clark), Geological Society of America.

86 Kirby, R.K., Hahn, T.A., and Rothrock, B.D. (1972) Thermal expansion, in *American Institute of Physics Handbook*, McGraw-Hill, New York.

87 Cartz, U. and Jorgenson, J.D. (1982) *Thermal Expansion*, vol. 7 (ed. O.C. Larsen), Plenum, New York, pp. 147–154.

88 Hazen, R.M. and Finger, L.W. (1982) *Comparative Crystal Chemistry*, John Wiley & Sons, Inc., New York.

89 Lager, G.A. and Meagher, E.P. (1978) *Am. Mineral.*, **63**, 365.

90 Winter, J.K., Okamura, F.P., and Ghose, S. (1979) *Am. Mineral.*, **64**, 409.

91 Knittle, E., Jeanloz, R., and Smith, G.L. (1986) *Nature*, **319**, 214.

92 Hazen, R.M. and Finger, L.W. (1987) *Phys. Chem. Minerals*, **14**, 426.

93 Angel, R.J. (2000) Equation of state, in *Reviews in Mineralogy and Geochemistry*, vol. 41 (eds R.M. Angel and R.T. Downs), Mineralogical Society of America, Washington, DC, pp. 35.

94 Anderson, O.L. (1995) *Equation of State of Solids for Geophysics and Ceramics Science*, Oxford University Press, Oxford, UK.

95 Martin, D.G. (1988) *J. Nucl. Mater.*, **152**, 94.

96 Badock, P.J., Spinder, W.E., and Baker, T.W. (1966) *J. Nucl. Mater.*, **18**, 305.

97 Rodriguez, P. and Sundaram, C.V. (1981) *J. Nucl. Mater.*, **100**, 227.

98 Tyagi, A.K., Ambekar, B.R., and Mathews, M.D. (2002) *J. Alloys Comp.*, **337**, 277.

99 Weber, A. and Ivers-Tiffee, E. (2004) *J. Power Sources*, **127**, 273.

100 Wandekar, R.V., Wani, B.N., and Bharadwaj, S.R. (2005) *Mater. Lett.*, **59**, 2799.

101 Wang, H., Tablet, C., Yang, W., and Caro, J. (2005) *Mater. Lett.*, **59**, 3750.

102 Yamanaka, S., Kurosaki, K., Maekawa, T., Matsuda, T., Kobayashi, S., and Uno, M. (2005) *J. Nucl. Mater.*, **344**, 61.

103 Yaremchenko, A.A., Kharton, V.V., Shaula, A.L., Patrakeev, M.V., and Marques, F.M.V. (2005) *J. Euro. Ceram. Soc.*, **25**, 2603.

104 Senyshyn, A., Vasylechko, L., Knapp, M., Bismayer, U., Berkowski, M., and Matkovskii, A. (2004) *J. Alloys Comp.*, **382**, 84.

105 Bayer, G. (1972) *J. Less-Common Met.*, **26**, 255.

106 Hazen, R.M., Finger, L.W., and Mariathasan, J.W.E. (1985) *J. Phys. Chem. Solids*, **46**, 253.

107 Hikishi, Y., Ota, T., Daimon, K., Hattori, T., and Mizuno, M. (1998) *J. Am. Ceram. Soc.*, **81**, 2216.

108 Li, H., Zhou, S., and Zhang, S. (2007) *J. Solid State Chem.*, **80**, 589.

109 Subbarao, E.C., Agrawal, D.K., McKinstry, H.A., Sallese, C.W., and Roy, R. (1990) *J. Am. Ceram. Soc.*, **73**, 1246.

110 Taylor, D. (1986) *J. Br. Ceram. Soc.*, **85**, 147.

111 Evans, J.S.O., Mary, T.A., Vogt, T., Subramanian, M.A., and Sleight, A.W. (1996) *Chem. Mater.*, **8**, 2809.

112 Sleight, A.W. (1995) *Endeavour*, **19**, 64.

113 Mary, T.A., Evans, J.S.O., Sleight, A.W., and Vogt, T. (1996) *Science*, **272**, 90.

114 Kowach, G.R. (2000) *J. Cryst. Growth*, **212**, 167.

115 Sleight, A.W. (1998) *Inorg. Chem.*, **37**, 2854.

116 Dove, M.T., Heine, V., and Hammonds, K.D. (1995) *Mineral. Mag.*, **59**, 629.

117 Tucker, M.G., Goodwin, A.L., Dove, M.T., Keen, D.A., Wells, S.A., and Evans, J.S.O. (2005) *Phys. Rev. Lett.*, **95**, 255501.

118 (a) Cao, D., Bridges, F., Kowach, G.R., and Ramirez, A.P. (2002) *Phys. Rev. Lett.*, **89**, 215902; (b) Cao, D., Bridges, F., Kowach, G.R., and Ramirez, A.P. (2003) *Phys. Rev. B*, **68**, 014303.

119 Hancock, J.N., Turpen, C., Schlesinger, Z., Kowach, G.R., and Ramirez, A.P. (2004) *Phys. Rev. Lett.*, **93**, 225501.

120 Mittal, R., Chaplot, S.L., Schober, H., and Mary, T.A. (2001) *Phys. Rev. Lett.*, **86**, 4692.

121 Mittal, R. and Chaplot, S.L. (1999) *Phys. Rev. B*, **60**, 7234.

122 Ramirez, A.P. and Kowach, G.R. (1998) *Phys. Rev. Lett.*, **80**, 4903.

123 Ravindran, T.R., Arora, A.K., and Mary, T.A. (2000) *Phys. Rev. Lett.*, **84**, 3879.

124 Lind, C., Vanderveer, D.G., Wilkinseon, A.P., Chen, J., Vaughan, M.T., and Weiner, D.J. (2001) *Chem. Mater.*, **13**, 487.

125 Korthuis, V., Khosrovani, N., Sleight, A.W., Roberts, N., Dupree, R., and Wareen, W.W. Jr. (1995) *Chem. Mater.*, **7**, 412.

126 Evans, J.S.O., Mary, T.A., and Sleight, A.W. (1997) *J. Solid State Chem.*, **133**, 580.

127 Tao, J.Z. and Sleight, A.W. (2003) *J. Solid State Chem.*, **173**, 442.

128 Evans, J.S.O., Mary, T.A., and Sleight, A.W. (1998) *J. Solid State Chem.*, **137**, 148.

129 Evans, J.S.O. and Mary, T.A. (2000) *Int. J. Inorg. Mater.*, **2**, 143.

130 Tyagi, A.K., Achary, S.N., and Mathews, M.D. (2002) *J. Alloys Comp.*, **339**, 207.

131 Evans, J.S.O., Mary, T.A., and Sleight, A.W. (1998) *Physica B*, **241–243**, 311.

132 Evans, J.S.O. (1999) *J. Chem. Soc., Dalton Trans.*, 3317.

133 Lind, C., Wilkinson, A.P., Hu, Z., Short, S., and Jorgensen, J.D. (1998) *Chem. Mater.*, **10**, 2335.

134 Woodcock, D.A., Lightfoot, P., and Ritter, C. (2000) *J. Solid State Chem.*, **149**, 92.

135 Forster, P.M., Yokochi, A., and Sleight, A.W. (1998) *J. Solid State Chem.*, **140**, 157.

136 Achary, S.N., Mukherjee, G.D., Tyagi, A.K., and Vaidya, S.N. (2002) *J. Mater. Sci.*, **37**, 2501.

137 Tiano, W., Dapiaggi, M., and Artioli, G. (2003) *J. Appl. Crystallogr.*, **36**, 1461.

138 Schafer, W. and Kirfel, A. (2000) *Appl. Phys. A*, **74**, S1010.

139 Li, J., Sleight, A.W., Jones, C.Y., and Toby, B.H. (2005) *J. Solid State Chem.*, **178**, 285.

140 Li, J., Yokochi, A., Amos, T.G., and Sleight, A.W. (2002) *Chem. Mater.*, **14**, 2602.

141 Amos, T.G., Yokochi, A., and Sleight, A.W. (1998) *J. Solid State Chem.*, **141**, 303.

142 Amos, T.G. and Sleight, A.W. (2001) *J. Solid State Chem.*, **160**, 230.

143 Wallez, G., Launay, S., Souron, J.-P., Quarton, M., and Suard, E. (2003) *Chem. Mater.*, **15**, 3793.

144 Taylor, D. (1984) *J. Br. Ceram. Trans.*, **83**, 129.

145 Gillery, F.H. and Bush, E.A. (1959) *J. Am. Ceram. Soc.*, **42**, 175.

146 Schreyer, W. and Schairer, J.F. (1961) *J. Petrology*, **2**, 324.

147 Li, C.-T. and Pecor, D.R. (1968) *Z. Kristallogr.*, **126**, 46.

148 Alamo, J. and Roy, R. (1984) *J. Am. Ceram. Soc.*, **67**, 78.

149 Harshe, G. and Agrawal, D.K. (1994) *J. Am. Ceram. Soc.*, **77**, 1965.

150 Limaye, S.Y., Agrawal, D.K., and Roy, R. (1991) *J. Mater. Sci.*, **26**, 93.

151 Tantri, P.S., Geetha, K., Umarji, A.M., and Ramesha, S.K. (2000) *Bull. Mat. Sci.*, **23**, 49.

152 Woodcock, D.A., Lightfoot, P., and Ritter, C. (1998) *Chem. Commun.*, 107.

153 (a) Tachaufeser, P. and Parker, S.C. (1995) *J. Phys. Chem.*, **99**, 10600; (b) Woodcock, D.A. and Lightfoot, P. (1999) *J. Mater. Chem.*, **9**, 2907.

154 Lightfoot, P., Woodcock, D.A., Maple, M.J., Villaescusa, L.A., and Wright, P.A. (2001) *J. Mater. Chem.*, **11**, 212.

155 Attfield, M.P. and Sleight, A.W. (1998) *Chem. Commun.*, 601.

156 White, G.K. (1993) *Contemp. Phys.*, **34**, 193.

157 Rottger, K., Endriss, A., Ihringer, J., Doyle, S., and Khuss, W.F. (1994) *Acta Crystallogr.*, **B50**, 664.

158 Lawson, A.C., Roberts, J.A., Martinez, B., Ramos, M., Kotliarz, G., Trouw, F.W., Fitzsimmons, M.R., Hehlen, M.P., Lashley, J.C., Ledbetter, H., Mcqueeney, R.J., and Migliori, A. (2006) *Phil. Mag.*, **86**, 2713.

159 Lawson, A.C., Cort, B., Roberts, J.A., Bennett, B.I., Brun, T.O., Von Dreele, R.B. and Richardson, J.W. Jr. (1999) Lattice effects in the light actinides, in *Electron Correlation and Materials Properties* (eds A. Gonis, N. Kioussis, and M. Ciftan), Kluwer Academic/Plenum Publishers, New York, pp. 75–96.

160 Salvador, J.R., Guo, F., Hogan, T., and Kanatzidis, M.G. (2003) *Nature*, **425**, 702.

161 Barron, T.K.H. (1979) *J. Phys. C. Solid State Phys.*, **12**, L155.

162 Berthold, J.W., Jacobs, S.F., and Norton, M.A. (1977) *Metrologia*, **13**, 9.

163 Shirane, G. and Takeda, A. (1951) *J. Phys. Soc. Jpn.*, **7**, 1.

164 Shirane, G. and Hoshina, S. (1951) *J. Phys. Soc. Jpn.*, **6**, 265.
165 (a) Smolenskii, G.A. (1970) *J. Phys. Soc. Jpn.*, **285**, 26; (b) Patwe, S.J., Achary, S.N., Mathews, M.D., and Tyagi, A.K. (2005) *J. Alloys Comp.*, **390**, 100.
166 Agrawal, D.K., Halliyal, A., and Belsick, J. (1988) *Mater. Res. Bull.*, **23**, 159.
167 Achary, S.N., Chakraborty, K.R., Patwe, S.J., Shinde, A.B., Krishna, P.S.R., and Tyagi, A.K. (2006) *Mater. Res. Bull.*, **41**, 674.
168 Gateshki, M. and Igartua, J.M. (2004) *J. Phys.: Condens. Matter.*, **16**, 6639.
169 Gehring, P.M., Chen, W., Ye, Z.-G., and Shirane, G. (2004) *J. Phys.: Condens. Matter*, **16**, 7113.
170 Patil, R.N. and Subbarao, E.C. (1969) *J. Appl. Crystallogr.*, **2**, 281.
171 Williams, D., Partin, D.E., Lincoln, F.J., Kouvetakis, J., and O'Keeffe, M. (1997) *J. Solid State Chem.*, **134**, 164.
172 Goodwin, A.L. and Kepert, C.J. (2005) *Phys. Rev. B*, **71**, 140301(R).
173 Phillips, A.E., Goodwin, A.L., Halder, G.J., Southon, P.D., and Kepert, C.J. (2007) *Angew. Chem.*, **120**, 1418.
174 Margadonna, S., Prassides, K., and Fitch, A.N. (2004) *J. Am. Chem. Soc.*, **126**, 15390.
175 Goodwin, A.L., Calleja, M., Conterio, M.J., Dove, M.T., Evans, J.S.O., Keen, D.A., Peters, L., and Tucker, M.G. (2008) *Science*, **319**, 794.
176 Van Uitert, L.G., O'Bryon, H.M., Lines, M.E., Guggenheim, H.G., and Zydzik, G. (1977) *Mater. Res. Bull.*, **12**, 261.
177 Yamanaka, S., Maekawa, T., Muta, H., Matsuda, T., Kobayashi, S., and Kurosakia, K. (2004) *J. Solid State Chem.*, **177**, 3484.
178 Megaw, H.D. (1939) *Z. Kristallogr.*, **100**, 58.
179 Megaw, H.D. (1971) *Mater. Res. Bull.*, **6**, 1007.
180 Hazen, R.M. and Prewitt, C.T. (1977) *Am. Mineral.*, **62**, 309.
181 Cameron, P. (1973) *Am. Mineral.*, **58**, 594.
182 Brown, I.D. (1992) *Acta Crystallogr.*, **B48**, 553.
183 Brown, I.D. and Altermatt, D. (1992) *Acta Crystallogr.*, **B41**, 244.
184 Bresse, N. and O'Keeffe, M. (1991) *Acta Crystallogr.*, **B47**, 192.
185 Brown, I.D., Dabakowski, A., and Mccleary, A. (1997) *Acta Crystallogr.*, **B53**, 750.
186 Kitaigorodoski, A.I. (1973) Chapter 5, in *Molecular Crystals and Molecules*, Academic Press, New York.
187 Haines, J., Cambon, O., Fraysse, G., and van der Lee, A. (2005) *J. Phys.: Condens. Matter*, **17**, 4463.
188 Haines, J., Cambon, O., and Hull, S. (2003) *Z. Kristallogr.*, **218**, 193.
189 Haines, J., Cambon, O., Astier, R., Fertey, P., and Chateau, C. (2004) *Z. Kristallogr.*, **219**, 32.
190 Haines, J. and Cambon, O. (2004) *Z. Kristallogr.*, **219**, 314.
191 Megaw, H.D. (1968) *Acta Crystallogr.*, **A24**, 589.
192 Zhao, Y., Weidner, D.J., Parise, J.B., and Cox, D.E. (1993) *Phys. Earth Planet. Interiors*, **76**, 1.
193 Zhao, Y. and Weidner, D.J. (2004) *Phys. Chem. Miner.*, **18**, 294.
194 Haines, J., Cambon, O., Philippot, E., Chapon, L., and Hull, S. (2002) *J. Solid State Chem.*, **166**, 434.
195 Achary, S.N. and Tyagi, A.K. (2004) *J. Solid State Chem.*, **177**, 3918.
196 Achary, S.N., Jayakumar, O.D., Tyagi, A.K., and Kulshresththa, S.K. (2003) *J. Solid State Chem.*, **176**, 37.
197 (a) Robinson, K., Gibbs, G.V., and Ribe, P.H. (1971) *Science*, **172**, 567; (b) Bevan, D.J.M. and Greis, O. (1980) *J. Strahle, Acta Crystallogr.*, **A36**, 889; (c) Withers, R., Wallenberg, R., Bevan, D.J.M., Thomson, J.G., and Hyde, B.G. (1989) *J. Less-Common Met.*, **156**, 17.
198 Mathews, M.D., Ambekar, B.R., and Tyagi, A.K. (2000) *J. Nucl. Mater.*, **280**, 246.
199 Yamashita, T., Nitani, N., Tsuji, T., and Inagaki, H. (1997) *J. Nucl. Mater.*, **245**, 72.
200 Fahey, J.A., Turcotte, R.P., and Chikalla, T.D. (1974) *Inorg. Nucl. Chem. Lett.*, **10**, 549.
201 Sudakov, L.V., Kapshukov, I.I., and Solntsev, V.M. (1973) *Atomnaya Energiya*, **35**, 128.
202 Taylor, D. (1984) *J. Br. Ceram. Trans.*, **83**, 32.
203 Tyagi, A.K. and Mathews, M.D. (2000) *J. Nucl. Mater*, **278**, 123.

204 Chidambaram, R. (1990) Proceedings of the Indo–Japan Seminar on Thoria Utilization, Bombay, Indian Nuclear Society, and Atomic Energy Society of Japan, December (eds M. Srinivasan and I. Kimura).

205 Kleykamp, H. (1999) *J. Nucl. Mater.*, **275**, 1.

206 Grover, V. and Tyagi, A.K. (2004) *J. Solid State Chem.*, **177**, 4197.

207 Grover, V. and Tyagi, A.K. (2002) *J. Nucl. Mater.*, **305**, 83.

208 Chavan, S.V. and Tayagi, A.K. (2006) *Mater. Sci. Eng. A.*, **433**, 203.

209 Mathews, M.D., Ambekar, B.R., and Tyagi, A.K. (2001) *J. Nucl. Mater.*, **288**, 83.

210 Tyagi, A.K., Ambekar, B.R., and Mathews, M.D. (2002) *J. Alloys Comp.*, **337**, 277.

211 Grover, V. and Tyagi, A.K. (2005) *Ceram. Int.*, **3**, 769.

212 Grover, V., Sengupta, P., and Tyagi, A.K. (2007) *Mater. Sci.Eng. B*, **138**, 246.

213 Tyagi, A.K. and Mathews, M.D. (2000) *J. Nucl. Mater.*, **78**, 123.

214 Tyagi, A.K., Mathews, M.D., Ambekar, B.R., and Ramachandran, R. (2004) *Thermochim. Acta*, **421**, 69.

215 Freshley, M.D. and Mattys, H.M. (1962) General Electric Report HW-76559, p. 116.

216 Bakker, K., Cordfunke, E.H.P., Konings, R.J.M., and Schram, R.P.C. (1997) *J. Nucl. Mater.*, **250**, 1.

217 Lambertson, W.A., Mueller, M.H., and Gunzel, F.H. (1953) *J. Am. Ceram. Soc.*, **36**, 397.

218 Kanno, M., Kokubo, S., and Furuya, H. (1982) *J. Nucl. Sci. Technol.*, **19**, 956.

219 Christensen, J.A. (1962) General Electric Report HW-76559, p. 117.

220 Latta, R.E., Duderstadt, E.C., and Fryxell, R.E. (1970) *J. Nucl. Mater.*, **35**, 347.

221 Baldock, P.J., Spinder, W.E., and Baker, T.W. (1966) *J. Nucl. Mater.*, **18**, 305.

222 Shackelford, J.W. and Alexander, W. (eds) (2009) *CRC Materials Science and Engineering Handbook*, 3rd edn, CRS Press, Washington, DC.

223 Chavan, S.V. and Tyagi, A.K. (2005) *Mater. Sci. Eng. A*, **404**, 57.

224 Mathews, M.D., Ambekar, B.R., and Tyagi, A.K. (2005) *J. Nucl. Mater.*, **341**, 19.

225 Mandal, B.P., Grover, V., and Tyagi, A.K. (2006) *Mater. Sci. Eng. A*, **430**, 120.

226 Grover, V. and Tyagi, A.K. (2004) *Mater. Res. Bull.*, **39**, 859.

227 Chavan, S.V., Mathews, M.D., and Tyagi, A.K. (2004) *J. Am. Ceram. Soc.*, **87**, 1977.

228 Mathews, M.D., Ambekar, B.R., and Tyagi, A.K. (2005) *J. Alloys Comp.*, **386**, 234.

229 Mathews, M.D., Ambekar, B.R., and Tyagi, A.K. (2005) *J. Nucl. Mater.*, **341**, 19.

230 Chavan, S.V., Mathews, M.D., and Tyagi, A.K. (2005) *Mater. Res. Bull.*, **40**, 1558.

231 Chavan, S.V. and Tyagi, A.K. (2005) *Mater. Sci. Eng. A*, **404**, 57.

232 Mathews, M.D., Ambekar, B.R., and Tyagi, A.K. (2006) *Ceram. Int.*, **32**, 609.

233 Grover, V. and Tyagi, A.K. (2006) *J. Am. Ceram. Soc.*, **89**, 2917.

234 Tyagi, A.K., Mathews, M.D., and Ramchandran, R. (2002) *J. Nucl. Mater.*, **294**, 198.

235 Chavan, S.V., Patwe, S.J., and Tyagi, A.K. (2003) *J. Alloys Comp.*, **360**, 189.

236 Chavan, S.V. and Tyagi, A.K. (2002) *Thermochim. Acta*, **390**, 79.

237 Hummel, F.A. (1949) *J. Am. Ceram. Soc.*, **32**, 320.

238 Perloff, A. (1956) *J. Am. Ceram. Soc.*, **39**, 83.

239 Beck, W.R. (1949) *J. Am. Ceram. Soc.*, **32**, 147.

240 Schaffer, E. and Roy, R. (1956) *J. Am. Ceram. Soc.*, **39**, 330.

241 Burger, M.J. (1948) *Am. Mineral.*, **33**, 751.

242 Mooney, R.C.L. (1956) *Acta Crystallogr.*, **9**, 728.

243 Achary, S.N., Mishra, R., Jayakumar, O.D., Kulshreshtha, S.K., and Tyagi, A.K. (2007) *J. Solid State Chem.*, **180**, 84–91.

244 Kulshreshtha, S.K., Jayakumar, O.D., and Sudarshan, V. (2004) *J. Phys. Chem. Solid*, **65**, 1141.

245 Ng, H.N. and Calvo, C. (1977) *Can. J. Phys.*, **55**, 677.

246 Wright, A.F. and Leadbetter, A.J. (1975) *Philos. Mag.*, **31**, 1391.

247 Keeffe, M.O. and Hyde, B.G. (1976) *Acta Crystallogr.*, **B32**, 2923.

248 Hatch, D.M., Ghose, S., and Bjorkstam, J.L. (1994) *Phys. Chem. Miner.*, **21**, 67.

249 Philippot, E., Armand, P., Yot, P., Cambon, O., Goiffon, A., McIntyre, G.J.,

and Bordet, P. (1999) *J. Solid State Chem.*, **146**, 114.

250 Haines, J., Chateau, C., Leger, J.M., Bogicevic, C., Hull, S., Klug, D.D., and Tse, J.S. (2003) *Phy. Rev. Lett.*, **91**, 015503.

251 Glazer, A.M. (1972) *Acta Crystallogr.*, **B28**, 3384.

252 Megaw, H.D. (1968) *Acta Crystallogr.*, **A24**, 589.

253 Howard, C.J., Kennedy, B.J., and Woodward, P.M. (2003) *Acta Crystallogr.*, **B59**, 463.

254 Nassau, K., Levindtein, H.J., and Loiacono, G.M. (1965) *J. Phys. Chem. Solids*, **26**, 1805.

255 Abrahams, S.C. and Berstein, J.L. (1966) *J. Chem. Phys.*, **45**, 2475.

256 Marinkovic, B.A., Jardim, P.M., de Avillez, R.R., and Rizzo, F. (2005) *Solid State Sci.*, **7**, 1377.

257 Wu, M.M., Xiao, X.L., Hu, Z.B., Liu, Y.T., and Chen, D.F. (2009) *Solid State Sci.*,**11**, 325.

258 Ari, M., Jardim, P.M., Marinkovic, B.A., Rizzo, F., and Ferreira, F.F. (2008) *J. Solid State Chem.*, **181**, 1472.

259 Gates, S.D. and Lind, C. (2007) *J. Solid State Chem.*, **180**, 3510.

260 Cheng, Y.Z., Wu, M.M., Peng, J., Xiao, X.L., Li, Z.X., Hu, Z.B., Kiyanagi, R., Fieramosca, J.S., Short, S., and Jorgensen, J. (2007) *Solid State Sci.*, **9**, 693.

261 Sumithra, S. and Umarji, A.M. (2006) *Solid State Sci.*, **8**, 1453.

262 Forster, P.M. and Sleight, A.W. (1999) *Int. J. Inorg. Mater.*, **1**, 123.

263 Sleight, A.W. and Bixner, L.H. (1973) *J. Solid State Chem.*, **7**, 172.

264 Harrison, W.T.A., Chowdhry, U., Machiels, C.J., Sleight, A.W., and Cheetham, A.K. (1985) *J. Solid State Chem.*, **60**, 101.

265 Battle, P.D., Cheetham, A.K., Harrison, W.T.A., Pollard, N.J., and Faber, J. Jr. (1985) *J. Solid State Chem.*, **58**, 221.

266 Harrison, W.T.A., Cheetham, A.K., and Faber, J. Jr. (1988) *J. Solid State Chem.*, **76**, 328.

267 Harrison, W.T.A. (1995) *Mater. Res. Bull.*, **30**, 1325.

268 Chen, H.Y. (1979) *Mater. Res. Bull.*, **14**, 1583.

269 Sumithra, S., Tyagi, A.K., and Umarji, A.M. (2005) *Mater. Sci. Eng. B*, **116**, 14.

270 Sumithra, S. and Umarji, A.M. (2004) *Solid State Sci.*, **6**, 1313.

271 Peng, J., Wu, M.M., Wang, H., Hao, Y.M., Hu, Z., Yu, Z.X., Chen, D.F., Ryoji Kiyanagi, J.S., Fieramosca, S., Short, J., and Jorgensen (2008) *J. Alloys Comp.*, **453**, 49.

272 Wu, M.M., Cheng, Y.Z., Peng, J., Xiao, X.L., Chen, D.F., Kiyanagi, R.J., Fieramosca, S., Short, S., Jorgensen, J., and Hu, Z.B. (2007) *Mater. Res. Bull.*, **42**, 2090.

273 Martinek, C. and Hummel, F.A. (1968) *J. Am. Ceram. Soc.*, **51**, 227.

274 Kennedy, C.A., White, M.A., Wilkinson, A.P., and Varga, T. (2007) *Appl. Phys. Lett.*, **90**, 151906.

275 Auray, M., Quarton, M., and Tarte, P. (1986) *Acta Crystallogr.*, **C42**, 257.

276 Auray, M., Quarton, M., and Tarte, P. (1987) *Powder Diffr.*, **2**, 36.

277 Freundlich, W. and Thoret, J. (1967) *C. R. Acad. Sci. Ser. C*, **265**, 96.

278 Trunov, V.K. and Kovba, L.M. (1967) *Russ. J. Inorg. Chem.*, **12**, 1703.

279 Chang, L.L.Y., Scroger, M.G., and Phillips, B. (1967) *J. Am. Ceram. Soc.*, **50**, 211.

280 Klevtsova, R.F., Glinskaya, L.A., Zolotova, E.S., and Klevtsov, P.V. (1989) *Sov. Phys. Dokl.*, **34**, 185.

281 Evans, J.S.O., Hu, Z., Jorgensen, J.D., Argyriou, D.N., Short, S., and Sleight, A.W. (1997) *Science*, **275**, 61.

282 Grzechnik, A., Crichton, W.A., Syassen, K., Adler, P., and Mezouar, M. (2001) *Chem. Mater.*, **13**, 4255.

283 Carlson, S. and Krogh Andersen, A.M. (2000) *Phy. Rev. B*, **61**, 11209.

284 Krogh Andersen, A.M. and Carlson, S. (2001) *Acta Crystallogr. B*, **57**, 20.

285 Muthu, D.V.S., Chen, B., Wrobel, J.M., Krogh Andersen, A.M., Carlson, S., and Kruger, M.B. (2002) *Phy. Rev. B*, **65**, 64101.

286 Mukherjee, G.D., Karandikar, A.S., Vijayakumar, V., Godwal, B.K., Achary, S.N., Tyagi, A.K., Lausi, A., and Busetto, E. (2008) *J. Phys. Chem. Solid*, **69**, 35.

287 Achary, S.N., Mukherjee, G.D., Tyagi, A.K., and Godwal, B.K. (2003) *Powder Diffr.*, **18**, 144.

288 Achary, S.N., Mukherjee, G.D., Tyagi, A.K., and Godwal, B.K. (2002) *Phys. Rev. B*, **66**, 184106.

289 Mittal, R., Chaplot, S.L., Lalla, N.P., and Mishra, R.K. (1999) *J. Appl. Crystallogr.*, **32**, 1010.

290 Allen, S., Ward, R.J., Hampson, M.R., Gover, R.K.B., and Evans, J.S.O. (2004) *Acta Crystallogr. B.*, **60**, 32.

291 Depero, L.E. and Sangalettis, L. (1997) *J. Solid State Chem.*, **129**, 82.

292 Muller, O. and Roy, R. (1974) *The Major Ternary Strctural Families, Crystal Chemistry of Non-Metallic Materials*, vol. 4, Springer-Verlag, Berlin.

293 Scott, H.P., Williams, Q., and Knittlem, E. (2002) *Phys. Rev. Lett.*, **88**, 015506.

294 Mursic, Z., Vogt, T., Boysen, H., and Frey, F. (1992) *J. Appl. Crystallogr.*, **25**, 519.

295 Grover, V. and Tyagi, A.K. (2005) *J. Alloys Comp.*, **39**, 112.

296 Nyman, H. and Hyde, B.G. (1984) *Acta Crystallogr.*, **B40**, 441.

297 Zalkin, A. and Templeton, D.H. (1964) *J. Chem. Phys.*, **40**, 501.

298 Deshpande, V.T. and Suryanarayana, S.V. (1969) *J. Phys. Chem. Solids*, **30**, 2484.

299 Achary, S.N., Patwe, S.J., Mathews, M.D., and Tyagi, A.K. (2006) *J. Phys. Chem. Solids*, **67**, 774.

300 Reddy, N.R.S. and Murthy, K.S. (1983) *J. Less-Common Met.*, **90**, L7.

301 Reddy, N.R., Murthy, K.S., and Rao, K.V.K. (1981) *J. Mater. Sci.*, **16**, 1422.

302 Reddy, C.V.V., Murthy, K.S., and Kistaiah, P. (1988) *Solid State Commun.*, **67**, 545.

303 Reddy, C.V.V., Kistaiah, P., and Murthy, K.S. (1985) *J. Phys. D., Appl. Phys.*, **18**, L27.

304 Reddy, C.V.V., Kistaiah, P., and Murthy, K.S. (1995) *J. Alloys Comp.*, **218**, L4.

305 Zhao, S., Zhang, H., Wang, J., Kong, H., Cheng, X., Liu, J., Li, J., Lin, Y., Hu, X., Xu, X., Wang, X., Shao, Z., and Jiang, M. (2004) *Opt. Mater.*, **26**, 319.

306 Patwe, S.J., Achary, S.N., and Tyagi, A.K. (2009) *Am. Mineral.*, **94**, 98.

307 Schopper, H.C., Urban, W., and Ebel, H. (1972) *Solid State Commun.*, **11**, 955.

308 Kahle, H.G., Schopper, H.C., Urban, W., and Wiechner, W. (1970) *Phys. Status Solidi*, **38**, 815.

7
Electronic Structure and High-Pressure Behavior of Solids

Carlos Moysés Araújo and Rajeev Ahuja

7.1
Introduction

The relation between the electronic structure and the crystallographic atomic arrangement is one of the fundamental questions in physics, geophysics, and chemistry. Since the discovery of the atomic nature of matter and its periodic structure, this has remained as one of the main questions regarding the very foundation of solid systems. Needless to say, this has also bearings on physical and chemical properties of matter, where again the relation between structure and performance is of direct interest. Solids have been mainly studied at ambient conditions, that is, at room temperature and pressure. However, it was realized early that there is also a fundamental relation between volume and structure, and that this dependence could most fruitfully be studied by means of high-pressure experimental techniques. From a theoretical point of view, this is an ideal type of experiment, since only the volume is changed, which is a very clean variation of the external conditions.

Here, we describe the electronic structure and crystallographic phase transformations of simple systems (elemental compounds) under high external pressure by means of first-principles theory. This is done for a number of examples covering most of periodic table groups. We will show that compounds made of single type of element can already display a rich and complex phase diagram due to peculiar pressure-induced electron band structure modifications. The chapter is organized as follows: in Sections 7.2 and 7.3, basic concepts of the general theoretical background are reviewed. The subsequent chapters present the results for each specific system and we finalize with an overview in Section 7.10.

7.2
First-Principles Theory

The main goal of first-principles theory of solids consists in solving the time-independent Schrödinger equation

Thermodynamic Properties of Solids: Experiment and Modeling
Edited by S. L. Chaplot, R. Mittal, and N. Choudhury

ISBN: 928-3-527-40812-2

$$H_t \Psi_{\mathbf{k},n}(\mathbf{r}_1, \ldots, \mathbf{r}_n, \mathbf{R}_1, \ldots, \mathbf{R}_t) = E_n(\mathbf{k}) \Psi_{\mathbf{k},n}(\mathbf{r}_1, \ldots, \mathbf{r}_n, \mathbf{R}_1, \ldots, \mathbf{R}_t) \tag{7.1}$$

to obtain the many-body eigenfunctions, $\Psi_{\mathbf{k},n}(\mathbf{r}_1, \mathbf{r}_2, \ldots \mathbf{r}_n, \mathbf{R}_1, \mathbf{R}_2 \ldots \mathbf{R}_t)$, which represent the stationary state of isolated system, and the corresponding eigenvalues, $E_n(k)$, which form the electron band structure of solids. Although this equation is written in an elegant and simple form, to find its solution is a rather difficult task requiring approximations that are usually done in three different levels. The first approximation level, which is called the Born–Oppenheimer approximation, considers the nuclei frozen with the electrons in instantaneous equilibrium with them. Thus, the nuclei are treated as an external potential applied to the electron cloud so that the electronic and ionic problems are decoupled. The other two approximation levels are, actually, implemented within a new formulation of the many-body problem, which is termed density-functional theory (DFT). This theory has been shown to be very successful to describe the underlying physics of solids. In the following, we elaborate briefly on its foundation and on some of its implementation methods for solids.

7.2.1
Density-Functional Theory: Hohenberg–Kohn Theorems and Kohn–Sham Equation

The DFT is based on the theorems formulated by Hohenberg and Kohn (HK) [1]:

Theorem 7.1
For any system of interacting particles in an external potential $V_{\text{ext}}(\mathbf{r})$, there is a one-to-one correspondence between the potential and the ground-state particle density $\varrho_0(\mathbf{r})$. The ground-state expectation value of any observable is, thus, a unique functional of the ground-state particle density $\varrho_0(\mathbf{r})$

$$\langle \psi | A | \psi \rangle = A[\varrho_0(\mathbf{r})]. \tag{7.2}$$

Theorem 7.2
For any external potential applied to an interacting particle system, it is possible to define a universal total energy functional of the particle density, which is written as

$$E[\varrho(\mathbf{r})] = E_{\text{HK}}[\varrho(\mathbf{r})] + \int V_{\text{ext}}(\mathbf{r})\varrho(\mathbf{r})d\mathbf{r}, \tag{7.3}$$

where the term $E_{\text{HK}}[\varrho(\mathbf{r})]$ includes all internal energies of the interacting particle systems. The global minimum of the functional in (7.3) is the exact ground-state total energy of the system E_0 and the particle density that minimizes this functional is the exact ground-state density $\varrho_0(\mathbf{r})$, that is,

$$\frac{\delta}{\delta\varrho} E\left[\varrho(\mathbf{r})\right]\big|_{\varrho=\varrho_0} = 0, \tag{7.4}$$

with

$$E_0 = E[\varrho_0(\mathbf{r})]. \tag{7.5}$$

The first theorem establishes that all observable quantities can be written as a functional of the particle density. It means that the density has as much information as the wavefunction does. In this way, instead of representing the system state by a wavefunction in a multidimensional space, one can represent it by the particle density that lives in a three-dimensional space. The second theorem states the universality of the energy functional and the variational principle, which provides a clue in finding a way to replace the original formulation of the many-body problem (7.1) by something that is more easily treatable. However, it can only be used if the functional E_{HK} is known (or a good approximation for it).

Kohn and Sham (KS) [2] have developed a framework within which the HK theorems can be applied for practical problems. The main idea consists in replacing the interacting many-body problem by a corresponding noninteracting particle system in an appropriate external potential. In the KS ansatz, the total energy functional can be written as

$$E[\varrho(\mathbf{r})] = T_0[\varrho] + E_H[\varrho(\mathbf{r})] + E_{\mathrm{ext}}[\varrho(\mathbf{r})] + E_{\mathrm{xc}}[\varrho(\mathbf{r})] + E_{\mathrm{II}}, \tag{7.6}$$

which is the so-called KS functional. The first, second, and third terms in Equation 7.6 are the functionals for the kinetic energy of a noninteracting electron gas, the classical Coulomb contribution (Hartree term) for the electron–electron interaction, and the external potential contribution due to nuclei and any other external potential, respectively. All many-body effects of exchange and correlation are incorporated into the term $E_{\mathrm{xc}}[\varrho(\mathbf{r})]$, which is called the exchange-correlation functional. The minimization of the KS functional with respect to the density $\varrho(\mathbf{r})$ leads to the following one-particle Schrödinger-like equation

$$\left[-\frac{1}{2}\nabla^2 + V_{\mathrm{ext}} + \int \frac{\varrho(r')}{|\mathbf{r}-\mathbf{r}'|}\,\mathrm{d}\mathbf{r}' + \frac{\delta E_{\mathrm{xc}}[\varrho]}{\delta\varrho(\mathbf{r})}\right]\psi_q(\mathbf{r}) = \varepsilon_q\psi_q(\mathbf{r}), \tag{7.7}$$

where ε_q are the KS eigenvalues, $\psi_q(\mathbf{r})$ are the KS orbitals, and $V_{\mathrm{eff}}(\mathbf{r})$ is the effective potential. The exact ground-state density, $\varrho(\mathbf{r})$, of a N-electron system is given by

$$\varrho(\mathbf{r}) = \sum_q |\psi_q(\mathbf{r})|^2. \tag{7.8}$$

Equations 7.7 and 7.8 must be solved in a self-consistent scheme so that the exact ground-state density and total energy of the many-body electron problem can be determined.

7.2.2
Exchange-Correlation Functional

The task of determining the exchange-correlation functional $E_{\mathrm{xc}}[\varrho(\mathbf{r})]$ enters on the second approximation step mentioned above and is the main challenge in the implementation of the KS scheme. The local density approximation (LDA) has been the most commonly used [8]. Here, the only information needed is the exchange-correlation energy of the homogeneous electron gas as a function of density [3]. The LDA is expected to work well for systems with a slowly varying density, as for instance the nearly free electron metals. Surprisingly, it also appears to be successful for many other systems including semiconductors and insulators. However, further improvements are still required. A natural way to improve the LDA is to consider the exchange-correlation energy depending not only on the density $\varrho(\mathbf{r})$ but also on its gradient $\nabla\varrho(\mathbf{r})$. This is implemented through an approach called generalized gradient approximation (GGA). The most widely used parametrizations for the GGA-functional are those obtained by Becke (B88) [4], by Perdew and Wang (PW91) [5], and by Perdew, Burke, and Enzerhof (PBE) [6]. More recent developments of the exchange-correlation functionals include, besides the electron density and its gradient, the KS orbital kinetic energy [7]. These new functionals are termed meta-GGA.

7.2.3 Plane Wave Methods

To solve the single-particle equation, the KS orbitals are expanded in basis functions. In solids, the plane waves (PWs) are the natural choice due to Bloch theorem. However, the representation of rapidly oscillating core states make the direct implementation of PW basis set computationally inviable. One approach to circumvent this problem is to represent the valence electron–ion interaction via pseudopotentials, which is fitted to free atom energy level. This pseudopotential-PW method is very accurate to describe sp-bonded solids [8]. However, to deal with the more localized d- and f-orbitals and to incorporate properly the effect of core states other approaches are required. One way is to split the basis set into two distinct classes where the basis functions close to the nucleus are more similar to atomic orbitals and in the interstitial region they are PW functions. This is the idea behind the augmented plane wave method (APW) [8]. In this method, the crystal lattice is divided into two main parts, which are termed muffin-tin (MT) and interstitial (I) regions. As originally developed, the APW basis functions are energy dependent what could lead to unphysical results. This problem was later solved by the linearized augmented plane wave (LAPW) method where the basis function are approximated to be energy-independent through an expansions around some reference energy levels [9]. Another PW-based method that is being widely used is the projector augmented wave (PAW) method [10]. This is an all-electron frozen core method, which combines the features of both the ultrasoft pseudopotentials and APW methods. Following the latter, the wavefunctions have a dual representation. Within spheres centered at each atomic position (the augmentation region), they are expanded by partial waves and outside they are expanded into plane waves or some other convenient basis set.

7.2.4 Linearized Muffin-Tin Orbitals Method

Besides the PW-based methods, there are other approaches to solve KS equation as for instance the linearized muffin-tin orbitals (LMTO) method [11], in which the real space is split into the muffin-tin and interstitial regions as in the LAPW method but no PWs are used as basis function. This method has been successfully employed in the investigation of materials physics under high pressure. LMTO was developed with the aim of not only providing a new electronic structure method but also bringing physical interpretation of the electronic structure in the terms of a minimal basis of orbitals. In fact, this method has led to new concepts, such as *canonical bands*, which has shown to be valuable to identify the underlying physics of crystallographic phase transitions. The implementation of LMTO is usually based on a number approximations as for instance the atomic sphere approximation (ASA) where the interstitial volume is reduced to zero and that is more appropriate to describe close-packed solids. When no geometrical approximations are applied to the symmetry of the potential, the method is called a full potential method (FPLMTO). This approach is significantly more time consuming but it allows the investigation of open structure solids.

7.2.5 Hellman–Feynman Theorem and Geometry Optimization

In this section, geometry optimization is briefly discussed, that is, the search for the spatial equilibrium configuration in which the atoms are arranged in the ground state. In the solid state, besides the atomic positions, the shape and volume of the unit cell must be optimized as well.

An atom feeling a net force moves in the direction of the force so that the total energy is minimized. The equilibrium configuration is reached when all such forces are equal to zero, or more realistic, when they are within some convergence criterion. These forces are calculated by using the force theorem or, as it is usually called, the Hellmann–Feynman theorem [8]. This theorem can be understood as follows. The forces due to the atomic displacements can be written as

$$\mathbf{F}_l = -\frac{\partial E}{\partial \mathbf{R}_l}, \tag{7.9}$$

where E is the system total energy,

$$E = \frac{\langle \Psi | H | \Psi \rangle}{\langle \Psi | \Psi \rangle}. \tag{7.10}$$

Thus, assuming $\langle \Psi | \Psi \rangle = 1$ and substituting (7.10) into (7.9), we get

$$\mathbf{F}_l = -\left\langle \Psi \left| \frac{\partial H}{\partial \mathbf{R}_l} \right| \Psi \right\rangle - \left\langle \frac{\partial \Psi}{\partial \mathbf{R}_l} | H | \Psi \right\rangle - \left\langle \Psi | H | \frac{\partial \Psi}{\partial \mathbf{R}_l} \right\rangle. \tag{7.11}$$

At the exact ground-state solution, the energy is extremal with respect to all possible variations of the wavefunction, and as a consequence the last two terms on the right-hand side of (7.11) vanish. Therefore, the forces are determined exclusively by the terms explicitly dependent upon atomic positions and it can be written as

$$\mathbf{F}_l = -\frac{\partial \langle \Psi | H | \Psi \rangle}{\partial \mathbf{R}_l} = -\left\langle \Psi | \frac{\partial H}{\partial \mathbf{R}_l} | \Psi \right\rangle. \tag{7.12}$$

Thus, one can keep $|\Psi\rangle$ at their ground-state values and calculate the partial derivative of the total energy with respect to the ionic positions only.

There are two main factors that can affect the use of the force theorem. One is the errors due to nonself-consistency and another is the explicit dependence of the basis functions upon the ionic positions. The latter gives rise to the so-called Pulay forces. These factors must be treated explicitly in order to avoid additional errors. A good discussion can be found in Ref. [8]. For a plane wave basis set, the Pulay forces are zero because the basis functions do not depend on the atomic position. However, in this case, one still needs to be careful if the volume and shape of a unit cell are being optimized. It is necessary to guarantee that the plane wave basis set is complete. In practice, this is done by using a large energy cutoff.

7.3
Structural Phase Transition from First Principles

To investigate the hydrostatic pressure effects, the volume of the unit cell (V) is varied, the internal degrees of freedom are reoptimized and the total energy recalculated producing a set of energy (E) – volume (V) values, which are fitted to an equation of states (EOSs). Then, from the first derivative of this equation with respect to volume, we obtain the pressure–volume $P(V)$ relation and so the enthalpy $H(P)$ at 0 K. For light elements, we need to go beyond the static lattice model and include the zero-point vibration energy in the 0 K enthalpy function.

As the applied hydrostatic pressure increases, the solids can undergo structural transitions. To investigate this phenomenon, a standard procedure is to calculate $H(P)$ curves for different candidate structures and the stabilized structure at a given pressure will be the one with the lowest enthalpy. Here, we have a problem of searching for crystal structures in a multidimensional configurational space. Many methods have been developed to achieve this goal where we could mention, for instance, the random search method [12], the data mine method [13], the evolutionary method [14], and so on. All these are searching methods based on 0 K total energy calculation. However, a more general approach to study equilibrium states of solids should include temperature effect. This can be achieved by evaluating the phonon dispersion curves and then the vibration free energy for different pressures within the quasiharmonic approximation. There are two main approaches to calculate the phonon modes from first principles, namely, frozen phonon and linear response method [8, 15]. The latter is derived from a second-order perturbation theory whereas the former make use of a direct method to calculate the derivatives of the total energy with respect to perturbations, by carrying out full self-consistent calculations for different values of perturbations and obtaining the derivatives by finite difference formulas.

Another way to investigate structural stability at finite temperature is through molecular dynamics (MD) simulations, which have experienced significant advances in the last two decades; thanks to developments in both theoretical methods and computational facilities. In the classical MD, which was the only possible approach in the past but still widely employed, the forces are obtained through effective potentials between the nuclei (parameterized pair potentials) that attempt to include the effects of electrons. More recently, *ab initio* (or first-principles-based) MD was made possible. In this case, the forces are determined directly from the solution of quantum mechanical problem of electrons without any adjustment of parameters. However, the nuclei are still treated as classical objects with their dynamics governed by Newton's equation. There are two main methods available to implement *ab initio* MD [8]. The first divides the problem in two parts: (a) the motion of the nuclei and (b) the self-consistent solution of Kohn–Sham equations for the electrons. This methodology is termed Born–Oppenheimer molecular dynamics. The alternative approach is Car–Parinello MD where the dynamics of nuclei and the quantum electronic problem are solved within the same algorithm.

These methods appear to be very powerful to study the thermodynamics of the structural transformation and to predict new crystallographic atomic arrangement at high pressure. However, to reveal the underlying physics of this phase transition, we need to inspect the electronic structure of the competing phases as a function of pressure. In the next sections, we will discuss some examples throughout the periodic table for which DFT has been able to provide the fundamental understanding of the phase transitions.

7.4 Alkali Metals

Alkali metals are often used as examples of so-called simple metals, where the electronic structure to a large extent can be compared with the highly simplified electron gas model. However, they are also of technological interest primarily as liquid coolants for nuclear reactors. These metals have high compressibilities and low melting points, so their phase diagrams have been determined over a considerable range of volume compressions. Unexpectedly, these high-pressure studies show complex phase diagrams with many crystal structures.

Among them, cesium (Cs) is the most-studied alkali metal and exhibits an unusual sequence of phase transitions under pressure. Cs is also the only alkali metal that has been demonstrated to become superconducting under high pressure [16]. At ambient conditions, Cs is stable in the high-symmetry body-centered cubic (bcc) structure, which transforms into a face-centered cubic (fcc) phase at approximately 2.3 GPa [17, 18]. The fcc phase undergoes a structural phase transition to a tetragonal structure (Cs IV) at approximately 4.3 GPa, which is stable up to 10 GPa, where it transforms into an orthorhombic structure (Cs V phase) with space group *Cmca* and 16 atoms in the unit cell [19, 20]. The coordination numbers change from 12 $\rightarrow$ 8 10/11 $\rightarrow$ 12 when going from fcc $\rightarrow$ Cs IV $\rightarrow$ Cs V $\rightarrow$ Cs VI as the pressure is increased.

First principles have been able to catch the correct crystallographic transformation sequence in Cs, namely, fcc $\rightarrow$ Cs IV $\rightarrow$ CsV $\rightarrow$ Cs VI; and this sequence can be explained by a simple canonical band energy model and population of d-states [21].

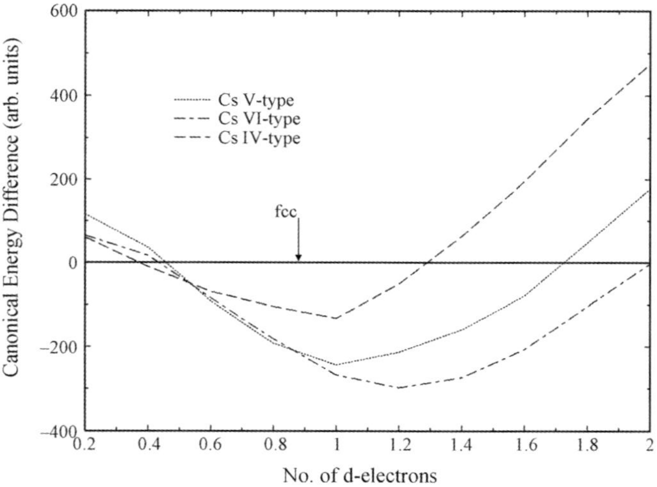

Figure 7.1 Energy differences obtained from canonical d-bands as a function of the d-band filling for Cs VI (dotted-dashed line), Cs IV (dashed line), and Cs V (dotted line) structure. The fcc (solid line) phase is used as the reference level. From Ref. [21].

This is illustrated by the canonical energy difference as a function of the number of d-electrons presented in Figure 7.1. One can see that for d-band filling up to 0.4 the eigenvalue sum of the canonical bands stabilizes the fcc structure, for d-band filling between 0.4 and 0.55, it stabilizes Cs IV; for 0.55–0.8 interval, it stabilizes Cs V; and beyond 0.8 and up to 2.0, it shows the stability of Cs VI (dhcp) phase. Thus, the d-bands, which are almost empty at ambient conditions, start to be filled out, as pressure is applied, by electrons that are transferred from s-like states. Once the s- to d-transition is completed, that is, when the 6s electron states is completely transformed to the d-band, the hcp structure become stable. It should be noticed that in the periodic table, another metal with one d-electron (yttrium) is also stable in the hcp structure. These results suggest that different crystal structures, as for instance Cs V, can be designed just by changing d-band population. This could be achieved by forming alloys, for instance.

There is another crystallographic phase of Cs (Cs III), which has caught a lot of attention both experimentally and theoretically [22]. This phase was first believed to be related to an isostructural transition. However, more recent experimental and theoretical investigations have demonstrated that Cs III phase possesses a complex type of structure with an orthorhombic symmetry belonging to the space group $C222_1$, which contains 84 atoms in the unit cell. This type of structure has also been observed in Rb III phase, which contains 52 atoms in the unit cell [23, 24]. To understand the stabilization of this phase, let us first consider the variation of d-band occupancy as a function of volume as calculated by full potential LMTO Figure 7.2. As can be seen, the d-band occupation is very similar for both elements in the region where Cs III and Rb III are stable, which is when V/V_0 is in the range 0.418–0.430 for Cs and 0.310–0.334 for Rb. One important result is that, in the case of Cs, at

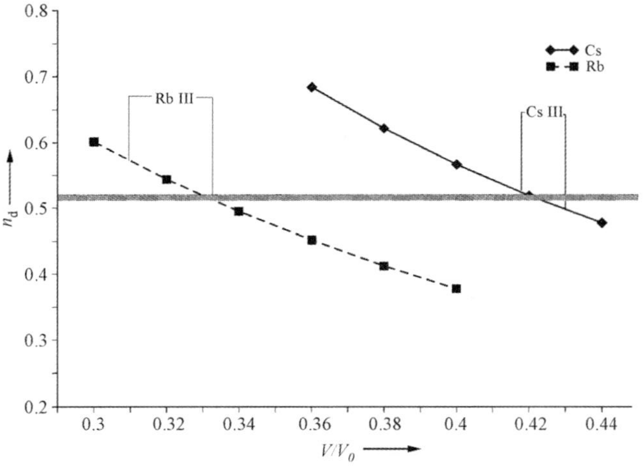

Figure 7.2 Occupation number for Cs 5d and Rb 4d electrons as a function of volume. From Ref. [23].

$V/V_0 = 0.42$, all other phases (namely, fcc, Cs IV, and Cs V) have the same self-consistent FPLMTO total energy, that is, they are degenerated. This volume corresponds to a d-band occupation of 0.52 (see Figure 7.1). Now, we have a remarkable agreement between FPLMTO and the canonical band model because this is the exact common crossing point for all three curves in Figure 7.1. The structures involved in this energy crossing display a variety of coordination numbers (12 → 8 → 10/11 → 12). This allows the stabilization of the complex Cs III phase, which possesses coordination numbers for individual atoms between 8 and 11. Thus, Cs III takes full advantage of the other near-degenerated phases. However, this can happens only in a narrow pressure interval.

7.5
Alkaline Earth Metals

The behavior of alkaline earth metals is strongly influenced by occupation of d-bands in a similar way as alkali metals shown above. Here, we describe such effects in calcium (Ca) and in less extent in strontium (Sr), which are elements in the middle of this group so that they may combine features of their lighter and heavier congeners.

The phase diagram of Ca is quite unique. With increasing hydrostatic pressure, in a range from 0 to 40 GPa, it undergoes the following sequence of structural phase transitions [25–27]: fcc → bcc → simple cubic (sc); that is, the number of nearest neighbors (NN) decreases starting from 12NN (fcc), via a structure with 8NN (fcc), and then with only 6NN (sc). First-principles calculations display very good agreement with the experimental values for the transition pressures and volume collapses [27]. The underlying physics of these phase transitions are understood from arguments based on the d-electron density distribution of valence electrons as a function of volume. The loss of coordination is compensated by a larger atomic wavefunction overlap as shown, from FPLMTO calculations, in Ref. [27].

Let us describe in more detail the physical mechanisms for the stabilization of the simple cubic structure under pressure in Ca. The reduced volumes causes the d-states to become increasingly populated, from being essentially empty at ambient conditions (at the experimental volume under ambient conditions, V_0) to become the dominant state at around $V/V_0 = 0.3$. Furthermore, the bandwidth of the d-states is demonstrated to scale as $(1/V)^{(5/3)}$ so that it becomes increasingly important for decreasing volumes. This effect favors the sc phase over the bcc phase. A further decrease in the volume stabilizes again the bcc structure mainly due to Born–Mayer repulsion of the 3p core states. In fact, the basic mechanism for the structural behavior of Ca under pressure is determined by an interplay between Madelung, d-band, and Born–Mayer contributions. This has been illustrated by fitting the calculated bcc–sc energy difference (the spd-valence curve in Figure 7.3) with the following function [27]:

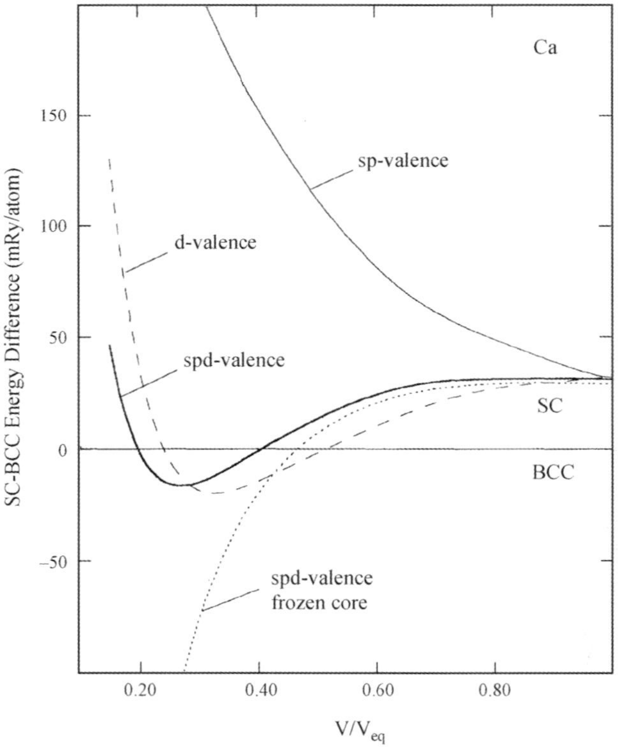

Figure 7.3 Energy difference between the bcc and sc crystal structures for Ca as a function of volume (V/V_{eq}, V_{eq} = equilibrium volume) for different choices of the basis set. The bee structure is used as the zero energy reference level. From Ref. [27].

$$\Delta E = \frac{A}{\nu^{1/3}} + \frac{B}{\nu^4} - \frac{F(\nu)}{\nu^{5/3}} C \tag{7.13}$$

where $\nu = V/V_{eq}$ and

$$F(\nu) = \frac{1}{2}(1-\nu)\left(1 - \frac{(1-\nu)}{10}\right). \tag{7.14}$$

In Equation 7.13, the first, second, and third terms represent the change in the Madelung, d-band energy, and Born–Mayer contribution, respectively. For Ca, the fitting parameters are found to be $A = 31.0$, $B = 0.12$, and $C = 27.0$. Besides Ca, magnesium (Mg) is also predicted to be able to undergo a phase transition to sc structure but at higher transition pressure of 6.6 Mbar. This is because unoccupied d-states in Mg lies at higher energies and therefore higher pressures are required to start their population. For Be, the d-orbitals are lying at very high energy so that the sc structure is not stable at any volume.

With increasing pressure for levels above 113 GPa, Ca undergoes two new structural phase transitions as reported in recent experiments [28]. These crystal structures, which could not be resolved directly from the analysis of the diffraction patterns, were determined from *ab initio* metadynamics simulations [29]. One of them, denoted Ca IV, forms a tetragonal structure, which consists of two helical chains along the *c*-axis, while the other one is an orthorhombic lattice of four zigzag chains. At even higher pressures, >122 GPa, first-principles calculations predicted that Ca V transforms into a stable Sc II type incommensurate complex [30]. This structure is composed by two tetragonal sublattices, the host (i) and the guest (ii), with the same a-lattice parameter but different values for c-lattice parameter with the ratio $\gamma = c_1/c_2$ being an irrational number. In the *ab initio* investigation, the incommensurate structure is modeled by a supercell containing the host and the guest cells but forming a commensurate structure with a rational γ ratio. This approach is based on the fact that the total energy is a continuous function of the structure parameters so that the energy of an analogue commensurate structure with γ ratio close to the incommensurate one will approach the energy of the true incommensurate structure. Therefore, the total energy as a function of γ ratio curve must display a minimum value for the true incommensurate structure. This method appears to be very good to assess the parameters of the incommensurate phase from first-principles calculations.

Another alkaline earth element displaying a complex phase diagram is Sr, which comes just bellow Ca in the periodic table. At ambient conditions, Sr stabilizes in an fcc structure, which transforms into a bcc structure at approximately 26 GPa. Subsequently, it undergoes the following sequence of structural transformations: bcc $\rightarrow$ Sr III $\rightarrow$ Sr IV $\rightarrow$ Sr V, with the transition pressures at approximately 26, 35, and 46 GPa, respectively [26]. Sr IV structure is shown to be monoclinic belonging to space group Ia with 12 atoms in the unit cell whereas Sr V is a kind of host–guest incommensurate structure. The Sr III phase is still not clearly resolved. The main candidate is tetragonal β-tin structure. However, first-principles-based studies have demonstrated that either it is not the correct structure or it exists as an alternative path (metastable structure) for the transition from bcc to Sr IV phase [31]. It could also be a coexisting structure with an unsolved smooth phase. All these crystallographic transformations are driven by an electron transfer process from sp-valence states to d-like states as for other elements such as Ca.

The results described so far, show that simple metals such as Ca and Sr displays quite a complex behavior under high pressure with a rich phase diagram. These discoveries have been made possible thanks to the advances on high-pressure techniques and on first-principles theory.

7.6
Transition Metals

In transition metals, the most important parameter controlling the stability of a given crystal structure is the d-state occupation number. This is due to the fact that the density of states projected on d-states for the bcc, fcc, and hexagonal close-packed structure show characteristic element independent shapes indicating that the one-electron contribution to the total energy dictates the crystal–structure stability. Therefore, by using the structure constant, which is a property related only to a given crystal structure and not to the specific material (or potential), Duthie and Pettifor [32] as well as Skriver [33] have been able to explain the well-known crystal structure sequence of the transition metals, that is, hcp $\rightarrow$ bcc $\rightarrow$ hcp $\rightarrow$ fcc. Their results show that the crystal structure of a certain metal is expected to be modified if an increase of its d-states occupation occurs. Thus, under high pressure where the occupation of d-bands are modified the transition metal must undergo crystallographic phase transformations. For example, Zr (hcp) is expected to become more similar to its neighbor to the right in the series, Nb (bcc), and consequently the bcc structure should be stabilized, as it is subjected to external pressure. This has been confirmed experimentally and demonstrated theoretically [34–36].

In fact, most of high-pressure experimental studies can be explained on the basis of the work by Duthie and Pettifor and Skriver. However, at very high pressures, other contributions become relevant requesting more advanced theoretical models. We show in Figure 7.4 the change in the total energy as a function of the cell volume with reference to fcc structure for Y, Ti, Rh, Pd, and Pt. For Y and Ti the fcc structure has been used instead of hcp, which has been shown to be a good approximation. One can notice that all studied metals are stable in the fcc structure at large volumes and that they all become bcc at sufficiently low volumes or high pressures. In the case of Y and Ti, this can be understood from the above-mentioned band filling effect since a reduced volume corresponds to an increased d occupation and thus Y and Ti are expected to eventually stabilize in the bcc structure. However, for Rh, Pd, and Pt, the stabilization of the bcc structure is unexpected for the following reason. For d-band occupation larger than 7, the fcc structure should be stable and for these metals an external pressure will not move the d-band occupation out of this range. The results of Figure 7.4 thus represent a breakdown of the arguments based on the filling of a canonical d-band. It should be pointed out that the density of states at these high

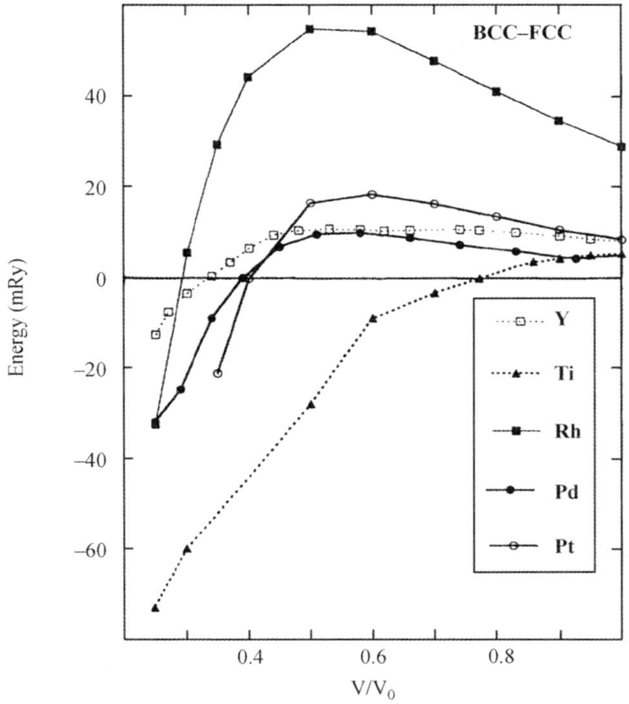

Figure 7.4 Calculated total energy difference between the bcc and fcc structures for Y, Ti, Rh, Pd, and Pt as a function of volume. The energy of the fcc phase is used as the reference level. From Ref. [37].

compressions no longer resemble the well-known canonical d-shape emphasizing that the standard concept of canonical d-bands is no longer applicable. At this high compression, the pseudo-p-bands starts to broaden into bands, which will hybridize with the d valence states. The influence of p–d hybridization is in fact very significant leading to stabilization of structures that are far from being stable at ambient conditions [37].

Below, we describe the behavior of transition metals in the Chromium group, which includes Cr, Mo, and W. These elements display pronounced similarities in their crystallographic behavior. All of them stabilize in a bcc structure at ambient conditions and undergo structural phase transformations to hcp and fcc with volume collapses in the range from 30 to 50% of the experimental volume. The band filling (not the Madelung) is the contribution to the total energy that dictates the crystal structure in these systems. At extraordinary high-pressure, Cr, Mo, and W transform back into the bcc structure and again this is explained by the p–d hybridization discussed above [38].

Another transition metal that plays an important role in the human society and deserves our attention in this text is gold (Au). This element is exceptionally stable to chemical reactions and to extreme pressures and temperatures [39]. For many years, the equation of state of Au has been used as standard for high-pressure studies [40, 41]. However, this status is now under debate since the experimental confirmation of the fcc–hcp phase transition [42], which was previously predicted from first-principles theory [43]. This crystallographic modification takes place at approximately 230 GPa and the transition pressure increases with increasing temperature. Such phenomenon is ascribed to d → sp electron transfer, which is connected to relativistic effects. Similar elements such as Cu, Ag, and Pt are not expected to undergo similar phase transitions. For Pt, this is explained by the fact that its d-band is not completely filled compared to d-band of Au, which, according to canonical band model, stabilizes the fcc structure. In the case of Cu and Ag, the relativistic effects are less pronounced (since they are lighter than Au) so that the d → sp electron transfer is lower. These results shown that, in the sense of structural stability at high pressure, Au is less stable than copper, silver, and platinum.

As a last example of the transition metal subsection, we analyze the behavior of iron (Fe) under high pressure. Iron is considered to be the main constituent of the Earth's core and therefore a great deal of effort has been devoted to understand its properties under high pressure and high temperature [44–49]. At ambient conditions, Fe assumes a hcp structure, which is stable for a wide range of pressure and temperature. The results from shock-wave experiments suggested that iron undergoes a structural phase transition to a bcc phase [50]. First-principles calculations at 0 K have not been able to reproduce this result [51, 52]. The energy difference between hcp and bcc is found to be large and furthermore the bcc phase is found to be mechanically unstable. Actually, the bcc phase is also dynamically unstable at low T. However, quasi-*ab initio* molecular dynamics simulations have shown that indeed Fe transforms into a bcc structure before melting showing that high temperature is required in order to promote this phase transition [49]. This could be connected to eventual anharmonic effects at high *T*, which is still a topic of study. These results indicate that Fe in the Earth's inner core must be stable in the bcc phase.

7.7
Group III Elements

The group-IIIA elements aluminum (Al), gallium (Ga), and indium (In) display a quite surprising behavior under high pressure, which has become a very active topic of experimental and theoretical investigations [53–56]. The lightest element, Al, adopts the high-symmetry fcc structure already at ambient conditions and no structural transformation is observed up to pressures of approximately 220 GPa. Gallium transforms into the bct-In phase (body-centered tetragonal structure), which is called Ga III phase, at a pressure of approximately 14 GPa. Increasing further, the pressure Ga undergoes a new crystallographic modification to the fcc structure at approximately 120 GPa. This structural phase transition from the lower symmetry bct to the high-symmetry fcc structure is not observed in In up to the highest applied pressure of 67 GPa and first-principles calculations at 0 K predicted that this phase transition may take place at extremely high pressure of approximately 800 GPa [56].

Such results indicate that these group IIIA elements follow a trend that is exactly the opposite behavior as regards the corresponding states rule.

The competition between the high-symmetry fcc and low-symmetry bct structure, for Al, Ga, and In, can be understood from the analysis of the s-p hybridization (or mixing) of the valence band. The reader should see the results presented in Figure 3 of Ref. [56]. The degree of s-p mixing is shown to follow the same trend as the change of the band energy as a function of tetragonal distortion of the fcc structure. Al displays a maximum of such s-p mixing at c/a ratio of $\sqrt{2}$ corresponding to fcc structure already at equilibrium volume. For Ga and In, this is achieved only at higher pressures. In the case of Ga, the maximum s-p mixing at c/a ratio of $\sqrt{2}$ occurs exactly at the transition pressure whereas for In this happens before the transition pressure is reached. This delay for the stabilization of the fcc-In is due to formation of van Hove singularity on the density of states as s-p mixing gap evolves, which indicates structure instability. This peak is moved away from the Fermi energy by tetragonal distortion of fcc structure. When this part of the band is moved above, the Fermi energy of the fcc phase becomes stable what occurs at even higher pressure. This different evolution of s-p hybridization gap was demonstrated to be related to relativistic effects that are more significant in In compared to Ga.

The simplest group IIIA element, boron (B), is one of the most challenging systems for high-pressure study due to its structural complexity in a variety of polymorphic forms even in the ground state [57]. One of its polymorphs β-B eventually transforms from a nonmetal to a superconductor above 160 GPa with a transition temperature (T_c) of 6 K [58]. This is an important finding that has prompted a great deal of studies. However, this superconducting phase remains unknown. A good candidate is the α-Ga phase since it has been shown, from total-energy calculations, to be more stable than bct and fcc phases at high pressure. In a recent study, Ma *et al.* [59] have predicted from *ab initio* theory that α-Ga B is a good superconductor with strong anisotropy, which is a consequence of the two-dimensional nature of its electronic structure. The calculated electron–phonon coupling constant λ increases with pressure, which shows consistence with the experimental finding where the transition temperature increases with pressure.

7.8 Group IV Elements

In this section, we will focus on the high-pressure behavior of silicon (Si) and germanium (Ge), which are group IV elements of great technological importance in the electronics industry. Besides that, these materials display a rich phase diagram, which has attracted a lot of theoretical as well as experimental attention. At ambient conditions, Si is stable in the diamond-type structure [60] and at approximately 12 GPa it transforms to the β-Sn structure [61]. The β-Sn structure is stable up to 16 GPa above which Si changes to the primitive hexagonal (ph) structure [62, 63]. Actually, in the narrow pressure range 13–16 GPa an intermediate phase forms, which is

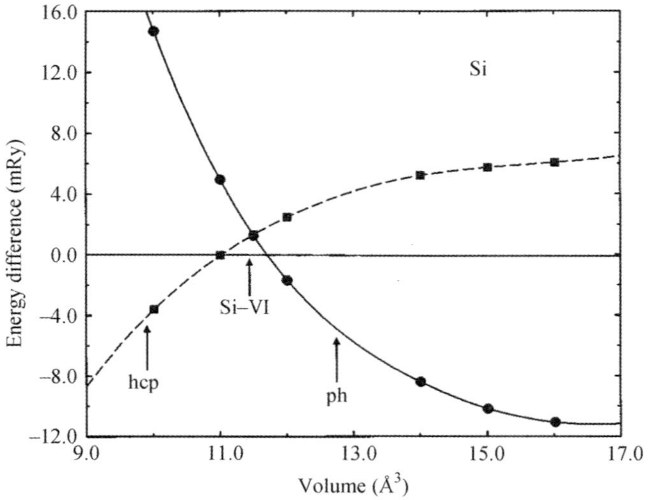

Figure 7.5 Energy difference between the ph, hcp, and Si VI crystal structures for Si as a function of volume. The Si VI structure is used as the reference level. From Ref. [64].

orthorhombic belonging to the *Imma* space group. Furthermore, Si becomes a superconductor in its high pressure, primitive hexagonal phase. Further increase in pressure induces other two crystallographic phase transitions, namely, ph → hcp → fcc at approximately 42 and 78 GPa, respectively. Beside those, a new phase has also been discovered, which is the so-called Si VI phase. FPLMTO method has been shown to be accurate enough to catch this phase transition even though Si VI exists in a very narrow pressure range [64]. In Figure 7.5, we show the energy difference between the ph, hcp, and Si VI crystal structures for Si as a function of volume. As can be seen, Si first transforms from ph to Si VI and then into hcp. The structure of Si VI is an orthorhombic structure belonging to space group *Cmca* and with coordination number of 10/11. Thus, with increasing pressure, Si increase the number of nearest neighbors in the sequence: 8NN (ph) → 10/11 (Si VI) → 12 (hcp). This is because the repulsive Born–Mayer term dominates the total energy in Si under high pressure so that at the lowest volumes the structures with better packing become stable.

Gemanium displays a sequence of pressure-induced phase transitions similar to Si. It adopts the diamond-type structure at ambient conditions and at approximately 10 GPa it transforms to the β-Sn structure [62]. This structure further transforms to the simple hexagonal (sh) structure at 75 GPa, which stays stable up to 102 GPa when a hcp structure stabilizes [65]. Between the β-Sn and sh structure at approximately 75 GPa, a new orthorhombic phase with the *Imma* space group was identified. Theoretical studies based on *ab initio* calculations have correctly reproduced the observed diamond to β-Sn phase transition as well as the phase sequence β-Sn → *Imma* → sh, as a function of increasing pressure. More recently, full potential LMTO calculations have demonstrated that the *Cmca* stabilizes in a narrow pressure interval between sh and hcp, similar to Si [66]. This is illustrated in Figure 7.6. Hence,

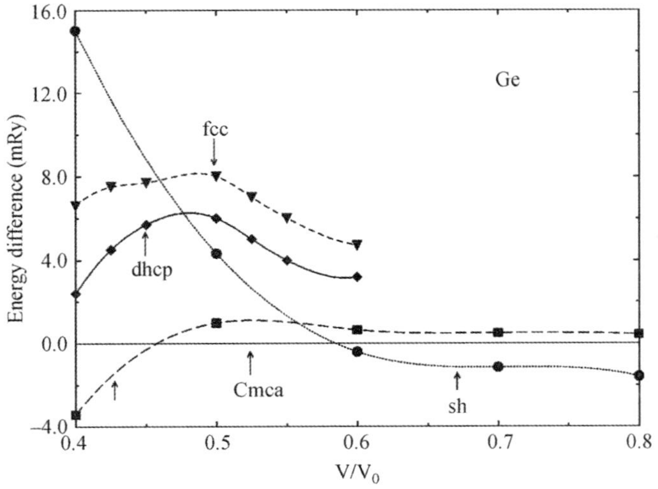

Figure 7.6 Energy difference between the sh, hcp, Cmca, dhcp, and fcc crystal structures for Ge as a function of volume. The *Cmca* structure is used as the energy reference level. Reprinted with permission of [R. Ahuja and B. Johannson, Journal of Applied Physics, Vol. 89, Issue 5, p. 2547]. Copyright American Institute of Physics 2001.

the repulsive Born–Mayer term also dominates the total energy in Ge stabilizing the close-packed structures at extremely high pressure.

7.9 Group V Elements

The group V elements are not considered to be of high technological importance except when combined with other elements as in the case of GaAs. Therefore, they have attracted less attention compared to other elements in the periodic table. Here, we will focus on the high-pressure behavior of phosphorus (P), which displays peculiar crystallographic properties. At ambient conditions, it forms a layered structure consisting of six-membered rings, which displays orthorhombic symmetry and is called A17 phase [67]. In this phase, phosphorus is a narrow band gap semiconductor with an energy gap of 0.3 eV [68]. A transition from the orthorhombic (A17) to the rhombohedral structure, A7 phase, occurs at 5.5 GPa and already at a slightly higher pressure of 11 GPa, it undergoes a new phase change to the simple cubic (sc) structure [69]. This phase is stable for a wide pressure range and only at around 137 GPa another crystallographic rearrangement takes place when a simple hexagonal (sh) structure is stabilized. High-pressure experimental studies on the stability of sc phase have revealed that the sc-sh transition occurs via an intermediate phase (P IV) whose crystal structure is still under debate. The diffraction measurements suggest that this structure might be the consequence of a monoclinic distortion along [1 1 0] direction in the sc structure. With increasing further the

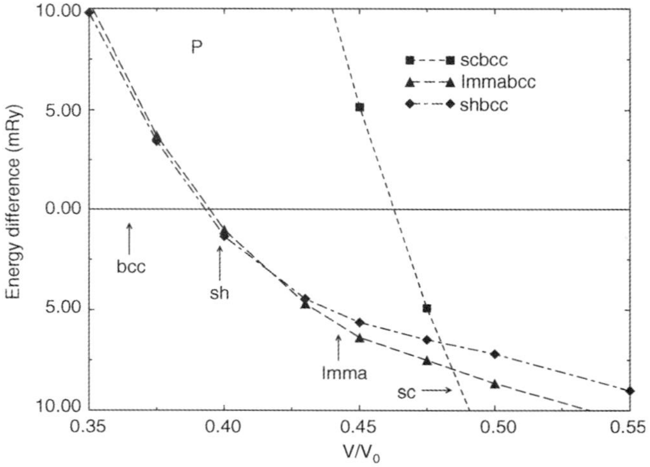

Figure 7.7 Total energy of phosphorus as a function of volume for different structures. The total energy of the bcc structure is taken as a reference level. From Ref. [70].

pressure, at around 262 GPa, sh is transformed into a bcc structure. The stabilization of this high-symmetry structure at very high pressure is in line with the phase diagram of other heavier group V elements.

Density-functional theory within local density approximation has been shown to be successful to describe underlying physics of P under pressure. First of all the semiconductor state with a narrow band gap in A17 phase is well reproduced from different implementations of DFT. The A17-A7 transition is found to be accompanied by a transition from the semiconductor to semi-metallic state whereas the higher pressure phases (sc, sh and bcc) display a more free electron-like density of states as can be seen in Figure 2 of Ref. [70]. Thus, the metallic behavior of phosphorus increases with increasing pressure. Possible structures to the intermediate P IV phase have also been investigated through FPLMTO calculations where the lowest energy configuration was found to assume an orthorhombic structure belonging to space group *Imma*. The total energy difference between different phases of P and its bcc phase as a function of volume is displayed in Figure 7.7, where the stabilization of *Imma* phase is illustrated. One can notice that sh and *Imma* structures are quite close in energy so that they may coexist for some pressure interval. The crystallographic sequence of Phosphorus under pressure can be understood from the variations of bandwidth as shown in Figure 2 of Ref. [70]. This is associated with the degree of hybridization between sp- and d-bands.

7.10 Overview

The advances on the first-principles calculation methods along with a significant development of computational facilities has made feasible the investigation of complex crystal structures improving our understanding of the solids behavior when subject to a wide range of pressure and temperature. This chapter shows some examples of elemental compounds in which the electronic structure theory successfully revealed the underlying physics of the pressure (or temperature in some cases)-induced crystallographic phase transitions. However, most of these studies still relied strongly on the experimental results to determine the structure sequences as pressure is increased. In the past 10 years, a great deal of effort has been devoted to relax this constrain, that is, to increase the prediction power of *ab initio* theory.

In fact, we are witnessing now an incredible improvement of *ab initio* theory capability to predict new structures thanks to the developments of search tools to explore the configurational space, which are based on genetic algorithm methods, random search methods, data mine approaches, topological modeling methods, and so on. These methods are based on 0 K calculations and vibrational contributions for the free energy is not included. However, with increasing temperature, more states become accessible on the energy landscape and the latter contribution must be added. To this end, two methods stand out, namely, the ergodicity search algorithm based on Monte Carlo simulations and the metadynamics approach. For getting a good overview about all these methods (at 0 K or finite temperature), the reader is recommended to see the review article by Woodley and Catlow [14].

Another important topic that has also experienced a significant advance is the investigation of dynamical stability of the different phases. In some cases, although the crystal structure is found to be the stable configuration, from thermodynamics viewpoint, it is not experimentally observed. This could be due to dynamical and/or mechanical instability. The latter is more straightforward to assess from first-principle calculations but the former request heavy calculations of the full phonon spectra. The development of frozen phonon methods, which makes use of a direct method to calculate the derivatives of the total energy with respect to perturbations, has allowed the calculation of phonon spectra of complexes systems. The dynamical stability of different phases is already taken into account in the metadynamics simulations, but it must be incorporated in the other search methods.

In spite of the significant advances already achieved by the scientific community, there still are many hurdles to be overcome in order to investigate more realistic situations when, for example, defects and interfaces play a role. To this end, we believe that the greatest progress will come from research efforts resulting from a close collaboration between theory and experiments.

References

1 Hohenberg, P. and Kohn, W. (1964) Inhomogeneous electron gas. *Phys. Rev.*, **136**, B864.

2 Kohn, W. and Sham, L. (1965) Self-consistent equations including exchange and correlation effects. *Phys. Rev.*, **140**, A1133.

3 Ceperley, D.M. and Alder, B.J. (1980) Ground-state of the electron-gas by a stochastic method. *Phys. Rev. Lett.*, **45**, 566–569.

4 Becke, A.D. (1988) Density-functional exchange-energy approximation with correct asymptotic-behavior. *Phys. Rev. A*, **38**, 3098–3100.

5 Perdew, J.P. and Wang, Y. (1992) Accurate and simple analytic representation of the

electron-gas correlation-energy. *Phys. Rev. B*, **45**, 13244–13249.
6 Perdew, J.P., Burke, K., and Ernzerhof, M. (1996) Generalized gradient approximation made simple. *Phys. Rev. Lett.*, **77**, 3865–3868.
7 Tao, J., Perdew, J.P., Staroverov, V.N., and Scuseria, G.E. (2003) Climbing the density functional ladder: nonempirical meta-generalized gradient approximation designed for molecules and solids. *Phys. Rev. Lett.*, **91**, 146401.
8 Martin, R.M. (2004) *Electronic Structure Basic Theory and Practical Methods*, Cambridge University Press; and the references therein.
9 Singh, D. (1994) *Plane Waves, Pseudopotentials and the LAPW Method*, Kluwer Academic.
10 Blöchl, P.E. (1994) Projector augmented-wave method. *Phys. Rev. B*, **50**, 17953–17979.
11 Skriver, H.L. (1984) *The LMTO Method*, Springer, Berlin.
12 Pickard, C.J. and Needs, R.J. (2007) When is H_2O not water? *J. Chem. Phys.*, **127**, 244503.
13 Curtarolo, S., Morgan, D., Persson, K., Rodgers, J., and Ceder, G. (2003) Predicting crystal structures with data mining of quantum calculations. *Phys. Rev. Lett.*, **91**, 135503.
14 Woodley, S.M. and Catlow, R. (2008) Crystal structure prediction from first principles. *Nat. Mater.*, **7**, 937–946; and the references therein.
15 Baroni, S., de Gironcoli, S., Corso, A.D., and Giannozzi, P. (2001) Phonons and related crystal properties from density-functional perturbation theory. *Rev. Mod. Phys.*, **73**, 515–562.
16 Wittig, J. (1970) Pressure-induced superconductivity in cesium and yttrium. *Phys. Rev. Lett.*, **24**, 812.
17 Hall, H.T., Merrill, L., and Barnett, J.D. (1964) High pressure polymorphism in cesium. *Science*, **146**, 1297.
18 Anderson, M.S., Gutman, E.J., Packard, J.R., and Swenson, C.A. (1969) Equation of state for cesium metal to 23 Kbar. *J. Phys. Chem. Solids*, **30**, 1587.
19 Takemura, K., Minomura, S., and Shimomura, O. (1982) X-ray-diffraction study of electronic transitions in cesium under high-pressure. *Phys. Rev. Lett.*, **49**, 1772–1775.
20 Schwarz, U., Takemura, K., Hanfland, M., and Syassen, K. (1998) Crystal structure of cesium-V. *Phys. Rev. Lett.*, **81**, 2711–2714.
21 Ahuja, R., Eriksson, O., and Johansson, B. (2000) Theoretical high-pressure studies of Cs metal. *Phys. Rev. B*, **63**, 014102.
22 Young, D.A. (1991) *Phase Diagrams of the Elements*, University of California Press, Berkeley.
23 Osorio-Guill_en, J.M., Ahuja, R., and Johansson, B. (2004) Structural phase transitions in heavy alkali metals under pressure. *ChemPhysChem*, **5**, 1411–1415.
24 See references in Ref. [23].
25 Skriver, H.L. (1982) Calculated structural phase-transitions in the alkaline-earth metals. *Phys. Rev. Lett.*, **49**, 1768–1772.
26 Olijnyk, H. and Holzapfel, W.B. (1984) Phase-transitions in alkaline-earth metals under pressure. *Phys. Lett. A*, **100**, 191–194.
27 Ahuja, R., Eriksson, O., Wills, J.M., and Johansson, B. (1995) Theoretical confirmation of the high pressure simple cubic phase in calcium. *Phys. Rev. Lett.*, **75**, 3473–3476.
28 Yabuuchi, T., Nakamoto, Y., Shimizu, K., and Kikegawa, T. (2005) New high-pressure phase of calcium. *J. Phys. Soc. Jpn*, **74**, 2391–2392.
29 Ishikawa, T. *et al.* (2008) Theoretical study of the structure of calcium in phases IV and V via *ab initio* metadynamics simulation. *Phys. Rev. B*, **77**, 020101(R).
30 Arapan, S., Mao, H., and Ahuja, R. (2008) Prediction of incommensurate crystal structure in Ca at high pressure. *Proc. Nat. Acad. Sci. USA*, **105**, 20627--20630.
31 Phusittrakool, A., Bovornratanaraks, T., Ahuja, R., and Pinsook, U. (2008) High pressure structural phase transitions in Sr from *ab initio* calculations. *Phys. Rev. B*, **77**, 174118.
32 Duthie, J.C. and Pettifor, D.G. (1977) Correlation between *d*-band occupancy and crystal-structure in rare-earths. *Phys. Rev. Lett.*, **38**, 564–567.
33 Skriver, H.L. (1985) Crystal-structure from one-electron theory. *Phys. Rev. B*, **31**, 1909–1923.

34 Sikka, S.K., Vohra, Y.K., and Chidambarum, R. (1982) Omega-phase in materials. *Prog. Mater. Sci.*, **27**, 245–310.

35 Xia, H., Parthasarathy, G., Luo, H., Vohra, Y.K., and Ruoff, A.L. (1990) Crystal structure of group-IVA metals at ultrahigh pressures. *Phys. Rev. B*, **42**, 6736–6738.

36 Ahuja, R., Wills, J.M., Johansson, B., and Eriksson, O. (1993) Crystal structures of Ti, Zr, and Hf under compression: theory. *Phys. Rev. B*, **48**, 16269–16279.

37 Ahuja, R., Söderlind, P., Trygg, J., Melsen, J., Wills, J.M., Johansson, B., and Eriksson, O. (1994) Influence of pseudocore valence-band hybridization on the crystal-structure phase stabilities of transition metals under extreme compressions. *Phys. Rev. B*, **50**, 14690–14693.

38 Söderlind, P., Ahuja, R., Eriksson, O., Johansson, B., and Wills, J.M. (1994) Theoretical predictions of structural phase transitions in Cr, Mo, and W. *Phys. Rev. B*, **49**, 9365–9371.

39 Hammer, B. and Norskov, J.K. (1995) Why gold is the noblest of all metals? *Nature*, **376**, 238–240.

40 Batani, D., Balducci, A., Beretta, D., Bernardinello, A., Lower, T., Koenig, M., Benuzzi, A., Faral, B., and Hall, T. (2000) Equation of state data for gold in the pressure range< 10 TPa. *Phys. Rev. B*, **61**, 9287–9294.

41 Dewaele, A., Loubeyre, P., and Mezouar, M. (2004) Equations of state of six metals above 94 GPa. *Phys. Rev. B*, **70**, 094112.

42 Dubrovinsky, L., Dubrovinskaia, N., Crichton, W.A., Mikhaylushkin, A.S., Simak, S.I., Abrikosov, I.A., de Almeida, J.S., Ahuja, R., Luo, W., and Johansson, B. (2007) Noblest of all metals is structurally unstable at high pressure. *Phys. Rev. Lett.*, **98**, 045503.

43 Ahuja, R., Rekhi, S., and Johansson, B. (2001) Theoretical prediction of a phase transition in gold. *Phys. Rev. B*, **63**, 212101.

44 Boehler, R. (1993) Temperatures in the Earth's core from melting-point measurements of iron at high-static pressures. *Nature*, **363**, 534–536.

45 Shen, G., Mao, H.K., Hemley, R.J., Duffy, T.S., and Rivers, M.L. (1998) Melting and crystal structure of iron at high pressures and temperatures. *Geophys. Res. Lett.*, **25**, 373–376.

46 Alfe, D., Gillan, M.J., and Price, G.D. (1999) The melting curve of iron at the pressures of the Earth's core from *ab initio* calculations. *Nature*, **401**, 462–464.

47 Laio, A., Bernard, S., Chiarotti, G.L., Scandolo, S., and Tosatti, E. (2000) Physics of iron at Earth's core conditions. *Science*, **287**, 1027–1030.

48 Steinle-Neumann, G., Stixrude, L., Cohen, R.E., and Gülseren, O. (2001) Elasticity of iron at the temperature of the Earth's inner core. *Nature*, **413**, 57–60.

49 Belonoshko, A.B., Ahuja, R., and Johansson, B. (2003) Stability of the body-centred-cubic phase of iron in the Earth's inner core. *Nature*, **424**, 1032–1034.

50 Brown, J.M. and McQueen, R.G. (1986) Grüneisen Parameter, and elasticity for shocked iron between 77 GPa and 400 GPa. *J. Geophys. Res*, **91**, 7485–7494.

51 Stixrude, L., Cohen, R.E., and Singh, D.J. (1994) Iron at high pressure: linearized-augmented-plane-wave computations in the generalized-gradient approximation. *Phys. Rev. B*, **50**, 6442–6445.

52 Söderlind, P., Moriarty, J.A., and Wills, J.M. (1996) First-principles theory of iron up to earth-core pressures: structural, vibrational, and elastic properties. *Phys. Rev. B*, **53**, 14063–14072.

53 Greene, R.G. *et al.* (1994) Al as a simple solid: high-pressure study to 220 GPa (2.2 Mbar). *Phys. Rev. Lett.*, **73**, 2075–2078.

54 Schulte, O. and Holzapfel, W.B. (1997) Effect of pressure on the atomic volume of Ga and Tl up to 68 GPa. *Phys. Rev. B*, **55**, 8122–8128.

55 Kenichi, T., Kazuaki, K., and Masao, A. (1998) High-pressure bct-fcc phase transition in Ga. *Phys. Rev. B*, **58**, 2482–2486.

56 Simak, S.I., Häussermann, U., Ahuja, R., Lidin, R., and Johansson, B. (2000) Gallium and indium under high pressure. *Phys. Rev. Lett.*, **85**, 142–145.

57 Donohue, J. (1974) *The Structures of the Elements*, John Wiley & Sons, Inc., New York.

58 Eremets, M.I., Struzhkin, V.V., Mao, H.K., and Hemley, R.J. (2001) Superconductivity in boron. *Science*, **293**, 272–274.

59 Ma, Y., Tse, J.S., Klug, D.D., and Ahuja, R. (2004) Electron–phonon coupling of α-Ga boron. *Phys. Rev. B*, **70**, 214107.

60 Minomura, S. and Drickamer, H.G. (1962) Pressure induced phase transitions in silicon, germanium and some 3–5 compounds. *J. Phys. Chem. Solids*, **23**, 451.

61 Jamieson, J.C. (1963) Crystal structure at high pressures of metallic modifications of silicon and germanium. *Science*, **139**, 762.

62 Olijnyk, H., Sikka, S.K., and Holzapfel, W.B. (1984) Structural-phase transitions in Si and Ge under pressures up to 50 GPa. *Phys. Lett. A*, **103**, 137–140.

63 Hu, J.Z. and Spain, I.L. (1984) Phases of silicon at high-pressure. *Solid State Commun.*, **51**, 263–266.

64 Ahuja, R., Eriksson, O., and Johansson, B. (1999) Theoretical high-pressure studies of silicon VI. *Phys. Rev. B*, **60**, 14475–14477.

65 Vohra, Y.K., Brister, K.E., Desgreniers, S., Ruoff, A.L., Chang, K.J., and Cohen, M.L. (1986) Phase transition studies of Germanium up to 1.25 Mbar. *Phys. Rev. Lett.*, **56**, 1944–1947.

66 Ahuja, R. and Johansson, B. (2001) Theoretical prediction of the *Cmca* phase in Ge under high pressure. *J. Appl. Phys.*, **89**, 2547.

67 Brown, A. and Rundqvist, S. (1965) Refinement of crystal structure of black phosphorous. *Acta Crystallogr.*, **19**, 684.

68 Keyes, R.W. (1953) The electrical properties of black phosphorus. *Phys. Rev.*, **92**, 580–584.

69 Kikegawa, T. and Iwasaki, H. (1983) An X-ray-diffraction study of lattice compression and phase-transition of crystalline phosphorous. *Acta Crystallogr. B*, **39**, 158–164.

70 Ahuja, R. (2003) Calculated high pressure crystal structure transformations for phosphorus. *Phys. Stat. Sol. B*, **235**, 282–287.

8
Ab Initio Lattice Dynamics and Thermodynamical Properties

Razvan Caracas and Xavier Gonze

8.1
Introduction

Many physical properties of an atomic crystalline lattice can be successfully determined from first principles within the static approximation at zero temperature. However, the crystal lattice is not just a rigid collection of atoms under symmetry constraints. On the contrary, it is in a continuous dynamical state, with atoms vibrating around their respective equilibrium positions. The description of as many of its physical properties, such as phase transitions, infrared or Raman spectra require the ability to describe the fluctuations of the nuclei positions with respect to their static equilibrium positions occurring during these vibrations. The treatment of such effects, once the properties of the static lattice system are established, can be done in a coherent framework, treating small deformations by way of perturbation theory inside density functional theory (DFT) [1, 2]. The corresponding formalism, called density-functional perturbation theory (DFPT) [3–14] has been already implemented in several software packages (e.g., see Refs [15–23]). The goal of this chapter is to present a general overview of the DFPT formalism to the readers to use and understand such programs.

Practical DFPT applications have already generated a large number of studies. In particular, the computation of selected vibrational frequencies, for example, for comparison with infrared and Raman data, is now routine, and can be performed for rather complicated crystal structures (e.g., see Ref. 24). Although more demanding in resources, both human and CPU time, the computation of the full vibrational spectrum is an invaluable source of information on the behavior of the crystal under different thermodynamic conditions. It allows determining whether a structure is stable against all possible small deformations (e.g., see Refs. [25, 26]), computing the temperature-dependent properties of this structure, such as thermal dilatation, temperature-dependent entropy, specific heat, internal energy, or free energy or allows determining whether a crystalline structure will present local instabilities. For high-pressure investigations, such analysis is crucial.

Thermodynamic Properties of Solids: Experiment and Modeling
Edited by S. L. Chaplot, R. Mittal, and N. Choudhury

ISBN: 928-3-527-40812-2

In the following, we will describe the theoretical basis that allow us to obtain starting from the first principles raw data, the specific quantities that can be linked to experimental results.

This chapter is organized as follows. In Section 8.2, we present the concept of interatomic force constants and dynamical matrices, and emphasize that they are second-order derivatives of the energy. In Section 8.3, we explain how we can compute within density-functional perturbation theory such as second-order derivatives. In Section 8.4, we focus on dielectric properties and on related infrared and Raman spectra. Then, in Section 8.5, we present the computation of thermodynamic quantities and Raman spectra of the polymeric high-pressure phase of nitrogen and we analyze the dynamical instabilities of ice X under pressure. The chapter ends with some discussions and perspectives.

8.2
Phonons

For the determination of the dynamical matrix, we start from a system of nuclei and electrons, in its electronic ground state. In the framework of the Born–Oppenheimer (BO) approximation (the adiabatic approximation, namely, stating that the electrons follow instantaneously the movement of the nuclei), the energy of such a system is a well-defined function of the position of the nuclei that can be computed on the basis of density functional theory, or any many-body approach to the electronic structure problem. We will refer to this energy as the Born–Oppenheimer ground-state energy of the system, to which the kinetic energy of the nuclei should be added to obtain a total energy. With respect to the periodic arrangement of nuclei, corresponding to the perfect classical crystal at zero temperature, small displacements around the equilibrium positions occur, and, in a classical viewpoint, evolve as a function of time. Consequently, the energy of the crystal can be expressed like a Taylor expansion as a function of the nuclear displacements [12, 14] as

$$
\begin{aligned}
E = E^{(0)} &+ \sum_{a\varkappa\alpha} \left(\frac{\partial E}{\partial \tau^{a}_{\varkappa\alpha}} \right) \Delta\tau^{a}_{\varkappa\alpha} + \sum_{a\varkappa\alpha} \sum_{b\varkappa'\beta} \left(\frac{\partial^2 E}{\partial \tau^{a}_{\varkappa\alpha} \partial \tau^{b}_{\varkappa'\beta}} \right) \Delta\tau^{a}_{\varkappa\alpha} \Delta\tau^{b}_{\varkappa'\beta} \\
&+ \sum_{a\varkappa\alpha} \sum_{b\varkappa'\beta} \sum_{ck''\gamma} \left(\frac{\partial^3 E}{\partial \tau^{a}_{\varkappa\alpha} \partial \tau^{b}_{\varkappa'\beta} \partial \tau^{c}_{\varkappa''\gamma}} \right) \Delta\tau^{a}_{\varkappa\alpha} \Delta\tau^{b}_{\varkappa'\beta} \Delta\tau^{c}_{\varkappa''\gamma} + \ldots
\end{aligned}
\tag{8.1}
$$

where $\Delta\tau^{a}_{\varkappa\alpha}$ is the displacement of the nucleus $\varkappa$ along direction α in the cell labeled a (with vector $\mathbf{R}_a$), from its equilibrium position $\tau_{\varkappa}$. The first term of the series, when all displacements vanish, correspond to the minimal energy in the static approximation of the lattice – $E^{(0)}$ (e.g., which is computed in a standard density-functional theory calculation). The second term of the series development corresponds to the forces on the atoms that vanish for the equilibrium configuration, at which energy is minimal and atoms lay in the bottom of the potential wells; hence, there are no linear terms in Equation 8.1. The third term of the series is called a "harmonic term" and the expansion of the energy truncated at second order is called the "harmonic approximation," the vibration of the nuclei being treated as a quasielastic lattice. Truncations of the energy expansion at higher orders are possible and the resulting equations correspond to anharmonic effects and to higher order properties and will be treated in some detail in Section 8.4.2. For the time being, let us remain within the harmonic approximation. Forces and energies are related through the principle of virtual works: the force exerted on one nucleus in a specific direction is the opposite of the derivative of the energy due to an infinitesimal change of position of this nucleus along that direction. When the nuclei are not at their equilibrium position, forces appear. In the harmonic approximation, they are linearly related to the displacement of every nucleus:

$$F^a_{\varkappa\alpha} = -\sum_{b\varkappa'\beta}\left(\frac{\partial^2 E}{\partial\tau^a_{\varkappa\alpha}\partial\tau^b_{\varkappa'\beta}}\right)\Delta\tau^b_{\varkappa'\beta}. \tag{8.2}$$

In order to describe the force on one nucleus a that arises because of the displacement of another nucleus b (which can be even itself), from Equation 8.2 one needs the matrix of *interatomic force constants* (IFCs):

$$C_{\varkappa\alpha,\varkappa'\beta}(a,b) = \left(\frac{\partial^2 E}{\partial\tau^a_{\varkappa\alpha}\partial\tau^b_{\varkappa'\beta}}\right). \tag{8.3}$$

The dynamics of the lattice obeys the classical mechanics law of Newton – forces equal mass times acceleration. The acceleration of each nucleus due to the forces acting upon it is then given as

$$F^a_{\varkappa\alpha}(t) = M_\varkappa\frac{\partial^2\tau^a_{\varkappa\alpha}(t)}{\partial t^2}. \tag{8.4}$$

In the harmonic approximation, the general solutions of the evolution equation, Equation 8.4, consist in a superposition of the so-called normal modes of vibrations, labeled by the index σ, with amplitude a_σ (to be determined by the initial conditions), that is,

$$\Delta\tau^a_{\varkappa\alpha}(t) = \sum_\sigma a_\sigma U^a_\sigma(\varkappa\alpha)e^{i\omega_\sigma t} + \text{c.c.} \tag{8.5}$$

where the normal mode angular frequency, ω_s, corresponds to a pattern of nuclear displacements U^a_σ. Both quantities are determined by the solution of a generalized eigenvalue equation, involving the interatomic force constants, as well as the masses of the nuclei

$$\sum_{\varkappa'\beta b} C_{\varkappa\alpha,\varkappa'\beta}(a,b)\,U^b_\sigma(\varkappa'\beta) = M_\varkappa\omega^2_\sigma U^a_\sigma(\varkappa\alpha) \tag{8.6}$$

To simplify the system of equations, we make use of the invariance of the structure with respect to the lattice periodicity. The interatomic force constant between the atom k from cell a and the atom k' from cell b is equal to that between atom k from cell

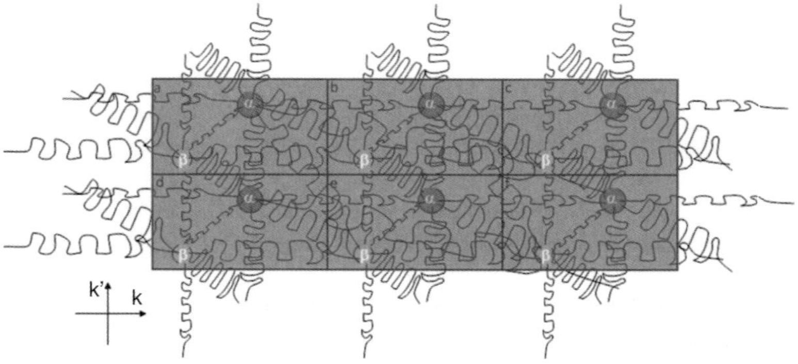

Figure 8.1 Interatomic force constants in the harmonic approximation.

a' and atom k' from cell b', if $a - b = a' - b' = q$ (Figure 8.1). Consequently, we can replace (a,b) by the wave vector $\mathbf{q}$, which characterizes the normal modes of vibration. These modes have the form of a Bloch wave, namely, the product of a wave vector-dependent phase factor that varies from cell to cell, by a wave vector-dependent periodic function:

$$U_{\sigma}^{a}(\varkappa\alpha) = e^{i\mathbf{q}\cdot \mathrm{R}_a} U_{m\mathbf{q}}(\varkappa\alpha) \tag{8.7}$$

In the framework of quantum mechanics, such patterns of displacements are quantized and are called phonons. Note that we replaced the index σ by the composite index $m\mathbf{q}$, where the dependence on the wave vector ($\mathbf{q}$) appears explicitly.

The periodic part of the Bloch wave fulfills a similarly generalized eigenvalue equation:

$$\sum_{\varkappa'\beta} D_{\varkappa\alpha,\varkappa'\beta}(\mathbf{q}) U_{m\mathbf{q}}(\varkappa'\beta) = M_{\varkappa}\omega_{m\mathbf{q}}^{2} U_{m\mathbf{q}}(\varkappa\alpha) \tag{8.8}$$

expressed in terms of the *dynamical matrices*, the Fourier transforms of the interatomic force constants

$$D_{\varkappa\alpha,\varkappa'\beta}(\mathbf{q}) = \sum_{B} C_{\varkappa\alpha,\varkappa'\beta}(0,b) e^{i\mathbf{q}\cdot \mathrm{R}_b} \tag{8.9}$$

The corresponding eigendisplacements are normalized such as

$$\sum_{\varkappa a} M_{\varkappa}[U_{m\mathbf{q}}(\varkappa\alpha)]^{*} U_{m\mathbf{q}}(\varkappa\alpha) = 1 \tag{8.10}$$

The dynamical matrix has dimension $3N_n \times 3N_n$, where N_n is the number of nuclei in the unit cell. By diagonalization, we obtain the eigenvalues, which are the square of the phonon frequencies, and its eigenvectors, which are the displacements corresponding to the atomic vibrations. Each eigenvector has $3N_n$ components, corresponding to the displacement of each of the N_n atoms along the three Cartesian directions. The phonons, also named phonon modes, can be degenerated.

The eigenvalues of the dynamical matrix correspond to the second derivative of the energy with respect to the corresponding atomic pattern of displacements.

One quick check of the dynamical stability of a given structure is the sign of the phonon frequencies: if all the phonons have positive frequencies for all wave vectors, then the structure is dynamically stable and hence at least thermodynamically metastable. If there are phonons with imaginary frequencies (namely, an eigenvalue of the dynamical matrix is negative) then the structure is dynamically unstable. In this case, the energy describes a double-well potential as a function of the amplitude of the atomic displacement corresponding to the unstable phonon eigenvector. The energy of the crystal is lowered if the atoms move according to this phonon, which usually results in a second-order phase transition [27].

The dispersion relations, giving the frequency of the phonons as a function of the wave vector $\mathbf{q}$, form "phonon bands." As for all waves propagating in a periodic medium, for example, as electronic waves, the wave vectors are restricted in a portion of the reciprocal space, the Brillouin zone, whose boundaries are Bragg planes. Traditionally, the phonon dispersion relations are represented for a selected set of (high symmetry) lines in the Brillouin zone.

The Γ point at $\mathbf{q} = \langle 0\ 0\ 0 \rangle$ is the center of the Brillouin zone. Phonons with that wave vector have a pattern of displacement with wavelength $1/0 = \infty$ along each of the three directions of the space. In other words, the displacement pattern is identical in all unit cells. By contrast, all the other points in the Brillouin zone correspond to the waves with a finite wavelength and a phase varying from cell to cell, that is, propagating waves. There are three phonon branches that go to zero frequency when $\mathbf{q}$ approaches 0. They are called the acoustic branches. In these vibrational modes, all the atoms of the structure move in phase, by the same amount along the same direction, such as the crystal is actually left invariant, due to the translational invariance of the energy. Such acoustic branches are found in all phonon spectra. The dispersion relations of the acoustic branches around Γ can be related to the elastic constants tensor. The computation of the dynamical matrix at one specific $\mathbf{q}$ wave vector is now a routine task, if performed in the framework of DFPT (see next section). Still, it takes a nonnegligible amount of computer time. Because of this, it is not considered as a good strategy to perform band structure calculations, like for the one of diamond, based on hundreds of single $\mathbf{q}$ wave vector direct evaluations of dynamical matrices.

Special techniques have been set up to interpolate the phonon dispersion relations throughout the whole Brillouin zone, from the knowledge of selected dynamical matrices. Such techniques, building upon the knowledge of asymptotic behavior of the interatomic force constants, are based on Fourier transforms, and specific treatment of the dipole–dipole interaction [7, 12].

In the adiabatic and harmonic approximations, the basic equations defining interatomic force constants, dynamical matrices, dispersion relations, and Bloch waves are exactly the same as in classical mechanics. However, the dynamics of the lattice should be described by a many-body nuclear time-dependent wavefunction, related to the static ground state by phonon creation operators. However, this difference with classical mechanics has negligible consequences, except at the level of the computation of the phonon contribution to thermodynamical quantities or the atomic temperature factors.

Next, we focus on DFPT that allows the efficient computation of the second-order derivatives of the energy with respect to arbitrary nuclei displacement, needed to obtain the dynamical matrices.

8.3 Density-Functional Perturbation Theory

In the density-functional theory [1, 2], we can derive the ground-state energy of the electronic system by minimization of the following functional:

$$E_{\mathrm{el}}\{\psi_\alpha\} = \sum_{\alpha}^{\mathrm{occ}} \langle \psi_\alpha | T + v_{\mathrm{ext}} | \psi_\alpha \rangle + E_{\mathrm{Hxc}}[n] \tag{8.11}$$

or alternatively

$$H|\psi_\alpha\} = \varepsilon_\alpha |\psi_\alpha\}, \quad \text{where} \quad H = T + v_{\mathrm{ext}} + v_{\mathrm{xc}} \tag{8.12}$$

where ψ_α's are the Kohn–Sham orbitals (to be varied until the minimum is found), T is the kinetic energy operator, v_{ext} is the external potential to the electronic system, including the one created by nuclei, E_{Hxc} is the Hartree and exchange-correlation energy functional of the electronic density $n(r)$, v_{xc} is the Hartree and exchange–correlation potential, H is the Hamiltonian operator and the summation runs over the occupied states α. In the following, we restrict ourselves to the insulating non-spin-polarized case.

The occupied Kohn–Sham orbitals are subject to the orthonormalization constraints,

$$\int \psi_\alpha^*(r)\, \psi_\beta(r)\, \mathrm{d}r = \langle \psi_\alpha | \psi_\beta \rangle = \delta_{\alpha\beta} \tag{8.13}$$

where α and β label occupied states. The density is generated as

$$n(r) = \sum_{\alpha}^{\mathrm{occ}} \psi_\alpha^*(r) \psi_\alpha(r) \tag{8.14}$$

The minimization of $E_{\mathrm{el}}\{\psi_\alpha\}$ under the orthonormality constraints Equation 8.12 can be achieved using the Lagrange multiplier method.

In the density-functional perturbation theory, one determines the derivatives of the DFT electronic energy with respect to different perturbations. This electronic energy is only a part of the Born–Oppenheimer energy: the nuclei–nuclei interaction energy must be added to it. However, the treatment of this additional contribution is much easier because it involves only computing the electrostatic repulsion between classical point charges (for example, [12]).

The perturbations treated in DFPT might be external applied fields (electric or magnetic), strains, as well as changes of potentials induced by nuclear displacements, or any type of perturbation of the equations that define the reference system. This powerful generic theory is able to deal with perturbations characterized by a non-zero, commensurate or incommensurate wave vector [28], with a workload similar to the one needed to deal with a periodic perturbation. Hence, it is particularly efficient for dealing with phonons.

8.3.1 Perturbation Expansion

The DFT equations have been defined for generic external potentials v_{ext}. We now choose a reference (unperturbed) external potential v_{ext} and expand the perturbed potential v_{ext} in terms of a small parameter λ, as follows [4, 8]:

$$v_{ext}(\lambda) = v_{ext}^{(0)} + \lambda v_{ext}^{(1)} + \lambda^2 v_{ext}^{(2)} + \lambda^3 v_{ext}^{(3)} \ldots \tag{8.15}$$

We are interested in the change of physical quantities, due to the perturbation of the external potential [29]. So, we expand the different perturbed quantities $X(\lambda)$ using the same form as for

$$X(\lambda) = X^{(0)} + \lambda X^{(1)} + \lambda^2 X^{(2)} + \lambda^3 X^{(3)} \ldots \tag{8.16}$$

where X can be the electronic energy E_{el}, the electronic wavefunctions $\psi_\alpha(r)$, the density $n(r)$, the electron eigenenergies $\varepsilon_{\alpha\beta}$, or the Hamiltonian H. For example, the lowest order expansion of Equation 8.16 is simply

$$H^{(0)}|\psi_\alpha^{(0)} >= \varepsilon_\alpha^{(0)}|\psi_\alpha^{(0)} > . \tag{8.17}$$

We suppose that all the zero-order quantities are known, as well as the change of external potential v_{ext} through all orders. In what follows, we suppose that the latter terms are the only applied perturbation, although the theory can be generalized to other forms of perturbation. For the computation of dynamical matrices and of phonons, only the first and second-order derivatives of the energy with respect to atomic displacements and electric fields are needed. For the computation of the Raman tensors intensities, the nonlinear optical coefficients, the anharmonicities and the infrared intensities we need to go at least to the third-order derivatives of the energy with respect to atomic displacements and electric fields. The DFPT allows building all these terms [3–5, 11].

Thanks to the variational property of the DFT electronic energy, the first-order derivative of the electronic energy can be evaluated without knowing any first-order quantity, except the change of external potential:

$$E_{el}^{(1)} = \sum_{\alpha}^{occ} \langle \psi_\alpha^{(0)} | v_{ext}^{(1)} | \psi_\alpha^{(0)} \rangle . \tag{8.18}$$

By contrast, the first-order change of wavefunctions, density, and Hamiltonian must be obtained self-consistently (or through a variational approach), in the same spirit as the self-consistent determination of the unperturbed wavefunctions, density, and Hamiltonian. Supposing the first-order changes of wavefunctions $\psi_\alpha^{(1)}(r)$ are known, then, the first-order change of density can be obtained as [3]

$$n^{(1)}(r) = \sum_\alpha^{\text{occ}} \psi_\alpha^{*(1)}(r)\psi_\alpha^{(0)}(r) + \psi_\alpha^{*(0)}(r)\psi_\alpha^{(1)}(r) \tag{8.19}$$

Based on this, one can compute the first-order change of Hamiltonian $H^{(1)}$, thanks to [3]

$$H^{(1)} = v_{\text{ext}}^{(1)} + v_{\text{Hxc}}^{(1)} = v_{\text{ext}}^{(1)} + \int \left.\frac{\delta^2 E_{\text{Hxc}}}{\delta n(r)\delta n(r')}\right|_{n^{(0)}} n^{(1)}(r')\,\mathrm{d}r'. \tag{8.20}$$

For the second-order derivative of the electronic energy, different expressions can be used, whose ingredients are zero- and first-order quantities only. There is a simple nonvariational expression

$$E_{\text{el}}^{(2)} = \sum_\alpha^{\text{occ}} \langle\psi_\alpha^{(0)}|v_{\text{ext}}^{(1)}|\psi_\alpha^{(1)}\rangle + \sum_\alpha^{\text{occ}} \langle\psi_\alpha^{(0)}|v_{\text{ext}}^{(2)}|\psi_\alpha^{(0)}\rangle \tag{8.21}$$

or a more complex, and more accurate variational expression [5, 6, 11]:

$$\begin{aligned} E_{\text{el}}^{(2)}\{\psi^{(0)};\psi^{(1)}\} = &\sum_\alpha^{\text{occ}} [\langle\psi_\alpha^{(1)}|H^{(0)} - \varepsilon^{(0)}|\psi_\alpha^{(1)}\rangle + (\langle\psi_\alpha^{(1)}|v_{\text{ext}}^{(1)}|\psi_\alpha^{(0)}\rangle + \langle\psi_\alpha^{(0)}|v_{\text{ext}}^{(1)}|\psi_\alpha^{(1)}\rangle) \\ &+ \langle\psi_\alpha^{(0)}|v_{\text{ext}}^{(2)}|\psi_\alpha^{(0)}\rangle] \\ &+ \frac{1}{2}\int\int \left.\frac{\delta^2 E_{\text{Hxc}}}{\delta n(r)\delta n(r')}\right|_{n^{(0)}} n^{(1)}(r)n^{(1)}(r')\mathrm{d}r\mathrm{d}r' \end{aligned} \tag{8.22}$$

where the first-order changes in wavefunctions $\psi_\alpha^{(1)}$ (herein after first-order wavefunctions) can be varied under the constraints

$$\langle\psi_\alpha^{(0)}|\psi_\beta^{(1)}\rangle = 0 \tag{8.23}$$

for all occupied states α and β.

In similar variational or nonvariational ways, one can build the third-order derivatives of the energy based on the knowledge of the first- and second-order derivatives of the energy and of wavefunctions.

Except for symmetry-breaking effects due to perturbations, the computer time needed to compute the self-consistent response to a given perturbation is comparable to the computer time needed to compute the self-consistent ground-state properties of the crystalline system, at fixed nuclei positions.

8.3.2
Response to Static Electric Fields

For the computation of the infrared response, one must treat not only changes in the potential due to the collective nuclei displacements but also changes associated with a homogeneous, low-frequency (compared to the typical electronic excitation energy) electric field. Two important problems arise when one attempts to deal with the response to such an electric field E_{mac}. The first problem comes from the fact that the potential energy of the electron, placed in such a field, is linear in space, and breaks the periodicity of the crystalline lattice:

$$v_{\mathrm{scr}}(r) = \sum_{\alpha} \varepsilon_{\mathrm{mac},\alpha} r_{\alpha}. \tag{8.24}$$

Second, this macroscopic electric field corresponds to a screened potential: the change of macroscopic electric field is the sum of an external change of field and an internal change of field, the latter being induced by the response of the electrons (the polarization of the material). In the theory of classical electromagnetism, the connection between the macroscopic displacement, electric, and polarization fields can be written as

$$D_{\mathrm{mac}}(r) = \varepsilon_{\mathrm{mac}}(r) + 4\pi P_{\mathrm{mac}}(r) \tag{8.25}$$

where $P_{\mathrm{mac}}(r)$ is related to the macroscopic charge density by

$$n_{\mathrm{mac}} = -\nabla P_{\mathrm{mac}}(r). \tag{8.26}$$

The long-wave method is commonly used to deal with the first problem: a potential linear in space is obtained as the limit for $\mathbf{q}$ tending to 0 of

$$v(\mathbf{r}) = \lim_{\mathbf{q}\to 0} \lambda \frac{2\sin \mathbf{q}\cdot\mathbf{r}}{|\mathbf{q}|} = \lim_{\mathbf{q}\to 0} \lambda \left(\frac{e^{i\mathbf{q}\cdot\mathbf{r}}}{i|\mathbf{q}|} - \frac{e^{-i\mathbf{q}\cdot\mathbf{r}}}{i|\mathbf{q}|} \right) \tag{8.27}$$

where $\mathbf{q}$ is in the direction of the homogeneous field. The detailed theoretical treatment of the response to an electric field, using the long-wave method, and treating the screening adequately (in order to solve the above-mentioned second problem) is given in Ref. [12]. It is found that an auxiliary quantity is needed: the derivative of the ground-state wavefunctions with respect to their wave vector, which can also be computed within DFPT. Once this quantity has been obtained, the computation of the response to a homogeneous electric field *per se* can be performed, also within DFPT.

8.3.3
Mixed Perturbations

The dynamical matrices or interatomic force constants are mixed second-order derivatives of the Born–Oppenheimer energy, corresponding to two different (groups of) nuclear displacements. Moreover, during the atomic vibrations, there is a coupling between phonons and electric field. For its computation, we need mixed derivatives of the energy with respect to both atomic displacements and homogeneous electric fields.

DFPT is able to deal straightforwardly with such mixed derivatives. We consider two or more simultaneous Hermitian perturbations, λ_{j_i} combined in a Taylor-like expansion of the following type (see Ref. [4, 12] for the notation):

$$v_{\text{ext}}(\lambda) = v_{\text{ext}}^{(0)} + \sum_{j_1} \lambda_{j_1} v_{\text{ext},j_1} + \sum_{j_1,j_2} \lambda_{j_1} \lambda_{j2} v_{\text{ext},j_1 j_2} + \sum_{j_1,j_2 j_3} \lambda_{j_1} \lambda_{j2} \lambda_{j3} v_{\text{ext},j_1 j_2 j_3} + \ldots \tag{8.28}$$

The mixed derivatives of the energy of the electronic system

$$E_{\text{el},j_1 j_2} = \frac{1}{2} \frac{\partial^2 E_{\text{el}}}{\partial \lambda_{j_1} \partial \lambda_{j_2}} \tag{8.29}$$

are obtained, respectively, from

$$E^{j_1 j_2} = \sum_{\alpha} \langle \psi_{\alpha}^{j_2} | v_{\text{ext}}^{j_1} | \psi_{\alpha}^{(0)} \rangle + \sum_{\alpha} \langle \psi_{\alpha}^{(0)} | v_{\text{ext}}^{j_1 j_2} | \psi_{\alpha}^{(0)} \rangle. \tag{8.30}$$

In the expression Equation 8.30, the first-order derivative of the wavefunctions with respect to the first perturbation $|\psi_{\alpha}^{j_1}\rangle$ are not needed, while the computation of $v_{\text{ext}}^{j_1}$ and $\langle \psi_{\alpha}^{(0)} | v_{ext}^{j_1 j_2} | \psi_{\alpha}^{(0)} \rangle$ takes little time. Similar expressions, which do not involve $|\psi_{\alpha}^{j_2}\rangle$ but $|\psi_{\alpha}^{j_1}\rangle$ are also available, as well as more accurate stationary expressions.

When three perturbations are considered (e.g., two electric fields and one atomic displacement in case of Raman tensors or three electric field in case of nonlinear optical coefficients) the latter expression becomes

$$E_{el}^{j_1 j_2 j_3} = \frac{1}{3} \frac{\partial^3 E_{el}}{\partial \lambda_{j_1} \partial \lambda_{j_2} \partial \lambda_{j_3}}. \tag{8.31}$$

Thus, the ability to compute the first-order responses (i.e., changes in wavefunctions and densities) to the basic perturbations described previously gives us also, as byproducts, mixed second-order derivatives of the electronic energy. Actually, even third-order mixed derivatives of the energy might be computed straightforwardly, thanks to the $2n + 1$ theorem of perturbation theory, within DFPT [4].

8.4
Infrared and Raman Spectra

Electromagnetic radiation or photons (in quantum theory) interact in several ways with a crystal. For an insulating solid, and considering electromagnetic frequencies in the infrared range, phonons are the predominant cause of features in the absorption or reflection spectra.

Only phonons with a very small wave vector $\mathbf{q} \to 0$ interact with photons. Indeed, the photons absorbed and emitted should have energy comparable to the one of phonons, that is, 0.3 eV at most. Such photons have a very large wavelength (larger than 3 mm), compared to the typical unit cell size, hence a very small wave vector. Even in the case of scattering of photons of higher energies (e.g., the Raman scattering, with photon energies around a few electronvolts), the involved phonons all have negligible wave vectors.

For this $\mathbf{q} \rightarrow 0$ limit symmetry-based group theory considerations allow to classify the phonons according to the irreducible representations of the crystal point group [28]. In particular, for crystals possessing a center of inversion, there are phonon modes whose corresponding representations are invariant with respect to inversion ("gerade" or "g" modes), or change sign under inversion ("ungerade" or "u" modes. Beyond phonon modes, many properties of crystals, such as dielectric tensors, elastic constants, thermal expansion coefficients, spontaneous polarization, might be classified according to irreducible representations. Group theory allows deducing transformation laws of these objects, and even predicts vanishing of some effects, or selected components of the response tensors. For example, in the case of the spontaneous polarization (a vector), one should examine the way a vector transforms under the operations of the point group. In particular, a vector always changes signs under inversion; hence, a crystal with inversion symmetry cannot have a spontaneous polarization.

Two experimental techniques are commonly used for phonon spectroscopy with electromagnetic fields: infrared reflectivity (or absorption), and Raman scattering. They are quite complementary because the involved phonon often belongs to different irreducible representations.

8.4.1 Infrared

The reflectivity of electromagnetic waves normal to the surface, having their electric field with direction $\mathbf{q}$ along an optical axis of the crystal, is given in terms of the frequency-dependent dielectric permittivity $\varepsilon_{\bar{q}}$:

$$R(\omega) = \left| \frac{\varepsilon_{\bar{q}}{}^{1/2}(\omega) - 1}{\varepsilon_{\bar{q}}^{1/2}(\omega) + 1} \right|^2 . \tag{8.32}$$

More general expressions for the reflectivity, or for the absorption, may be found in classical textbooks [29]. The dielectric permittivity along the direction $\bar{q}$ is computed from the dielectric permittivity tensor

$$\varepsilon_{\bar{q}}(\omega) = \sum_{\alpha\beta} \bar{q}_\alpha \varepsilon_{\alpha\beta}(\omega) \bar{q}_\beta . \tag{8.33}$$

Let us examine the first principle approach to this tensor. In the infrared frequency regime, the dielectric permittivity tensor obtained in the harmonic approximation can be split in two parts, the electronic contribution $\varepsilon_{\alpha\beta}$, taken as frequency independent, and the phonon contribution.

For the evaluation of the coupling between phonons and a homogeneous electric field, we need to consider the forces created by an applied electric field, and the polarization created by nuclear displacements. The Born-effective charge tensor $Z^*_{\varkappa,\beta\alpha}$ is the change in polarization (per unit cell) created along the direction β due to the displacement along the direction α of the nuclei j, under the condition of zero electric field [3, 12, 30]. The same tensor corresponds to the energy derivative with respect to atomic displacements and electric fields or, equivalently, to the change in atomic force due to an electric field:

$$Z^*_{\varkappa,\beta\alpha} = \Omega_0 \frac{\partial P_{\mathrm{mac},\beta}}{\partial \tau_{\varkappa\alpha}(\mathbf{q}=\mathbf{0})} = \frac{\partial F_{\varkappa,\alpha}}{\partial \varepsilon_\beta} = -\frac{\partial E}{\partial \varepsilon_\beta \partial \tau_{\varkappa\alpha}(\mathbf{q}=\mathbf{0})}. \tag{8.34}$$

The Born effective charge tensors fulfill an important sum rule stemming from the fact that a global translation of a neutral crystal, as a whole, should not change its polarization. This sum rule implies that the charge neutrality is fulfilled at the level of the Born effective charges. For every direction α and β, one must have [30]

$$\sum_{\varkappa} Z^*_{\varkappa,\beta\alpha} = 0, \tag{8.35}$$

that is, the sum of the Born effective charges of all nuclei in one cell must vanish, element by element. In DFT computations, this sum rule will be broken because of (i) the incompleteness of the basis set used to represent wave functions, (ii) the discreteness of special point grids, or (iii) the discretization of the real space integral (needed for the evaluation of the exchange-correlation energies and potentials). Techniques to recover the Born effective charge neutrality sum rule are described in Ref. [12].

Then, based on the Born effective charge tensors for each nuclei, and the eigenvectors of the dynamical matrix at the Brillouin zone center, one can fully derive the photon-phonon interaction in the harmonic approximation. Following Refs [30, 31], the quantity

$$p_{m\alpha} = \sum_{\varkappa\beta} Z^*_{\varkappa,\beta\alpha} U_{mq=0}(\varkappa\beta) \tag{8.35}$$

that combines Born effective charges with the phonon eigendisplacements, $U_{mq=0}(\varkappa\beta)$ is referred to as the polarity of the phonon mode m. The three components p_{ma} form a vector $\mathbf{p}_m$, whose sign is arbitrary [32]. In terms of mode polarities, the dielectric tensor $\varepsilon_{\alpha\beta}(\omega)$ has a rather simple expression

$$\varepsilon_{\alpha\beta}(\omega) = \varepsilon^{\infty}_{\alpha\beta} + \frac{4\pi}{\Omega_0} \sum_m \frac{p_{m\alpha} p_{m\beta}}{\omega_m^2 - \omega^2} \tag{8.36}$$

and the dielectric permittivity along some direction $\mathbf{q}$, Equation 8.36, becomes

$$\varepsilon_{\bar{q}}(\omega) = \sum_{\alpha\beta} \bar{q}_\alpha \varepsilon^{\infty}_{\alpha\beta}(\omega) \bar{q}_\beta + \frac{4\pi}{\Omega_0} \sum_m \frac{(\mathbf{p}_m \cdot \mathbf{q})^2}{\omega_m^2 - \omega^2}. \tag{8.37}$$

The last two equations express $\varepsilon_{\alpha\beta}(\omega)$ and $\varepsilon\bar{q}(\omega)$ in terms of an electronic contribution $\varepsilon^{\infty}_{\alpha\beta}$ (approximated as frequency-independent in the infrared regime), and contributions from each possible phonon mode m at the Brillouin zone center. The phonon contributions have characteristic frequency dependence: there is a resonant behavior when the frequency of light matches the one of a phonon, in which case the denominator of Equation 8.37 vanishes. The latter equation shows that, if the vector $\mathbf{p}_m$ is perpendicular to $\mathbf{q}$, the direction of the electric field, the mode m does not contribute to the dielectric permittivity constant along $\mathbf{q}$. Then, for each mode m there is one direction along which the mode contributes to the dielectric permittivity constant, in which case it is referred to as longitudinal, while for the perpendicular directions, the mode will be referred to as transverse. In this way, we find the (well-known) distinction between the longitudinal optic (LO) modes and the transverse optic (TO) modes. This distinction will be the subject of further explanation later.

Alternatively, the same value of the dielectric permittivity tensor might be obtained in terms of the Born effective charge tensors and the zone-center dynamical matrix, see Equation 52 of Ref. 11, from which one deduces the following expression of the static $\omega = 0$ dielectric permittivity tensor,

$$\varepsilon^{0}_{\alpha\beta}(\omega) = \varepsilon^{\infty}_{\alpha\beta} + \frac{4\pi}{\Omega_0}\sum_{\varkappa\varkappa'}\sum_{\alpha'\beta'}\left(Z^{*}_{\varkappa,\alpha\alpha'}[\tilde{C}(\mathbf{q}=\mathbf{0})]^{-1}_{\varkappa\alpha',\varkappa'\beta'}Z^{*}_{\varkappa',\beta\beta'}\right). \tag{8.38}$$

This equation highlights that when the frequency is sufficiently small as to allow nuclei to relax to their equilibrium position under the applied field, their masses do not play a role anymore: the static $\omega = 0$ dielectric permittivity tensor is independent of the masses.

8.4.2 **Raman**

In a Raman experiment, the (polarized) incident light is scattered by the sample, and the energy as well as polarization of the outgoing light is measured. A Raman spectrum, presenting the energy of the outgoing photons, will consist of rather well-defined peaks, around an elastic peak (corresponding to outgoing photons that have the energy of the incident photons – the Rayleigh peak).

At the lowest order of the theory, the dominant mechanism is the absorption or emission of a phonon by a photon. A measure of the energy difference between the outgoing and incident photons gives the energy of the absorbing or emitting phonon. Thus, even more straightforwardly than the IR spectrum, a Raman spectrum relates to the energy of phonons at the Brillouin-zone center: when the zero of the frequency scale is set at the incident light frequency, the absolute value of the energy of the peaks corresponds to the energy of the phonons.

The Raman intensity of a specific mode is determined by the specific Raman scattering efficiency (S) via a prefactor. The Raman scattering efficiency depends on the frequency of the incoming photon, ω_0, on the frequency of the phonon, on temperature, T, and on the Raman tensor α_m. Its full expression is [33, 34]

$$\frac{dS}{dV} = \frac{(\omega_0 - \omega_m)^4}{c^4} |e_S \cdot \alpha \cdot e_0|^2 \frac{h}{2\omega_m}(n_m + 1). \tag{8.39}$$

The dependency on temperature is given by the boson factor as

$$n_m = \frac{1}{e^{h\omega_m/k_B T} - 1} \tag{8.40}$$

and the Raman tensor

$$\alpha_{ij}^m = \sqrt{\Omega_0} \sum_{\varkappa\beta} \frac{\partial \chi_{ij}^{\infty(1)}}{\partial \tau_{\varkappa\beta}} \eta_m(\varkappa\beta) = \sqrt{\Omega_0} \sum_{\varkappa\beta} \frac{\partial}{\partial \tau_{\varkappa\beta}} \left(\frac{\partial^2 E}{\partial \varepsilon_i \partial \varepsilon_j} \right) \varepsilon_i \varepsilon_j \eta_m(\varkappa\beta) \tag{8.41}$$

is the derivative of the macroscopic dielectric tensor with respect to the set of atomic displacements that correspond to the phonon eigenvector [31, 35, 36]. The middle term in Equation 8.39, $|e_S \cdot \alpha \cdot e_0|$, gives the coupling between the incoming phonon with polarization e_0, the crystal, characterized by a Raman tensor α and the scattered phonon with polarization e_S.

The Raman tensors are the key ingredients needed to calculate the Raman spectra. They can be computed either from finite differences, as the change of the dielectric tensor due to infinitesimal atomic displacements, or from perturbation theory, as the derivative of the energy with respect to three perturbations: two electric fields and one atomic displacement. For this, one needs to define an electronic (variational) enthalpy that takes into account both the electronic energy and its changes under an external electric field

$$F_{e+i}[\mathbf{R}_k, \varepsilon] = \min_{\psi_n}(E_{e+i}[\mathbf{R}_k, \psi_k] - \Omega_0 \varepsilon P[\psi_k]$$

that we develop, according to Equation 8.1, a Taylor expansion to third order [31]:

$$\begin{aligned} F_{e+i}[\lambda] = F_{e+i}{}^{(0)}[\lambda] + \sum_i \left(\frac{\partial F_{e+i}}{\partial \lambda_i} \right) \lambda_i + \frac{1}{2} \sum_{ij} \left(\frac{\partial^2 F_{e+i}}{\partial \lambda_i \partial \lambda_j} \right) \lambda_i \lambda_j \\ + \frac{1}{6} \sum_{ij} \left(\frac{\partial^3 F_{e+i}}{\partial \lambda_i \partial \lambda_j \partial \lambda_k} \right) \lambda_i \lambda_j \lambda_k + \ldots \end{aligned} \tag{8.42}$$

Considering as perturbations atomic displacements, τ, and electric fields, ε, then each of the different terms of the full expansion corresponds to a (measurable) physical property:

$$\begin{aligned} F_{e+i}[R_k, E] = {} & F_{e+i}[R_k, E] \\ & - \Omega_0 \sum_{\alpha} \underbrace{P_{\alpha}^{s}}_{\text{polarization}} \varepsilon_{\alpha} - \sum_{\alpha} \sum_{k} \underbrace{F_{\alpha}^{0}}_{\text{forces}} \tau_{\varkappa\alpha} \\ & - \Omega_0/2 \sum_{\alpha\beta} \underbrace{\chi_{\alpha\beta}^{\infty(1)}}_{\text{dielectric tensor}} \varepsilon_{\alpha} \varepsilon_{\beta} - \sum_{\alpha\beta} \sum_{\varkappa} \underbrace{Z_{\varkappa,\alpha\beta}^{*}}_{\text{dynamical charges}} \tau_{\varkappa\alpha} \varepsilon_{\beta} \\ & + 1/2 \sum_{\alpha\beta} \sum_{\varkappa'\varkappa} \underbrace{C_{\alpha\beta}(\varkappa, \varkappa')}_{\text{interatomic force constants}} \tau_{\varkappa\alpha} \tau_{\varkappa'\beta} \end{aligned}$$

$$-\Omega_0/3\sum_{\alpha\beta}\underbrace{\chi^{\infty(2)}_{\alpha\beta\gamma}\varepsilon_\alpha\varepsilon_\beta\varepsilon_\gamma}_{\text{nonlinear optical coefficients}} \quad -\Omega_0/2\sum_{\varkappa}\sum_{\alpha\beta}\underbrace{\frac{\partial\chi^{\infty(1)}_{\alpha\beta}}{\partial\tau_{\varkappa\gamma}}\varepsilon_\alpha\varepsilon_\beta\tau_{\varkappa\gamma}}_{\text{Raman coefficient}}$$

$$-1/2\sum_{\varkappa\varkappa'}\sum_{\alpha\beta}\frac{\partial Z^*_{\varkappa,\alpha\beta}}{\partial\tau_{\varkappa'\gamma}}\tau_{\varkappa\alpha}\tau_{\varkappa'\gamma}\varepsilon_\beta$$

$$+1/3\sum_{\varkappa\varkappa'\varkappa''}\sum_{\alpha\beta}\underbrace{\Xi(\varkappa,\varkappa',\varkappa'')\tau_{\varkappa\alpha}\tau_{\varkappa'\beta}\tau_{\varkappa''\gamma}}_{\text{anharmonicities}}+\ldots \tag{8.43}$$

By inverting this equation one can obtain the different physical properties of interest.

The terms of the Taylor expansion up to the second order are already computed using density functional perturbation theory as in Section 8.3. The third-order terms can be computed following the same approach, like in the $2n + 1$ theorem [4]. Of course, the complexity of the exact expression for each higher order term of the Kohn–Sham equations and consequently the complexity of the implementation increase with the order of the derivation. Currently, there are two available implementations, which are the nonlinear optical coefficients, that is, the derivative of the energy with respect to three electric fields, and the Raman tensors, that is, the derivative of the energy with respect to two electric field and one atomic displacements [31, 36]. The third term of the third-order derivative of the energy, that is, with respect to one electric field and two atomic displacements participates to the width of the infrared and Raman peaks, while the very last term from the expression yields anharmonicities, related to the phonon time of life, the line broadening of Raman and infrared spectra and thermal transport due to phonon scattering.

Concerning the Raman spectra, the Raman tensor is given by the Raman coefficient term. For noncentrosymmetric crystals, some modes are active in both Raman and infrared. In this case, the same formalism as above holds for the TO component, while a supplementary correction is needed for the modes in LO geometry. In this case, the Raman tensors become [31]

$$\left.\frac{\partial\chi^{\infty(1)}_{ij}}{\partial\tau_{\varkappa\beta}}\right|_{D=0} = \left.\frac{\partial\chi^{\infty(1)}_{ij}}{\partial\tau_{\varkappa\beta}}\right|_{E=0} - \frac{8\pi}{\Omega_0}\frac{\sum_l Z^*_{\varkappa\beta l}q_l}{\sum_{ll'} q_l\varepsilon^{\infty}_{ll'}q_{l'}}\sum_l \chi^{\infty(2)}_{ijl}q_l. \tag{8.44}$$

Quite often, experimental data are recorded using polarized or unpolarized lasers on powdered samples. If we neglect the surface effects on the Raman tensors, an approximation that is valid for meso- and macrocrystals, then the resulting Raman spectra can be obtained y performing averages over all possible orientation of the crystals and then summing up over the parallel and perpendicular laser polarizations. The intensities of the two polarized components of the powder spectra, parallel and perpendicular, and the resulting total powder spectra are [34, 37]

$$G_0 = \frac{(\alpha_{xx} + \alpha_{yy} + \alpha_{zz})^2}{3}$$

$$G_1 = \frac{(\alpha_{xy} - \alpha_{yz})^2 + (\alpha_{yz} - \alpha_{zx})^2 + (\alpha_{zx} - \alpha_{xy})^2}{2}$$

$$G_2 = \frac{(\alpha_{xy} + \alpha_{yz})^2 + (\alpha_{yz} + \alpha_{zx})^2 + (\alpha_{zx} + \alpha_{xy})^2}{2}$$

$$+ \frac{(\alpha_{xx} - \alpha_{yy})^2 + (\alpha_{yy} - \alpha_{zz})^2 + (\alpha_{zz} - \alpha_{xx})^2}{3} \tag{8.45}$$

$$\begin{aligned} I_{\parallel}^{\text{powder}} &= C(10G_0 + 4G_2) \\ I_{\perp}^{\text{powder}} &= C(5G_1 + 3G_2) \\ I_{\text{tot}}^{\text{powder}} &= I_{\parallel}^{\text{powder}} + I_{\perp}^{\text{powder}}. \end{aligned} \tag{8.46}$$

This formalism has been applied successfully, for example, in the ABINIT implementation to various systems [38–42].

8.5
Thermodynamical Properties

Now, we can compute the Helmholtz free energy, the internal energy, the constant-volume specific heat, and the entropy as functions of temperature. Such thermodynamic functions of a solid are determined mostly by the vibrational degrees of freedom of the lattice, since, generally speaking, the electronic degrees of freedom play a noticeable role only for metals at very low temperatures [43]. However, the complete knowledge of the phonon band structure, with sufficient accuracy, is required for the calculation of these thermodynamic functions. The formulas presented here neglect all anharmonic effects. For most solids, the harmonic approximation will be accurate for a temperature smaller than a significant fraction of the melting temperature or the temperature of the lowest solid–solid phase transition (e.g., about 500 K for quartz, that undergoes a phase transition above 800 K). On the other hand, the quantum effects are correctly included, unlike in an approach based on the classical dynamics of nuclei.

The above-mentioned thermodynamic functions require summations over the eigenstates of all phonons m at all wave vectors $\mathbf{q}$. This sum can be transformed into a one-dimensional integral over the phonon density of states, $g(\omega)$:

$$(3N_n) \int_0^{\omega_L} f(\omega) g(\omega) \mathrm{d}\omega \tag{8.47}$$

where N_n is the number of nuclei per unit cell, ω_L is the largest phonon frequency, and $g(\omega)\mathrm{d}\omega$ defines the phonon population in the range ω and $\omega + \mathrm{d}\omega$. The phonon density of states $g(\omega)$ can be normalized so that

$$\int_0^{\omega_L} g(\omega)d\omega = 1 \tag{8.48}$$

such as

$$g(\omega) = \frac{(2\pi)^3}{\Omega_0(3N_n)} \sum_m \delta(\omega - \omega_{mq})dq \tag{8.49}$$

Specifically, the phonon contribution to the Helmholtz free energy ΔF, the phonon contribution to the internal energy ΔE, as well as the constant-volume specific heat C_V, and the entropy S, at temperature T, evaluated for one unit cell, have the following expressions within the harmonic approximation [44], where $x = \hbar\omega/k_B T$ and k_B is the Boltzmann constant:

$$\begin{aligned}
\Delta F &= (3N_n)k_B T \int_0^{\omega_L} \{\ln(e^x - 1) - x/2\} g(\omega)d\omega; \\
\Delta E &= (3N_n) \int_0^{\omega_L} \frac{(e^x + 1)}{(e^x - 1)} \frac{\hbar\omega}{2} g(\omega)d\omega; \\
C_V &= (3N_n)k_B \int_0^{\omega_L} \left(\frac{x}{e^{x/2} - e^{-x/2}} \right)^2 g(\omega)d\omega; \\
S &= (3N_n)k_B \int_0^{\omega_L} \left[\frac{xe^x}{e^x - 1} - \ln(e^x - 1) \right] g(\omega)d\omega.
\end{aligned} \tag{8.50}$$

8.6 Examples and Applications

We will exemplify these theoretical developments with two cases: with the determination of the thermodynamic properties and of the Raman spectra of polymeric nitrogen – the cubic gauche phase of atomic nitrogen; and with the analysis of the dynamical instabilities of the ionic phase of water ice – ice X.

8.6.1 Polymeric Nitrogen

Crystalline nitrogen undergoes a series of phase transitions that compact its structure and reduce the degrees of freedom of molecules by increasing the order. At low pressures, in the α phase the free rotation of the molecules describes spheres, while in the β phase it describes disks in two dimensions. At higher pressure, the degree of freedom of the molecular motions is further reduced and partial disorder or complete order occurs, as exhibited by different crystalline structures such as the δ or ε [45, 46]. All these phases have molecular structures. From 10–20 GPa up to about 100 GPa, there are several crystalline phases obtained experimentally whose structures and stabilities are not yet fully understood. Above approximately 100 GPa, first principles calculations suggested that crystalline atomic polymeric structures should develop, where the integrality of the nitrogen molecules is lost. Several theoretical structures have been predicted [47–49], and the most stable one, named "cubic gauche," has been experimentally confirmed afterward experimentally [50, 51]. These experiments that lead to the formation of the cubic gauche structure passed through the amorphous state as an intermediate state.

The cubic gauche structure is cubic, with $I2_13$ space group, and the atoms positioned in the special 8a(x x x) Wyckhoff positions. Each nitrogen atom has three nearest neighbors at equal distance. This topology thus yields an atomic character to the structure, with the nitrogen atoms polymerizing in left-handed spiral chains parallel to the [1 1 1] directions. The chains delimit different void spaces with quasitrigonal cross sections in the (1 1 1) planes or quasihexagonal in the (1 1 0) planes. The strong covalent bonding and the three-dimensional framework topology of the structure determine the exceptional hardness of this material.

There are 12 optical phonon modes that according to group theory decompose in the zone center as A + E + 2T. The Tmodes are both Raman and infrared active, and the A and E modes are only Raman active. All the modes harden under pressure as shown in Figure 8.2.

The dispersion of the phonon bands is shown in Figure 8.3a and b at two pressures [39]. All the modes have positive frequencies, thus the cubic gauche structure of nitrogen is dynamically stable over a wide pressure range. The density of states, the integral of the phonon population as a function of energy levels is represented in panel c of the Figure 8.3. Based on this density of states we can derive the thermodynamic properties of the structure using equations (8.50). Some of these are illustrated in Figure 8.4.

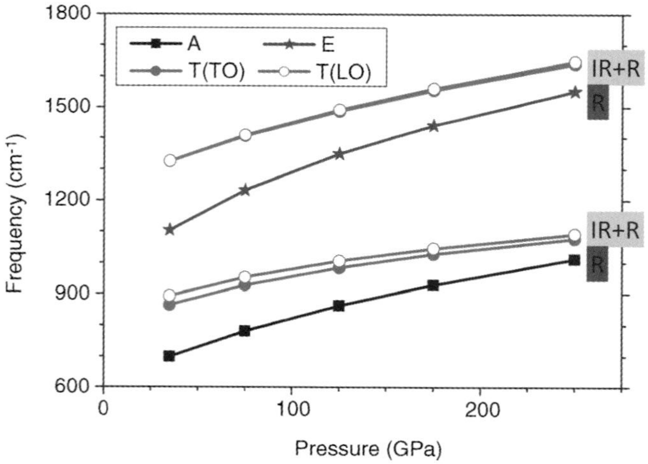

Figure 8.2 Variation of the phonons frequencies at gamma for the cubic gauche structure of nitrogen. The T, A, and E modes are Raman active and the T modes are also infrared active.

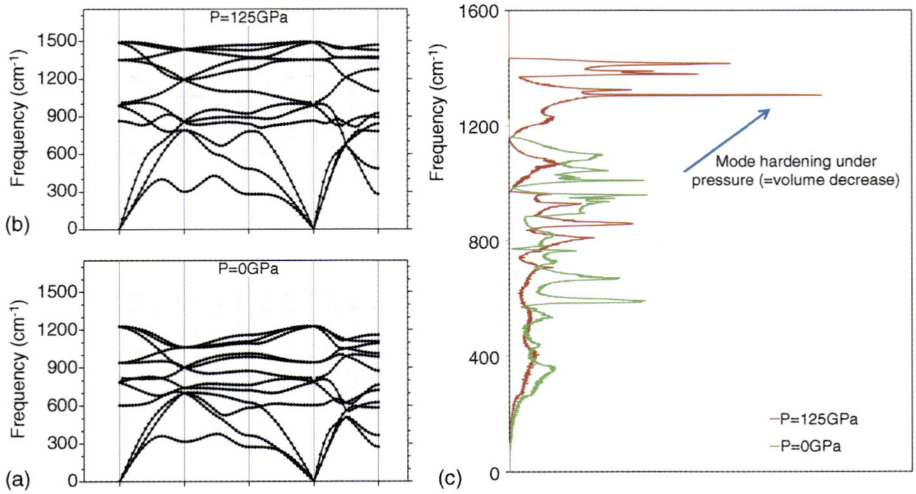

Figure 8.3 Phonon band dispersion in the cubic gauche structure of nitrogen at several pressures. All the phonons have positive frequencies hence the structure is dynamically stable.

Then we compute the Raman spectra for the cubic gauche structure of nitrogen [39]. As this structure is stable at high pressures, Raman spectroscopy is the usual tool used for identification. The theoretical Raman spectra are dominated by the A mode. The intensity of the other modes is less than 10% and they would be hardly observable in a measurement. Our theoretical prediction is in excellent agreement with the experimental measurements [39, 50], in terms of both peak position and intensity, as shown in Figure 8.5.

8.6.2 Ice X

The phase diagram of water ice is dominated by the ices VII, VIII, and X above approximately 3GPa [52, 53]. The structure of these three phases is similar and consists of a body-centered cubic lattice of oxygen atoms with the hydrogen atoms staying in-between. In ice VII and ice VIII the integrality of the molecules is preserved, the structure being thus "molecular." Ice VII is disordered the molecules pointing randomly in all Cartesian directions. The structure has cubic symmetry. Ice VIII is ordered with the molecules pointing only around the *c*-direction. The structure has tetragonal symmetry with a center of inversion. In ice X, the integrality of the molecules is lost. The hydrogen atoms lay between every two neighboring oxygen at equal distance. The structure is cubic and has an ionic character.

As much experimental effort has been dedicated to the study of the phase boundaries and the physical properties of these three ices, we have performed a dynamical analysis by computing the phonons at different pressures for the ice X [54]. Phonon band dispersion at three representative pressures is shown in Figure 8.6.

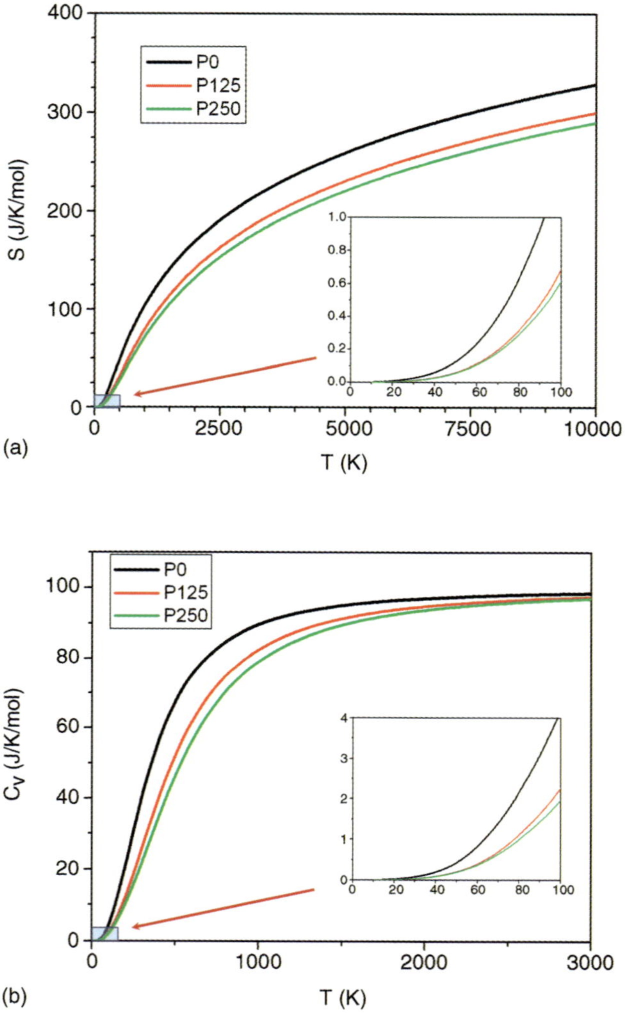

Figure 8.4 Temperature variation of selected thermodynamic properties of the cubic gauche structure of nitrogen at several pressures.

At low pressures, below approximately 114 GPa ice X exhibits an unstable phonon mode (Figure 8.6a). This mode is weakly dispersive: we retrieve it unstable around the same frequency all over the Brillouin zone. This behavior is an indication about the highly localized character in the direct space of the atomic displacement pattern. Indeed if we look at the corresponding vibration in the direct space we observe that this mode corresponds to the bouncing back and forth of the hydrogen atoms between their two oxygen neighbors (Figure 8.6b).

At low pressures, the distance between the two oxygen atoms is large and thus the potential opens up and forms the characteristic two-well shape of the unstable modes

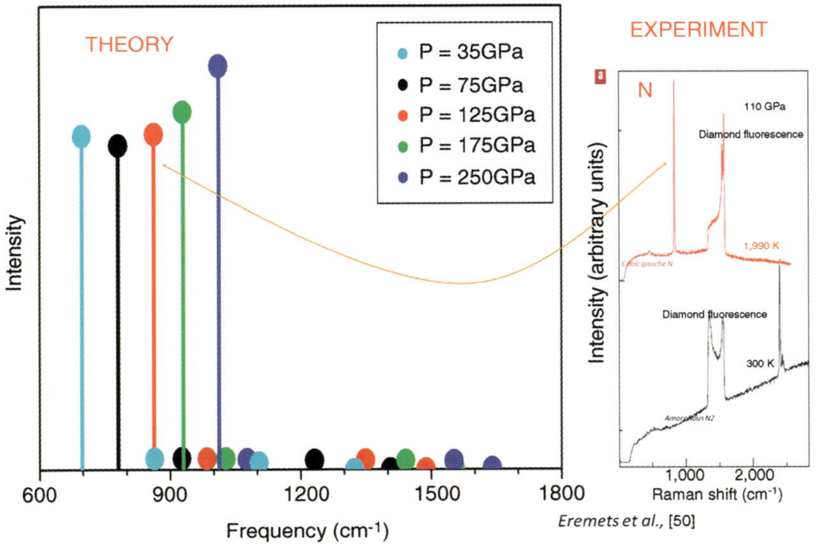

Figure 8.5 The Raman spectra of the cubic gauche phase of nitrogen are dominated by one strong peak with A symmetry that hardens under pressure. The theoretical calculations and the experimental measurements are in excellent agreement.

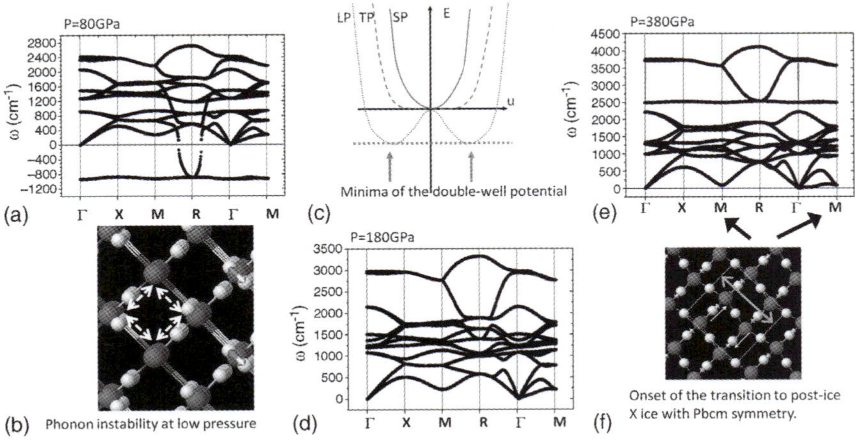

Figure 8.6 Analysis of the phonon dispersion bands in ice X as a function of pressure. At low pressures, below 100 GPa, there is one unstable phonon mode, represented by convention with negative frequencies (a) that corresponds to bouncing back and forth of the hydrogen atoms between their two oxygen neighbors (b). This displacement describes a double-well potential (c) that leads to the disordering of the structure. At intermediate pressures all phonons are stable (d) while at high pressure a new instability develops in M (e). This triggers a phase transition toward a new orthorhombic post-ice X phase (f).

(Figure 8.6c). The hydrogen atoms oscillate between these two minima, which are energetically equivalent positions. The structure resulting from the condensation of this unstable phonon mode is disordered ice X. There is experimental evidence [55] showing that the transition between ices VII or VIII and X goes through an intermediate phase, which is disordered.

As the pressure increases the distance between the oxygen atoms decreases and the double-well potential closes and transforms into a normal potential well with the minimum at the mid-distance between the two oxygen atoms. The hydrogen atoms are stabilized and vibrate around this minimum, situated at equal distance to both oxygen neighbors. This is the stable configuration of ice X, which can be found in a pressure range extending from about 114 GPa up to 430 GPa (Figure 8.6d).

Toward the upper part of this stability range another phonon softens, around the M point of the reciprocal space (Figure 8.6e). This phonon mode corresponds to displacements of both hydrogen and oxygen atoms parallel to the diagonal (1 1 0) planes. The result of this displacement is shear of the tetrahedra and a breaking of the symmetry. This is the phase transition to the post-ice X phase of ice, with Pnm symmetry (Figure 8.6f).

It is important to note that the analysis of the phonon modes in ice X allows us (i) to explain experimental observations from Raman spectroscopy in the region below 100 GPa that can be explained by the disordered ice X and (ii) to confirm previous *ab initio* molecular dynamics calculations [56] that predicted a phase transition to a post-ice X with Pbcm symmetry.

8.7 Conclusions

We have reviewed the basics of lattice dynamics from the perspective of the density-functional perturbation theory. We have presented the theory used for (i) the calculation of the phonons at the Brillouin-zone center, and connected this with infrared and Raman experiments; (ii) the calculations of the phonon band structures allowing to analyze the stability or instability of a phase; (iii) the determination of the intensity of the Raman peaks; and (iv) the computations of the thermodynamical properties (specific heat, internal energy, free energy, and entropy).

We have shown that the basic idea behind DFPT is expressing the energy of a lattice as a Taylor expansion function of one or more perturbations. The different terms of the expansion are evaluated analytically, in increasing derivation order. Each term or combination of terms is then related to specific physical properties. Here, we have discussed in detail only atomic displacements and electric field type perturbations, but other perturbations are possible, such as strain or magnetic field. Some of these properties are already available in several density-functional software packages, while others still wait to be coded. In these cases, finite differences can be used successfully on its own or on top of DFPT to derive nonlinear properties that have not yet been implemented.

It is possible to obtain interatomic force constants also by finite differences of the total energy and by force calculations, in large supercells. Eventually if one uses the same basis set for the representation of the electronic wavefunctions, with the same relevant parameters, the final results from both methods should be identical. However, the DFPT has the power and elegance of a pure analytical development that can also provide more physical answers with a smaller computational cost.

References

1 Hohenberg, P. and Kohn, W. (1964) Inhomogeneous electron gas. *Phys. Rev.*, **136**, B864–B871.

2 Kohn, W. and Sham, L.J. (1965) Self-consistent equations including exchange and correlation effects. *Phys. Rev.*, **140**, A1133–A1138.

3 Baroni, S., Giannozzi, P., and Testa, A. (1987) Green's-function approach to linear response in solids. *Phys. Rev. Lett.*, **58**, 1861–1864.

4 Gonze, X. and Vigneron, J.-P. (1989) Density-functional approach to nonlinear-response coefficients of solids. *Phys. Rev.*, **B49**, 13120–13128.

5 Gonze, X., Allan, D.C., and Teter, M.P. (1992) Dielectric tensor, effective charges and phonon in α-quartz by variational density-functional perturbation theory. *Phys. Rev. Lett.*, **68**, 3603–3606.

6 Savrasov, S.Yu. (1992) Linear response calculations of lattice dynamics using muffin-tin basis sets. *Phys. Rev. Lett.*, **69**, 2819–2822.

7 Gonze, X., Charlier, J.-C., Allan, D.C., and Teter, M.P. (1994) Interatomic force constants from first principles. The case of α-quartz. *Phys. Rev.*, **B50**, 13035–13038.

8 Gonze, X. (1995) Adiabatic density functional perturbation theory. *Phys. Rev.*, **A52**, 1096–1114.

9 Lee, C. and Gonze, X. (1995) Ab initio calculation of the thermodynamical properties and atomic temperature factors of SiO_2 α-quartz and stishovite. *Phys. Rev.*, **B51**, 8610–8613.

10 De Gironcoli, S. (1995) Lattice dynamics of metals from density-functional perturbation theory. *Phys. Rev.*, **B51**, 6773–6776.

11 Gonze, X. (1997) First-principle responses of solids to atomic displacements and homogeneous electric fields implementation of a conjugate-gradient algorithm. *Phys. Rev.*, **B55**, 10337–10354.

12 Gonze, X. and Lee, C. (1997) Dynamical matrices, Born effective charges, dielectric permittivity tensors, and interatomic force constants from density-functional perturbation theory. *Phys. Rev.*, **B55**, 10355–10368.

13 Putrino, A., Sebastiani, D., and Parrinello, M. (2000) Generalized variational density functional perturbation theory. *J. Chem. Phys.*, **113**, 7102–7109.

14 Baroni, S., de Gironcoli, S., Dal Corso, A., and Giannozzi, P. (2001) Phonons and related crystal properties from density-functional perturbation theory. *Rev. Mod. Phys.*, **73**, 515–562.

15 Scandolo, S., Giannozzi, P., Cavazzoni, C., de Gironcoli, S., Pasquarello, A., and Baroni, S. (2005) First-principles codes for computational crystallography in the Quantum-ESPRESSO package. *Z. Kristallogr.*, **220**, 574–579.

16 http://www.pwscf.org.

17 Gonze, X., Rignanese, G.-M., Verstraete, M., Beuken, J.-M., Pouillon, Y., Caracas, R., Jollet, F., Torrent, M., Zérah, G., Mikami, M., Ghosez, Ph., Veithen, M., Raty, J.-Y., Olevano, V., Bruneval, F., Reining, L., Godby, R., Onida, G., Hamann, D.R., and Allan, D.C. (2005) A brief introduction to the ABINIT software package. *Z. Kristallogr.*, **220**, 558–562.

18 http://www.abinit.org.

19 Gonze, X., Beuken, J.-M., Caracas, R., Detraux, F., Fuchs, M., Rignanese, G.-M., Sindic, L., Verstraete, M., Zerah, G., Jollet, F., Torrent, M., Roy, A., Mikami, M.,

Ghosez, Ph., Raty, J.-Y., and Allan, D.C. (2002) First-principle computation of material properties the ABINIT software project. *Comput. Mater. Sci.*, **25**, 478–492.

20 Hutter, J. and Iannuzzi, M. (2005) CPMD: Car–Parrinello molecular dynamics. *Z. Kristallogr.*, **220**, 549–551.

21 http://www.cpmd.org.

22 Savrasov, S.Y. (2005) Program LMTART for electronic structure calculations. *Z. Kristallogr.*, **220**, 555–557.

23 http://www.fkf.mpg.de/andersen/docs/interest.html.

24 Rignanese, G.-M., Detraux, F., Gonze, X., Bongiorno, A., and Pasquarello, A. (2002) Dielectric constants of Zr silicates a first-principles study. *Phys. Rev. Lett.*, **89**, 117601.1–117601.4.

25 Lee, C. and Gonze, X. (1995) The pressure-induced ferroelastic phase transition of SiO_2-stishovite. *J. Phys. Condens. Matt.*, **7**, 3693–3698.

26 Caracas, R. and Gonze, X. (2005) First-principles determination of the dynamical properties of Pb2MgTeO6 perovskites. *Phys. Rev.*, **B71**, 054101.

27 Landau, L. and Lifschits, E. (1960) *Electrodynamics of Continuous Media*, Pergamon Press.

28 Caracas, R. and Gonze, X. (2005) First-principle study of materials involved in incommensurate transitions. *Z. Kristallogr.*, **220**, 511–520.

29 Altmann, S.L. (1991) *Band theory of solids an introduction from the point of view of symmetry*, Oxford University Press, Oxford.

30 Ghosez, Ph., Michenaud, J.-P., and Gonze, X. (1998) Dynamical atomic charges: the case of ABO_3 compounds. *Phys. Rev.*, **B58**, 6224–6240.

31 Veithen, M., Gonze, X., and Ghosez, Ph. (2005) Non-linear optical susceptibilities, Raman efficiencies and electro-optic tensors from first-principles density functional perturbation theory. *Phys. Rev.*, **B71**, 125107.

32 Because of possible conflicting definitions for associated quantities, such as the mode-effective charges and the oscillator strengths, we use mode polarities in all our formulas, since these are not subject to such conflicting definitions.

33 Cardona, M. (1982) *Light Scattering in Solids II: Basic Concepts and Instrumentation*, vol. **50**, Topics in Applied Physics (eds M. Cardona and G. Güntherodt), Springer, New York, pp. 19–168.

34 Placzek, G. (1934) Rayleigh-Streuung und Raman-effekt, *Handbuch der Radiologie*, vol. **6** (ed. G. Marx), Akademische, Frankfurt-Main, Germany, pp. 205–374.

35 Baroni, S. and Resta, R. (1986) Ab initio calculation of the low-frequency Raman cross section in silicon. *Phys. Rev. B*, **33**, 5969–5971.

36 Lazzeri, M. and Mauri, F. (2003) First-principles calculation of vibrational Raman spectra in large systems: signature of small rings in crystalline SiO_2. *Phys. Rev. Lett.*, **90**, 36–401.

37 Prosandeev, S.A., Waghmare, U., Levin, I., and Maslar, J. (2005) First-order Raman spectra of AB0 1/2B00 1/2O3 double perovskites. *Phys. Rev. B*, **71**, 214–307.

38 Caracas, R. and Cohen, R.E. (2005) Effect of chemistry on the stability and elasticity of the perovskite and post-perovskite phases in the $MgSiO_3$–$FeSiO_3$–Al_2O_3 system and implications for the lowermost mantle. *Geophys. Res. Lett.*, **32**, L16310. doi:

39 Caracas, R. (2007) Raman spectra and lattice dynamics of cubic gauche nitrogen. *J. Chem. Phys.*, **127**, 144510.

40 Caracas, R. and Cohen, R.E. (2007) Post-perovskite phase in selected sesquioxides from density-functional calculations. *Phys. Rev. B*, **76**, 184101.

41 Hermet, P., Veithen, M., and Ghosez, Ph. (2007) First-principles calculations of the nonlinear optical susceptibilities and Raman scattering spectra of lithium niobate. *J. Phys. Condens. Matt.*, **19**, 456202.

42 Hermet, P., Goffinet, M., Kreisel, J., and Ghosez, Ph. (2007) Raman and infrared spectra of multiferroic bismuth ferrite from first-principles. *Phys. Rev. B*, **75**, 220102(R).

43 Landau, L. and Lifschits, E. (1960) *Electrodynamics of Continuous Media*, Pergamon Press.

44 Maradudin, A.A., Montroll, E.W., Weiss, G.H., and Ipatova, I.P. (1971) *Theory of Lattice Dynamics in the Harmonic*

Approximation in Solid State Physics (eds H.E. Ehrenreich, F. Seitz, and D. Turnbull), Academic Press, New York.

45 Bini, R., Ulivi, L., Kreutz, J., and Jodl, H. (2000) High pressure phases of solid nitrogen by Raman and infrared spectroscopy. *J. Chem. Phys.*, **112**, 8522.

46 Tassini, L., Gorelli, F., and Ulivi, F. (2005) High temperature structures and orientational disorder in compressed solid nitrogen. *J. Chem. Phys.*, **122**, 074701.

47 McMahan, A.K. and LeSar, P. (1985) Pressure dissociation of solid nitrogen under 1 Mbar. *Phys. Rev. Lett.*, **54**, 1929–1932.

48 Martin, R.M. and Needs, R.J. (1986) Theoretical study of the molecular-to-nonmolecular transformation of nitrogen at high pressures. *Phys. Rev. B*, **34**, 5082–5092.

49 Caracas, R. and Hemley, R.J. (2007) New structures of dense nitrogen: pathways to the polymeric phase. *Chem. Phys. Lett.*, **442**, 65–70.

50 Eremets, M., Gavriliuk, A.G., Trojan, I.A., Dzvienko, D.A., and Boehler, R. (2004) Single-bonded cubic form of nitrogen. *Nat. Mater.*, **3**, 558.

51 Eremets, M., Gavriliuk, A.G., and Trojan, I.A. (2007) Single-crystalline polymeric nitrogen. *Appl. Phys. Lett.*, **90**, 171904.

52 Pruzan, Ph., Chervin, J.C., Wolanin, E., Canny, B., Gauthier, M., and Hanfland, M. (2003) Phase diagram of ice in the VII–VIII–X domain. Vibrational and structural data for strongly compressed ice VIII. *J. Raman Spectrosc.*, **34**, 591.

53 Hemley, R.J. (2000) Effects of high pressure on molecules. *Annu. Rev. Phys. Chem.*, **51**, 763–800.

54 Caracas, R. (2008) Dynamical instabilities of ice X. *Phys. Rev. Lett.*, **101**, 085502.

55 Goncharov, A.F., Struzhkin, V.V., Somayazulu, M.S., Hemley, R.J., and Mao, H.K. (1996) Compression of ice to 210 gigapascals: infrared evidence for a symmetric hydrogen-bonded phase. *Science*, **273**, 218.

56 Benoit, M., Bernasconi, M., Focher, P., and Parrinello, M. (1996) New high-pressure phase of ice. *Phys. Rev. Lett.*, **76**, 2934.

Index

Thermodynamic Properties of Solids: Experiment and Modeling
Edited by S. L. Chaplot, R. Mittal, and N. Choudhury

ISBN: 928-3-527-40812-2

i

l

m